SYSTEM RELIABILITY
Concepts and Applications

SYSTEM RELIABILITY

Concepts and Applications

Klaas B. Klaassen and Jack C. L. van Peppen

IBM Almaden Research Center

Edward Arnold
A division of Hodder & Stoughton
LONDON NEW YORK MELBOURNE AUCKLAND

© 1989 VSSD

First published in Great Britain 1989

Distributed in the USA by Routledge, Chapman and Hall, Inc.
29 West 35th Street, New York, NY 10001

British Library Cataloguing in Publication Data

Klaassen, K. B.
 System reliability: concepts and applications.
 1. Computer systems. Reliability
 I. Title II. van Peppen, J. C. L.
 004

ISBN 0-340-50142-1

All rights reserved. No part of this publication may be reproduced or transmitted in any form or by any means, electronically or mechanically, including photocopying, recording or any information storage or retrieval system, without either prior permission in writing from the publisher or a licence permitting restricted copying. In the United Kingdom such licences are issued by the Copyright Licensing Agency: 33–34 Alfred Place, London WC1E 7DP

Text set in Times and Univers by VSSD, Delft, The Netherlands.
Printed and bound in Great Britain for Edward Arnold, the educational, academic and medical publishing division of Hodder and Stoughton Limited, 41 Bedford Square, London WC1B 3DQ by J. W. Arrowsmith Ltd, Bristol

Preface

The mind-boggling rate of industrial expansion of the past few decades has produced innumerable technical devices and systems on which we rely in our daily life for modern convenience, safety, and sometimes even preservation of human lives. These modern artifacts cover a broad spectrum ranging from a relatively simple electronic watch to very complex transportation systems such as airplanes or spacecraft. Often, one is not even aware of the use of particular systems (part of our electrical energy is generated by nuclear reactors) until one is most unpleasantly reminded (Chernobyl disaster).

It is a proven fact that all these technical systems are *producible*, in other words: One can at least make them work at the time of first use. A higher order requirement, however, is that they remain serviceable throughout their expected useful life; i.e. that they are *reliable*. The consequences of an unreliable functioning of these systems may vary from inconvenience, extra costs, environmental damage, to even death. Such inability to perform reliably may not only arise from the product itself (usually manifested in hardware or software failures), but also from human errors. Take for instance the (pilot) error where an aircraft is put down on the runway extremely hard. As the cover picture shows, this can result in a cracked fusilage and the dragging of the entire tail section over the runway until the aircraft comes to a complete stop (Eastern Airlines, Florida, Dec. 28, 1987).

Only recently has the reliability aspect of our industrial activity been increasingly emphasised. The U.S. automobile industry, after having lost out almost completely to the Japanese and their more reliable cars, has only lately improved the reliability of its products drastically. Of course, producibility, yield, and quality are of eminent importance for an industrial product, but a healthy reliability over the entire planned life span of the product is at least of equal importance. The hesitation of many manufacturers in accepting a high reliability as one of the product design goals can be explained by the extra cost associated with the reliability program and by the intangible nature of product reliability to the customer. The customer (at least initially) does not know that one system is more reliable than another, and also does not know that the price difference between the two is more than warranted if one takes into account the later savings on "inconveniences" such as repair costs, aggravation, loss of production, accidents, environmental damage, etc.. Judging from the large number of unreliable systems around today, not everybody recognizes the principle underlying reliability engineering: *"Invest now, save later"*.

A secondary cause for the hesitation to regard reliability as an important product specification is the lack of training product design engineers receive in this field. To fill this gap for graduating engineers, Dr. K.B. Klaassen started a lecture series on Reliability Engineering at the Electrical Engineering Department of the Delft University of Technology in the Netherlands. Due to the great student interest in this topic (judging from the large enrollment figures), good lecture notes became a necessity. These notes found such a receptive audience outside the University that it was decided to publish them in the form of a book, first in the Dutch language and later, when the authors had joined IBM's Almaden Research Center in San Jose, California, also in English.

This book is composed of nine chapters. At the end of each chapter the reader finds a number of problems designed to rehearse the subject matter of that specific chapter. To aid in solving the problem, the end of the book provides not only answers to these exercises, but also a detailed explanation of the solutions. Throughout the text of the book, practical examples are provided, taken from the various applications of reliability engineering such as: electronics, control engineering, avionics, power engineering etc.

Chapter 1 discusses the definition of reliability and the various associated aspects. It reviews the reasons for reliability improvement, dwells briefly on the probabilistic versus deterministic approach to reliability engineering, and gives the most important ways in which the reliability of a system may be increased.
Chapter 2 is devoted to the deterministic approach to reliability engineering, which is often indicated as the "physics of failure" approach. It deals with several degradation models, gives examples of important physical failure mechanisms, and explains the use of screening techniques for removing the potentially weak components.
From Chapter 3 on, the book focuses on probabilistic reliability engineering. *Chapter 3* covers the nomenclature, definitions and, mathematical relationships of all essential probabilistic reliability, availability, and maintainability parameters.
Chapter 4 deals with all frequently encountered failure probability distributions. It also covers reliability testing, confidence levels, and accelerated testing.
Chapter 5 is dedicated to probabilistic reliability models, in particular the catastrophic failure model, the stress-strength model, and the Markov model.
Chapter 6 discusses the effect of system structure on the reliability of non-maintained systems. It deals with series, parallel, m-out-of-n, and majority voting systems. This chapter also acquaints the reader with various techniques for reliability analysis and reliability optimization.
Chapter 7 deals with maintained systems. It introduces various forms of maintenance and their effect on the system's availability. The effect of redundancy combined with maintenance is also discussed. The chapter closes with a look at the problem of spare-parts provisioning.
Chapter 8 deals with system evaluation techniques such as Fault Tree Analysis (FTA) ans Failure Mode Effect and Criticality Analysis (FMECA). It also introduces the concepts of risk and safety.
Finally, *Chapter 9* is dedicated to software reliability. It discusses how to write reliable programs, how to test software for reliability, and gives an effective software failure model.

Acknowledgements
The authors wish to express their gratitude to the many students of the Delft University whose comments made this a better book. In addition, they wish to acknowledge the efforts of Mr. J.C. van Dijk in coordinating the drafts of this book between San Jose and Delft. They thank Mr. G. van Berkel for making most illustrations, and Mr. J.D.

Schipper for his assistance with the first editions of the lecture notes. Last but not least, they thank Dr. A. Bossche for his remarks and suggestions to improve the didactic quality of this book.

Spring 1989   Klaas B. Klaassen
San Jose, California   Jack C.L. van Peppen

Contents

1 INTRODUCTION	11
1.1 Definitions	12
1.2 Need for reliability engineering	15
1.3 Statistical versus deterministic approach	16
1.4 Methods for increasing reliability	19
Problems	22
2 DETERMINISTIC RELIABILITY	23
2.1 Arrhenius' model	23
2.2 Failure mechanisms	26
2.3 Screening	29
Problems	30
3 STATISTICAL RELIABILITY	32
3.1 Nomenclature	32
3.2 Operational reliability quantities	34
3.2.1. Derived quantities	36
Problems	39
4 STATISTICAL FAILURE OF COMPONENTS	41
4.1 Failure distributions	41
4.1.1 Negative-exponential distribution	44
4.1.2 Normal distribution	48
4.1.3 Lognormal distribution	50
4.1.4 Weibull distribution	51
4.1.5 Gamma distribution	53
4.2 Life distribution measurements	57
4.2.1 Failure distribution from life tests	58
4.2.2 Confidence level of life tests	58
4.2.3 Accelerated life tests	61
Problems	62
5 RELIABILITY MODELS	63
5.1 Catastrophic failure model	63
5.2 Stress-strength model	68
5.3 Markov model	73
Problems	83

6	NON-MAINTAINED SYSTEMS	85
	6.1 Introduction	85
	6.2 Series systems	86
	6.3 Redundancy	89
	6.4 Parallel systems	92
	6.4.1 Dependent failures	97
	6.5 m-out-of-n systems	101
	6.6 Majority voting systems	103
	6.7 Mixed systems	106
	6.8 Optimisation	107
	6.9 Analysis methods	111
	6.9.1 Network reduction method	111
	6.9.2 Tie set and cut set method	112
	6.9.3 Decomposition method	114
	6.9.4 State-space method	115
	Problems	119
7	MAINTAINED SYSTEMS	126
	7.1 Introduction	127
	7.2 Systems with preventive maintenance	129
	7.2.1 Scheduled maintenance	129
	7.2.2 Condition-based maintenance	132
	7.3 Systems with corrective maintenance	135
	7.3.1 Replacement	136
	7.3.2 Repair	138
	7.3.3 Repairable systems without redundancy	141
	7.3.4 Repairable systems with redundancy	150
	7.3.5 Shared-repair facilities	156
	7.3.6 Inhomogeneous systems	159
	7.4 Maintenance aspects	164
	7.4.1 Maintenance strategies	165
	7.4.2 Spare parts supplies	166
	Problems	167
8	EVALUATION METHODS	172
	8.1 Introduction	172
	8.2 Causal evaluation	173
	8.2.1 FMEC Analysis	174
	8.3 Anti-causal evaluation	177
	8.3.1 Fault tree analysis	178
	8.4 Risk and safety	187
	Problems	191

9	RELIABILITY OF COMPUTER SOFTWARE	196
	9.1 Introduction	196
	9.2 Writing reliable software	198
	9.3 Reliability testing	200
	9.4 Failure models for software	201
Problems		204
SOLUTIONS TO PROBLEMS		205
	1 Introduction	205
	2 Deterministic Reliability	206
	3 Statistical Reliability	207
	4 Statistically Failing System Components	209
	5 Reliability Models	210
	6 Non-Maintained Systems	212
	7 Maintained Systems	223
	8 Evaluation Methods	234
	9 Reliability of Computer Software	242
APPENDIX		245
	A.1 Applied Laplace Transforms	245
	A.2 The Central Limit Theorem	245
	A.3 Most Commonly Used Symbols	246
LITERATURE		247
INDEX		251

1
Introduction

The field of reliability engineering covers a large and extremely varied area of applied science; for that reason it is impossible to do justice to the many aspects of reliability engineering in one book. The total province of reliability engineering can be divided roughly as follows:
- *Reliability theory:* The mathematical approach of solving reliability problems by statistical and stochastic means, for example: Estimation theory, renewal theory, queueing theory, logistics, etc.
- *Measuring, testing and certifying reliability:* Measuring the achieved reliability of a product on the basis of experiments (tests), performed on only part of the products (sample), during a relatively short time (accelerated tests) which are discontinued before the entire sample has failed (truncated tests) and determining the statistical confidence of these measurements.
- *Reliability analysis:* Collection of failure data, reduction and archiving of these data for use in future designs. The occurring failures can be analysed physically (physics of failure), but also statistically (statistical failure analysis). The information gathered about causes of failure, failure mechanisms, and ways in which components fail is then used in the design phase to avoid such failures in the future.
- *Design for reliability:* Increasing the inherent reliability of a product by such means as: Using special, highly reliable components (hi-rel components), decreasing the loading level of the components (derating), reviewing the designs at certain intervals (design reviews), adapting the product to user and environment (human engineering, fail-safe methods), making a product well maintainable (modular design, standardisation), and using extra parallel components (hardware redundancy) or extra parallel calculations or operations (software redundancy).
- *Management and organisation:* Creating and maintaining an (industrial) organisation suited for the design, development, production, and maintenance of reliable products. The development of the necessary administrative and logistic support. Furthermore, training programmes, inspection, test and maintenance procedures, as well as cost-benefit analyses of the applied reliability measures are usually included in this category.

Of the above subjects, the management aspect will not be discussed in this book. The theory of reliability will be treated and elucidated by means of a number of examples from areas such as energy technology, avionics, electronics, control engineering, computer technology, and everyday life. Further, a number of topics from the reliability analysis area will be discussed. A number of design techniques will be evaluated. The importance of the choice of proper maintenance techniques will also be treated.

12   *Introduction*

N.B.: The concept *'reliability'* is often confused with another concept: *'quality'*. The quality of a component, product or service (generally speaking a 'system') is determined by the degree to which the properties of that system are within predetermined and specified tolerances. If there are no specifications with regard to the expected life in a system specification, and hence the quality only pertains to the state of the system at the time of delivery by the producer to the consumer, the fraction of the total number of systems that meets the specification is expressing the *conformity* of that system. If there are also specifications with regard to the life of a system, and therefore the properties of the supplied system are also of recognised importance after the time of delivery, the fraction of the total number of supplied products that still functions in accordance with the specifications at a time t after the time of delivery t_0 is expressing the *reliability* of that system.

The following section defines exactly what is understood by the reliability of a system.

1.1 Definitions

In this book the *reliability* of a system shall be the *probability* that this *system* uninterruptedly performs certain (accurately) *specified functions* during a stated interval of a *life variable*, on the condition that the system is used within a certain *specified environment*. This general definition contains six elements which will be explained briefly below:

- *Reliability*: This is a statistical probability which is usually denoted as $R(t)$. It is often confused with the concept 'quality'. Both concepts originated in the area of quality control, from which reliability engineering later emerged as a separate field of specialisation.
- *Probability*: One should distinguish predicted, or *a priori* reliability, which is defined as a sheer likelihood, and proved, or *a posteriori* reliability, which is a retrospective certainty, and is defined as the fraction of surviving systems. For a future design one can only predict; afterwards, in a case history for example, one has certainty.
- *System*: A system encompasses a collection of elements (components, units, modules) between which there is a mutual interaction (interconnection) which can be separated from the environment of that system (system boundaries). The mutual interaction between the elements of a system realises the system function, which can, in general, be divided into a number of specified attributes or properties.
  The designation 'system' does not only imply technical systems such as equipment, installations, and machines, but also non-technical systems such as biological organisms, organisations, and services. For convenience we will restrict our examples to technical systems
- *Specified function*: The purpose of a certain system is reflected by the system functions, which in turn consist of one or more specified properties or attributes. In systems with signals continually varying between certain limits (analog systems) a system function (for example amplification) can be separated into a number of properties (e.g. voltage amplification 100, bandwidth 2 MHz) which are subjected to tolerances (voltage amplification 100 ± 5%, bandwidth > 2 MHz). In Table 1.1 the

Name	Instrumentation Amplifier	
Manufacturer	XXX Corporation	
Model Number	3456-B	
All specifications traceable to US Bureau of Standards		
Function	Voltage Amplification	
Specifications	Gain	100 ± 5%
	Frequency Range (−3 dB)	DC −2 MHz
	Noise (referred to input)	< 1.5 nV/$\sqrt{Hz}$
	Input Impedance	> 1 Mohm
	Output Impedance	< 0.1 ohm
	Nonlinearity (input < 1 V)	< 10^{-3}
	Max. Output Current	> 100 mA (short circuit protected)
	Required Line Power	< 42 VA
Environment	Temperature Range	
	Operational	0 °C to 50 °C
	Storage	−40 °C to 75 °C
	Humidity Range	< 95%, no condensation
	Altitude	
	Operational	< 4.5 km
	Mechanical Shocks	< 50 m/s^2
	Line Voltage Range	120 V +5%, −10%
	Line Frequency Range	48 Hz to 440 Hz
Reliability	Mean time to failure (no maintenance)	5 years

Table 1.1 An example of a system (instrumentation amplifier) with a certain function (voltage amplification) which is specified. The environment in which the instrument should be used is also specified. The reliability specification is given as the expected average life.

specifications of an analog measurement instrument are given. If one or more specified properties exceed the tolerance intervals the system is no longer reliable; it has failed. In the case of analog systems (here the amplifier) the system may still be able to function, but outside the tolerances. In systems operating with binary signals (digital systems) one usually sees that a certain function (for example, access to a background memory) or a property of it (the ability to store information) ceases completely, i.e. can no longer be used, after a failure has occurred. Therefore, the temptation to continue using a failed system is not as strong here.

- *Life variable*: The elapsed time in almost all cases will be the life variable. This may be calendar time, but also accumulated user time (operation time). The time that the system is not in use must be accounted for, however, if it contributes to a shortening of the system's life. The total time is then $t = t_o + at_{oo}$, in which t_o is the operation time and t_{oo} the time that the system is not in use. The coefficient 'a' which indicates the severity of 'non use' is almost always smaller than 1. However, there are cases in which systems out of operation have, per unit of time, a greater mortality then when in operation. Just think of electrolytic capacitors, effects of condensation in systems that are not in use, and think also of people with a task too light or no task at all who more often make mistakes from plain boredom. Besides time, the life variable may also be the number of times a system is switched on and off (relay), the number of load changes (fatigue fractures in airplane wings, landing-gears, jet turbine blades, etc.), or it may be the distance travelled (cars).

- *Specified environment*: Every system is placed in a certain environment. All elements that are not part of the system belong to this environment, thus most of the time also the user and the rest of the installation of which the system in question is a part. If a system is put in the wrong environment (i.e. outside the specified environment parameter ranges), either on purpose or inadvertently, the system may fail or age more quickly. Examples are an environment that is too hot or too wet, a supply voltage that is too high, input signals that are too large, or a load that is too great or too small (applying full throttle while the car's gear is in neutral). This so-called *misuse* of a system outside the specified environment cannot be accurately forecast by the designer and must therefore be excluded in the reliability definition.

N.B.: In practice most systems fail due to misuse, either by the user or by the designer who wrongly applies the components in the system; so most systems fail because of human error.

In the above, the definition of reliability has been explained in detail. It turns out that no statement about the reliability of the system can be made without an explicit, clearly formulated description of the *system* under observation, the *system functions*, and the *allowed environment*. For example, what is the reliability of a human being? Is a human outside the specifications if he or she has a headache?

In technical systems, but also in services and the like, it is therefore of major importance to describe these matters as exactly as possible, also with regard to later legal and financial consequences (legal liability for and warranty on products etc.).

We shall later see that it is important to distinguish between systems that are maintained and systems that are not. By *maintenance* we understand any human intervention which keeps the system operational or returns it to an operational state. If a system is maintainable but *de facto* is not maintained due to neglect, for instance, that system belongs actually to the second above-mentioned group of systems without maintenance. Rather than use the term 'maintainable', which indicates a degree of freedom, we shall use the term 'maintained'. We shall therefore call the two categories mentioned above 'maintained' and 'non-maintained' systems.

The concept reliability only pertains to non-maintained systems, since in the considered interval of the life variable the system has to function correctly without interruption, so no failures may occur. Repairs are not allowed here.

Definitions 15

For that reason we introduce a more general concept: *Availability*. For non-maintained systems the availability is equal to the reliability. We shall discuss the availability of a maintained system in more detail in Chapter 7, where we deal with maintained systems.

Reliability engineering can now be defined as the whole of mathematical, organisational, and other applied scientific technologies, methods, and strategies to achieve a reliable product and determine its degree of reliability.

1.2 Need for reliability engineering

The necessity to practise reliability engineering is obvious from the relation between the elements of the reliability definition given in the previous section. The size of the system, the intricacy of the specified functions, the length of the useful interval of the life variable, and the degree of hostility of the system's environment all influence the reliability.

It will be clear that the tendency to larger systems, i.e. systems with larger numbers of components, would decrease the reliability if the development of more reliable system components and structures does not keep in step. There are many such systems with a large *quantitative complexity,* such as energy distribution networks, telecommunication systems, digital computer networks, and space probes.

In addition, there is a tendency towards the use of more complex system functions, that is, more functions to be performed by a single system, the functions are more involved (which is expressed in more specified properties), and the allowed tolerances become smaller. This increase in *qualitative complexity* also causes the reliability to drop if no adequate countermeasures are taken. We may think of: Multi-function measuring equipment with a higher, required accuracy, automatic landing systems for airplanes, process control equipment, and so on.

Further, the correct functioning of a system over a longer interval of the life variable is increasingly important as we become more dependent on such systems (energy generation systems, pacemakers and the like). These so-called critical systems require a high reliability, often over long periods (e.g. 25 years for telecommunication systems). A source of concern in pacemakers, for instance, is the energy source, since circuit failures in pacemakers occur with a probability of less than $140 \cdot 10^{-9}$ per hour. In Figure 1.1 the reliability of a number of different energy sources for pacemakers is shown.

Besides this, our technical systems are more and more put to use in hostile environments; they have to be suitable for a wider variety of environments. Just think of applications in the process industry (heat, humidity, chemical substances), mobile applications in aircraft, ships, and vehicles (mechanical vibrations, shocks, badly defined power supply voltages, high electromagnetic interference level).

All in all, these are sufficient reasons for reliability engineering to be so much in the limelight these days. Add to that the emphasis on reliability in situations where no maintenance is possible, because of an isolated location (unmanned arctic weather stations, remote space probes, underwater amplification stations in transatlantic cables, etc.). Even if maintenance were possible, it is often better (more cost-effective) to increase the initial reliability of a system because of the high costs associated with that system being down for repairs. Despite the higher initial costs, the life cycle cost may turn out to

be lower. This is called the 'invest now, save later' principle of reliability.
Also the socio-ethical aspects of products with a reliability that is too low cannot be underestimated. These low-reliability disposable products lead to a waste of labour, energy, and raw materials that are becoming more and more scarce.

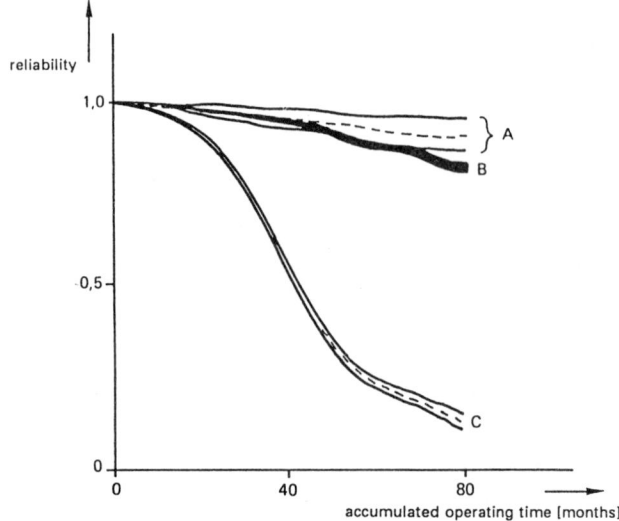

Figure 1.1 The reliability of various energy sources for pacemakers.
a) nuclear batteries (140 elements);
b) lithium batteries (5600 elements);
c) mercury-zinc batteries (2000 elements).

1.3 Statistical versus deterministic approach

As we have seen in Section 1.1 one has to distinguish between *a priori*, or predicted reliability, and *a posteriori*, or proven reliability.
In the statistical, predicting approach of the reliability problem the designer will try to make a judgement about the expected reliability of a future system on the basis of information about the field behaviour of previously produced components and on the basis of the results of (artificially accelerated) reliability measurements of current components. Taking this approach can give rise to a number of problems.
With today's fast development of technology, future products will hardly contain components of which the reliability history is known. So, in general, we do not have access to statistical data for the calculation of system reliability. Even if we do use components developed in the past, with a known reliability history, the components to be used will fairly certainly have been manufactured at another time. Usually the production process has been adjusted in the meantime. From investigations it has become clear that these, at first sight, small adjustments may have great consequences for the reliability. The components produced later no longer fulfill the previously proven reliability (non-homogeneity in time of the production line).

An alternative would be to measure the reliability of components by accelerating the ageing process. Here too, a number of problems may occur.

How large is the applied acceleration factor exactly? Are the component parameters which are stressed to produce accelerated ageing really representative of the actual ageing of the components in the field, i.e. during practical use? Are perhaps other failure mechanisms also triggered which would result in too low a predicted reliability? Are certain failure mechanisms occurring during practical use not excited at all, in the test or are they excited with an acceleration factor which deviates from the intended acceleration?

Another problem is that we usually are not able to measure 100% of the components; for example because the components surviving the test have a considerably shortened remaining life expectancy. We will therefore have to make our judgement based on a sample out of the total population of components. If the production is not sufficiently homogeneous, a small sample will result in an inaccurate assessment of the reliability of the entire collection (take for instance the production non-homogeneity within one batch or between batches).

All in all, the conclusion is that the statisticians hand us very fine algorithms for sampling and testing, which find widespread use in reliability engineering, but we have to work with an appalling lack of information. In practice one often has to make do with many best judgement estimates. The confidence level of the final results is then so low that one achieves little more than a rather uncertain assessment of the expected reliability of a future system. In this respect, it should be noted that the statistical methodology (when using estimated reliability data for many components) gives a far better estimate for the *ratio* of the reliabilities when we are comparing different design alternatives.

For the above reasons, an alternative to the statistical approach to the reliability problem, the deterministic approach, is also important. The deterministic approach entails the study of physical deterioration processes leading to failure in components. Important is what starts these processes, which environment accelerates them, how they lead to breakdown of a component, and how these processes can be stopped or slowed down. Based on the know-how of the (dominant) deterioration process (evaporation of a filament in an incandescent lamp, for example) and the rate of the degeneration (depending on the temperature of the filament) one can make a prediction about the life (*in casu* the number of burning-hours until the filament opens up).

As an example of the deterministic approach to a reliability problem, we shall briefly discuss a study of failure mechanisms in light bulbs.

Light bulbs are made for a certain mains voltage V (for example $V = V_{rms} = 220$ volt), so that the dissipated power P ($P = V^2/R_{hot}$) has a certain value (e.g. $P = 100$ watt). This determines, among other things, the length and cross sectional area of the tungsten filament. Usually the filament is spiralled (sometimes even twice) to increase the heat production (temperature increase per watt) and thus the light production (lumen/watt). After switching on, the filament reaches in ca. 10–20 ms a final temperature of about 2500 to 2600 °C (4500 to 4700 °F). The accompanying rapid expansion (and contraction when switching off) may result in thermal fatigue of the filament. The life variable of this failure mechanism is clearly the number of on-off cycles of the lamp. If the filament is left

18 Introduction

on, the dominant failure mechanism is the evaporation of the filament. Here the life variable is the number of burning-hours. However, one is faced with a paradox here: *a uniformly evaporating filament, supplied with a constant voltage, cannot fail by evaporation!* This is because the filament's resistance will increase more and more as the filament evaporates, thereby reducing the dissipated power and consequently the filament's temperature. In turn, this lower temperature will slow down the evaporation process more and more. The lamp's real cause of failure is a local, greatly increased evaporation, for example at the location of a crack in the filament or at a narrow site caused by the surface roughness of the drawn filament. At this site the cross section is smaller and the dissipation, and therefore also the temperature, is higher. This causes the evaporation to be much faster here. In Figure 1.2 it is shown how the life t decreases accordingly as the temperature T_{hs} of a hot spot rises higher above the temperature T_w of the rest of the filament.

N.B.: Small differences in the diameter and thus in the temperature have large consequences! Therefore, we have to conclude that the quality control of the filament during production is of decisive importance for the later life of the lamp.

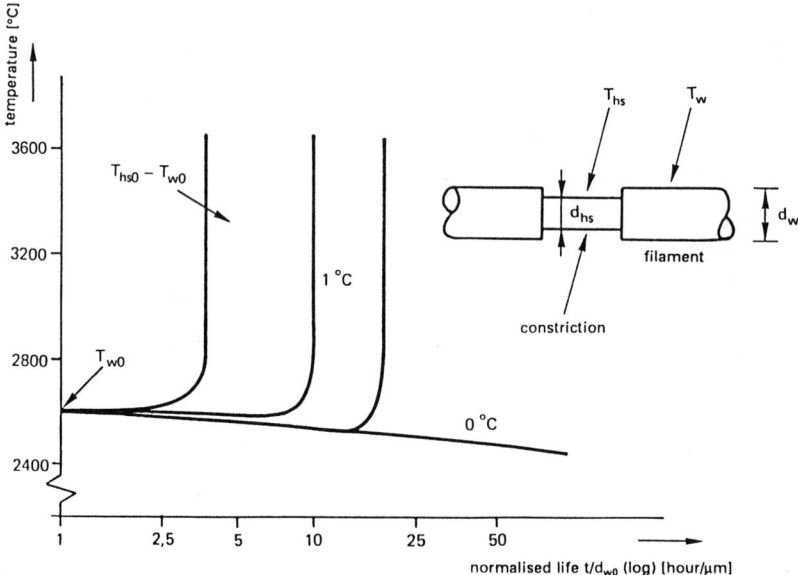

Figure 1.2 *Normalised life t/d$_{w0}$ of a filament of a light bulb as a result of an initial hot spot caused by a constriction. The initial diameter of the filament is d$_{w0}$, the initial hot spot temperature is T$_{hs0}$, the initial filament temperature is T$_{w0}$. A minor difference in temperature has great consequences!*

Figure 1.3a shows the temperature profile along a filament that is left on continuously. It shows four instances, viz. 0, 30, 60 and 95 % of the life t_0 of the filament. We clearly see the development of a hot spot. Finally, Figure 1.3b indicates how the temperature profile (designated T) correlates with the measured diameter profile of the filament (designated D). Concludingly, the following remark. Many light bulbs die prematurely from yet another cause: Mains voltage spikes. After all, a mains voltage above nominal is an accelerating factor for both above-mentioned failure mechanisms. It does not introduce a new failure process; it just accelerates existing failure processes.

The information often neglected in accelerated testing conducted on a purely statistical basis, is the failure mechanism of the defective components. This failure mechanism will probably also be present in the other components, but has not caused a failure within the duration of the test under the applied test conditions. In the field this may be different.

The questions arising if one would only follow the deterministic approach (i.e. the physics of failure approach) are, among others: Can statistical fluctuations in the production process cause some components to fail, for example, by a failure mechanism which is not probable in most components? Small cracks, for example, may be created in the filament of some light bulbs by fluctuations in the drawing process, where an ordinary filament has a life which is limited by the surface roughness of the filament.

Another question is whether many of these physics of failure studies are not aimed too much at the 'typical' component. The entire production process, which also produces 'atypical' specimens, gets too little attention. Precisely these atypical specimens may later dominate the failure behaviour of the total population.

One cannot limit oneself to the statistical or the deterministic approach alone. Both are one-sided: The statistician is not interested in the cause of the failure, the physicist is only interested in the 'typical' failure mechanisms.

This is the reason to use both approaches in mutual harmony to obtain accurate life expectancy information and a reliable product.

1.4 Methods for increasing reliability

There are several ways in which the inherent reliability of a system can be assured. The inherent reliability is the reliability intrinsic to the system that will indeed be realised in the field provided the system is not misused. In this section the most important measures that can be taken to secure a high inherent reliability will be discussed briefly below. Many of them will later be treated in more detail.

- The introduction of reliability in as early a phase of the system design as possible as one of the aims of that design. Figure 1.4 shows an example of how important a well-considered design is in this respect. This early introduction is necessary because, if the reliability is only introduced in a late phase where the design is final or nearly final, the only thing a designer can do is to resort to the use of reliable (and therefore expensive) components, or he can apply redundancy at the system level (which is very ineffective), or he can improve only the weakest link in the chain. These are all methods that are not very cost effective. We will return to this later.
- The choice of those technical means and technologies that can easily realise the required system functions without necessitating a *tour de force*. After the choice of a proper technology or a proper combination of technologies, one should be able to design the system with configuration necessitating the minimal quantitative and qualitative complexity. The aim of the design should be that the system functions are determined by only a few, reliable components and the design must be tolerant of variations with time in the properties of the other, less critical components.

20 *Introduction*

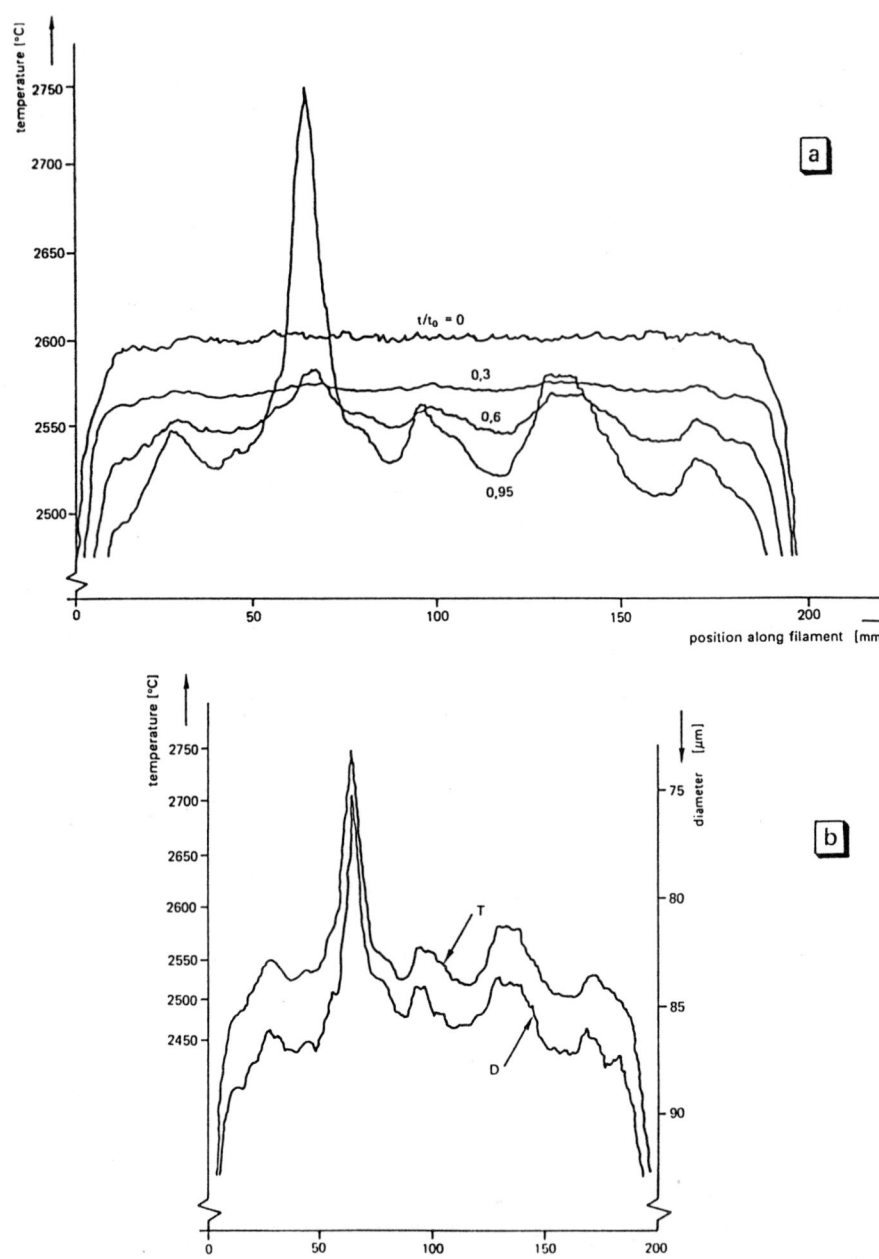

Figure 1.3 Temperature variation along a filament of an incandescent lamp.
a) Temperature profile at four different instants t/t_0 in the life (0, t_0) of a filament.
b) Correlation between the temperature profile (indicated by T) and the profile of the filament diameter (indicated by D) for the instant $t/t_0 = 0.95$.

Methods for increasing reliability 21

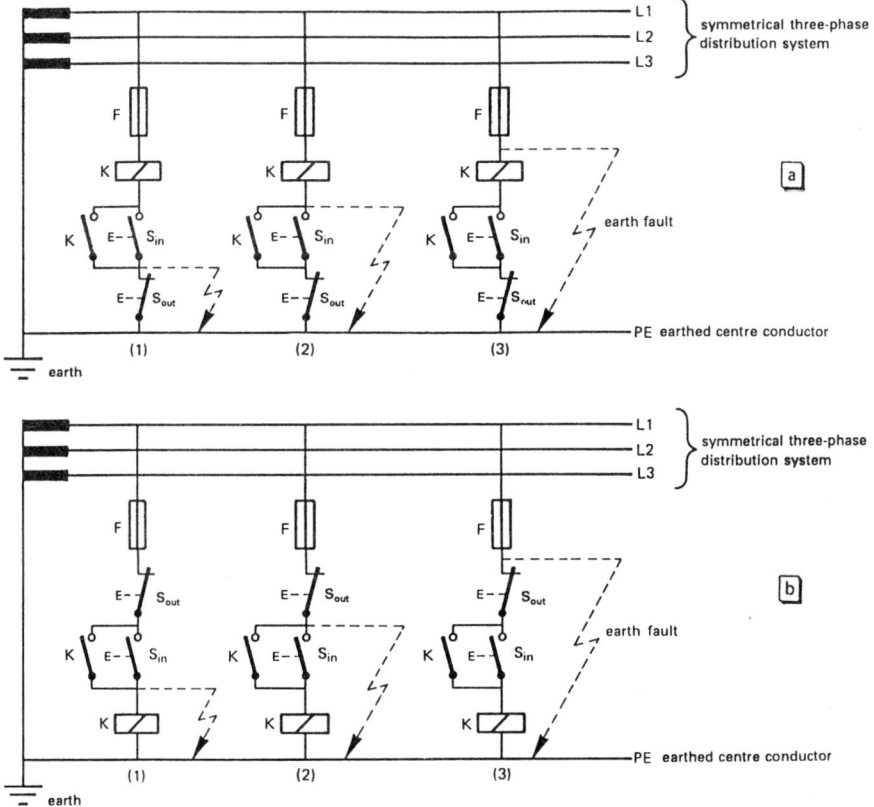

Figure 1.4 *Design errors in a earth symmetrical three-phase distribution system. In spite of short circuits to earth, the relay K must be able to be switched off, or (because the fuse F melts) must switch off. Earth faults may never switch on K. S_{on} is a normally open push contact for switching on, S_{off} a normally closed push contact for switching off.*
a) Wrong design. Before switching on, earth fault 2 can switch on K without a command. After switching on, earth faults 1 and 2 cause K to be unable to be switched off.
b) Good design. Both, before and after switching on, earth faults cannot lead to an undesired switching on or a refusal to switch off.

- The application of derating. Derating is the reduction of the operational and the environmental stress to which the components of a system are exposed. The components here are operated well below their maximum ratings by using more components to share the load or by utilising other stress derating measures. As long as the stress probability density function encountered in practical use and the strength probability density function of a component coming from the manufacturing line still overlap, derating will give an improvement. We will return to this matter in Section 5.2.
- The thorough testing of the system prototypes for unreliability and the interim inspection of the systems for flaws during production.
- The introduction of a burn-in period to trace early failures by running the system for a period of time, possibly under increased stress. This will be treated further in Section 2.3.

22 Introduction

- The conduction of life tests resulting in failure-rate data which can be used to adjust the initial design. More failure-rate data are obtained from the field in a later phase and should be reported back to the designer. These actual data are used for validity studies of the life tests which have been conducted and for use in later designs or design updates.
- The use of redundancy. Incorporating redundancy into a system is providing alternative means for the realisation of the required system functions when the primary means have failed. To avoid dependent errors, the redundant (sub)systems should preferably realise the required function in a different way than the primary (sub)systems. They should consist of different components and be made by different manufacturers. If the reliability of the primary (sub)systems is higher, the redundant system connected in parallel has a greater effect, i.e. the reliability of the combination is increased to a greater extent. Because of this, redundancy should be used in the system at a hierarchical level that is as low as possible, so preferably at component level (see Section 6.3).
- The introduction of preventive maintenance where this is possible. Preventive maintenance aims to avoid system breakdown. Because preventive maintenance is usually conducted according to a predetermined plan, the costs are lower than those of corrective maintenance (repairs). Also, the costs resulting form an unscheduled stop of the system due to a sudden breakdown are reduced. However, preventive maintenance is not useful in all systems. Moreover, some corrective maintenance will always be necessary (see Chapter 7).
- The establishment of an organisational structure aimed at designing, developing, producing, and maintaining a reliable product. The principal management aspects involved here are organisation, training, logistics and coordination of manpower and means, etc.

Problems

1.1. What are the essential elements in the reliability definition?

1.2. Why will a uniform filament of a light bulb with power supplied by a constant voltage source never fail?

1.3. What is your conclusion for a uniform filament in an incandescent lamp whose power is supplied by a constant current source?

1.4. Why is the allowed temperature range in Table 1.1 lower under operational conditions than under storage conditions?

1.5. What is meant by 'deterministic reliability engineering'?

1.6. Explain why a well thought through organisation is indispensable to realise a reliable product.

1.7. What kinds of environmental elements might have an influence on the ageing process of electronic integrated circuits?

2
Deterministic Reliability

As we have already seen, the deterministic approach to reliability is especially interested in the physical process that leads to failure. This process is called the *failure mechanism*. Eventually, it leads to the situation where a component does not function at all or where it functions outside the specified tolerances. The consequences of a failure mechanism that has led to the component exceeding its specified limits can be observed externally. The observed failure is called the *failure mode* of that component. A failure mechanism is usually activated and accelerated by a certain environmental quantity or combination of environmental quantities. Such quantities are called *stress quantities*. If we use the example with the light bulb from Section 1.3, one of the *failure mechanisms* is an increasing local evaporation caused by the creation of hot spots along the filament due to cracks or surface roughness. The externally observable *failure mode* is an open filament. So the light bulb fails in the open mode. A *stress quantity* for this failure mechanism is the applied voltage (mains voltage). The stress as a result of a slightly increased supply voltage is large: The mains voltage has only to be increased very little above the nominal value for a much shorter life.

2.1 Arrhenius' model

One of the most common and important stress quantities is the increase of a component's *temperature*. The temperature of a component is determined by the temperature of the environment (external stress) and the power dissipation in the component, which in combination with the heat resistance to ambient (*environment*) makes the internal temperature rise above the ambient temperature (*internal stress*). A temperature increase accelerates all sorts of physicochemical processes. It is often assumed that the failure process behaves as a chemical process with a certain reaction rate Q for which the following equation holds:

$$Q(T) = Q_0 \, e^{-E_A/kT},$$

here Q_0 is a constant, E_A the *activation energy* in electron-volt, k is Boltzmann's constant ($k = 8.6 \times 10^{-5}$ eV/K) and T the absolute temperature. This expression was determined experimentally by Arrhenius in 1880.

If it is assumed that the drift in the properties (parameters) of the component as a function of time t is proportional to:

$$t^n Q(T),$$

in which linear drift in time is a special case: $n = 1$, we can compare failure rates at two different temperatures T_1 and T_2. Suppose that with temperature T_1 the time needed to have a parameter of the component, which started at the original nominal value, drift away to the tolerance limit is t_1. This parameter drift is characterised by $Q(T_1)t^n$, so that the time t_2 needed for the same parameter drift at a temperature T_2 is found from:

$$Q(T_2)t_2^n = Q(T_1)t_1^n,$$

or more generally:

$$Q(T)t^n = K'$$

Here t is the life at temperature T and K is an arbitrary constant. Consequently:

$$\ln t = K + \frac{E_A}{nkT}.$$

Here the new constant $K = \ln(K'/Q_0)^{1/n}$. The importance of this relation lies in the fact that many failure mechanisms can be characterised in this way. Plotting the results of two experiments conducted at different temperatures T_1 and T_2 on logarithmic paper is a simple way of determining the effective activation energy E_A/n associated with a certain failure mechanism. This can easily be seen from:

$$\ln t_2 - \ln t_1 = \ln \frac{t_2}{t_1} = \frac{E_A}{nk}\left(\frac{1}{T_2} - \frac{1}{T_1}\right)$$

Once E_A/n is known, one can also calculate the *acceleration factor* t_2/t_1 for all other temperatures from the above equation. The expression shows that the exponent n of the drift in time t has the same effect as a change in the activation energy E_A. For that reason E_A/n is called the *effective activation energy*. In Figure 2.1 this is illustrated. This figure compares five failure processes with different effective activation energies E_A/n. Here the *stress quantity* is the environmental temperature to which the tested components are exposed. Along the logarithmic vertical axis is plotted how long a component lasts on average at that environmental temperature. This quantity is usually referred to as MTTF or Mean Time To Failure. Once it is known (or assumed) that components meet Arrhenius' model, then life tests at two different stress levels (in this case temperatures) are sufficient to determine the rest of the life-stress curve.

A model that is also used to determine the acceleration in the mean time to failure resulting from an increased temperature is given by the relation of Eyring:

$$Q(T) = \alpha T\, e^{-E_A/kT}.$$

This model is derived from quantum-mechanical considerations. If the (absolute) temperature variation is small, αT can be considered to be a constant Q_0. This gives us back the Arrhenius relation.

An important stressing method for components composed of several different materials is the temperature *step stress* method which abruptly varies the temperature back and forth between two predetermined values. There are expressions which relate the resulting fatigue (micro cracks), which gives rise to open electrical contact paths, to the number of temperature cycles and the size of the temperature step. If such an expression fits the measured failure data, one can take the same course as above for Arrhenius' model.

Arrhenius' model 25

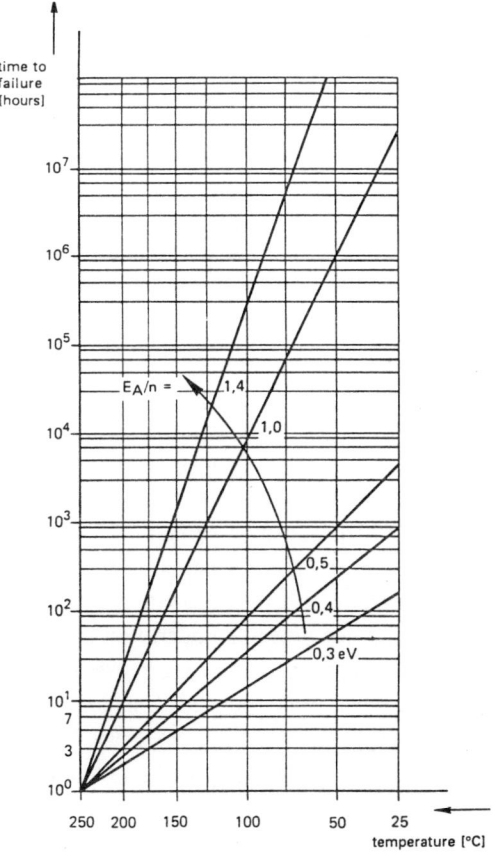

Figure 2.1 *Plot of Arrhenius' relation for failure processes having (five) different levels of endurance. The endurance is expressed in the effective activation energy E_A/n.*

Usually the result depends a great deal on the nature of the mechanical structure, the materials used, and the mechanical stress already present. The 'epoxy A' formerly used for plastic transistor casings was especially notorious for the many failures caused by breaking leads in the casing during thermal fatigue experiments.

The effect of *environmental humidity* as a stress quantity largely depends on two factors: The permeability of the casing or the coating of the components, and the effect of internal moisture in the component itself. Sometimes one can see moisture causing electrolytic corrosion between metallic conductors which are at different electrical potentials if they are embedded in a synthetic plastic casing. Practically all the available literature shows that if the component surface (e.g. a chip) is exposed to a relative humidity of more than 1% and if the electrical dissipation in the component is small (and therefore the reduction in relative humidity due to the hotter surface is small), problems will develop in a relatively short period of time. In systems with larger dissipation, problems may arise during those time intervals in which the component is not used (for example during transport or storage) when the relative humidity at the internal component surface is higher than under normal operation.

In components where high voltage gradients across an insulator develop (such as MOS-structures) an *externally applied voltage* usually has a shortening effect on the life of the component. It is often impossible to give a systematic relationship.

Switching on and off of the power supply to a component, switching on and off the voltage across it, or the current through it, results in rapid temperature changes caused by the internal dissipation in the component. The thermal resistance between the interior of the component and ambient determines the size of the temperature step. This rather resembles the stress caused by temperature cycling, though power off-on induced temperature variations are usually much faster and the highest temperature occurs internally in the component.

It has been known for quite some time that bipolar transistors irradiated with *ionising radiation* break down because of electric charges generated in the gas-filled casing or in the oxide layer of the semiconductor. These components can be restored to their original state by thermal annealing. As a remedy, the components are exposed to the expected dose, the usable components are selected and these are subsequently restored to their former state by a thermal treatment (radiation hardening).

More recently, there have been publications about the effect of alpha particles on the charge in the storage cells of charge-coupled devices and dynamic RAMs. Since the soft-error rate created by these charges depends linearly on the flux density, plotted on full logarithmic paper, this type of failure is rather easy to predict. By increasing the charge in the storage cells, the failure rate can be reduced considerably. For high-density storage devices this is usually not acceptable. The same could also be accomplished by increasing the size of the cells. Since the trend is just the opposite, namely smaller cell size and lower cell charge, another solution has been found by Hitachi and Motorola in the form of a protecting layer of polyamide with a thickness of ca. 10 μm which serves as a radiation filter against α-particles coming from the component casing.

2.2 Failure mechanisms

In deterministic reliability engineering one is first of all interested in the failure mechanisms which lead to failure in the field, how these mechanisms operate, and in which ways the underlying failure process can be prevented or slowed down. In principle, the latter can be effected by giving the component a changed physical structure and/or chemical composition, so that the failure process does not occur or occurs at a lower rate. Another possibility which can be used for non-dominant failure mechanisms is increasing the specific stress which excites this failure mechanism to such an extent that the surviving components are free of failures caused by this mechanism (e.g. burning-in weak components). This way of *screening components* has distinct disadvantages. The applied stress quantity may also affect the life of the remaining components, the production yield decreases and the product becomes more expensive because of the costs of this screening. Therefore, the emphasis here is on changing the product in such a way and to such an extent that its life meets the requirements. Only if this is not possible (or not possible any more) will one need to resort to constructing screens to sift out the so-called *early failures*.

To illustrate this, an example is given in Figure 2.2 of a growing crack in a mechanical construction caused by material fatigue. In a good design, mechanical stress concentrations should be avoided as much as possible. By testing the manufactured construction, possible crack growth in time can be measured. In Figure 2.2a the relation is shown between the number of load changes (load cycles) and the resulting crack length d for a sample with a 10 mm test slot. Figure 2.2b depicts the relation between the crack length and the remaining strength of this construction. A crack length larger than a certain minimum length a_d is detectable, i.e. instrumentation is available that would detect the existance of such a crack. The minimum required strength is F_k. If the crack propagates deeper than a_k the construction will break. On the basis of fracture mechanics one may therefore decide on the following periodical inspection. The first inspection is to be after $n_d + (n_k - n_d)/2$ load cycles. If no cracks are found ($d < a_a$) then the minimum life estimate is $n_k - n_d$ and the next inspection will have to take place after, for example, $(n_k - n_d)/2$ depending on the *safety margin* one wants to build into the inspection process.

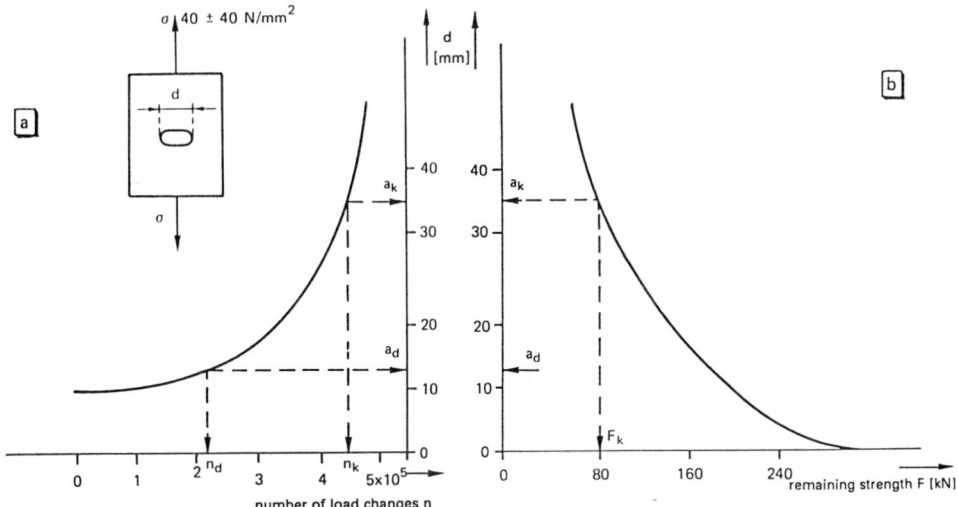

Figure 2.2 *Crack growth caused by fatigue as a result of load changes (in 2024-T3 aluminium alloy). Here n is the number of load changes, d the total crack length, and F the remaining strength. The construction is loaded with 80 kN. 4.5× 10⁵ load changes are just critical (nk). ak is the corresponding critical crack length, ad is the just detectable crack length. The first inspection for cracks can be done for n = nd + (nk − nd)/2. If no cracks are found, a minimum life of nk − nd is decided on.*

The study of failure processes that may occur in the components of technical systems is particularly strongly developed in microelectronics. Below, we shall therefore briefly discuss the physical processes behind a number of commonly occurring failure mechanisms in semiconductor components.

- *Corrosion*: In ICs the aluminium metallisation can be affected by corrosion. This corrosion is brought about by humidity, contamination, and electric potentials. Moisture seeps through the plastic casing of the component or migrates along the material boundary between the plastic and the connection pins. Contamination is caused by ions diffusing from the epoxy material to the metallisation especially there where water is present. The externally applied electric voltage causes an electric drift field for these ions that is proportional to the supply voltage and inversely proportional to the distance between the metallisation traces. This can start two corrosion processes, namely *anodic corrosion* and *cathodic corrosion*. Anodic corrosion does not depend on temperature but only on leakage current density:

$$Al \rightarrow Al^{3+} + 3e^-,$$

$$2\,Al^{3+} + 3\,H_2O \rightarrow Al_2O_3 + 6H^+.$$

The Al_2O_3 forms a protecting, non-conductive layer and halts further corrosion. For cathodic corrosion the process is:

$$2\,H_2O + 2\,e^- \rightarrow 2\,OH^- + H_2\uparrow,$$

$$2\,H_2O + 2\,Al + 2\,OH^- \rightarrow 2\,AlO_2^- + 3\,H_2\uparrow.$$

The second chemical reaction makes this corrosion temperature dependent with an activation energy of 0.5 eV. AlO_2^- is soluble in water and forms $Al(OH)_3$, which causes cracks in the SiO_2 passivation layer deposited on top of the metallisation. Passivation with silicon nitride layers gives a better protection against this corrosion.

■ *Electromigration*: If the current density in the metallisation is large enough and the temperature high, electromigration occurs. If the current density becomes larger than 1 mA/μm² (for aluminium), due to traces that are designed too small or currents that are chosen too high, this effect becomes noticeable and results in interruptions of the aluminium traces after some time. Trace cross-sectional areas can be too small at the location of scratches, and constrictions in the covering of steps where the trace steps to a lower or higher level.

The explanation for this electromigration process is as follows: A positive metal ion freed by thermal agitation from its potential well in the metal atomic lattice experiences two forces, namely a force as a result of the electric field which points against the (electron) current direction, and a force in the direction of the (electron) current caused by the impulse of the electrons 'colliding' with the activated metal ion (which is still more or less part of the conductor). Because of the shielding effect of the many electrons, the first force is small in comparison with the second force resulting from the 'electron wind'. These free metal ions have a better chance to occupy a free spot at the surface than the restrained metal ions around such a spot. The free metal ions will therefore be trapped by the 'dangling bonds' extending from such a free spot at the surface of the lattice. In this way there will be material transported by ions in the direction of the positive end of the conductor, which causes asperities and constrictions. The constrictions increase the current density and cause the process to go faster and faster. This can be expressed as below by Arrhenius' expression:

$$\ln t = K + \frac{E_A}{nkT}.$$

Experiments have shown that $K = -m \cdot \ln(AJ)$, in which m and A are constants and J is the current density. This results in a time to failure t:

$$t = (AJ)^{-m} \cdot e^{\frac{E_A}{nkT}}.$$

The effective activation energy E_A/n associated with electromigration is in the range: $0.5 \text{ eV} < E_A/n < 0.8 \text{ eV}$. The constants m and A depend on the size of the grains and the amount of contamination in the material.

- *Purple plague*: This failure mechanism arises because, with thermal compression bonding of the gold connection wires onto the aluminium bond flaps of an IC, the gold and aluminium diffuse after some time. This creates intermediate layers of diffusion products, namely purple $AuAl_2$ and white $AuAl$. However, these layers are stronger than the gold wire itself and thus do not create a brittle connection that might break off after some time. The reason that this interface gives way is the unequal diffusion rate of Al and Au and the influence of local contaminations on this diffusion rate. The contamination precipitates at the diffusion front and influences the local diffusion rate and gives rise to weak spots in the thermo-compression weld. This process is, of course, sensitive to temperature and has an effective activation energy $E_A/n \approx 1 \text{ eV}$.

2.3 Screening

The term screening will be used here for tests performed on all components with the purpose of removing the failed or potentially weak ones. The use of screening methods means therefore that one assumes the reliability of the components to be a given *fact* and (*ad hoc*) tries to improve it.

A rather obvious screening method is the automatic testing of all components for their specifications. Sometimes extra parameters can be measured which have proven to have a good correlation with the expected life of the component. Thus, for example, resistors with high excess noise and a 'large' non-linearity have been demonstrated to have a low life expectancy.

Beside the above quality testing, 'burn-in screens' can be used to make the weak components fail during the test. Only if the screening is ideal would all early failures be removed from the production batch. A burn-in screen is designed by accelerating the failure mechanisms coupled with the early failures, without affecting the life of the surviving components as much as possible.

In semiconductor components early failures are usually caused by production flaws such as bad wire-metallisation welds, scratches in the metallisation, bad soldering of the IC to the package base. Screens against these early failures are respectively mechanical shock tests, operating the IC with full supply voltage and possibly at a raised ambient temperature. An example of a strict screening test as used by a semiconductor manufacturer is shown in Figure 2.3.

30   *Deterministic Reliability*

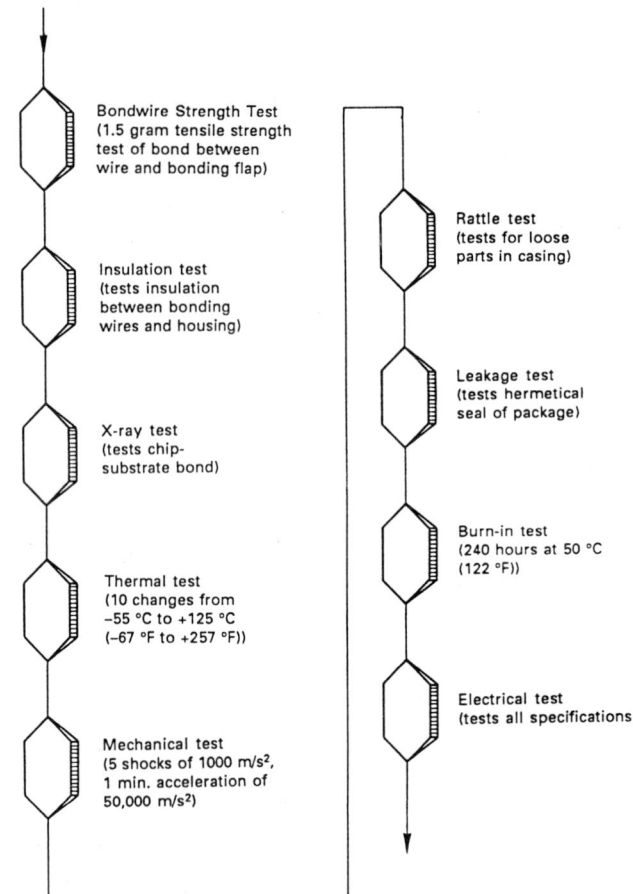

Figure 2.3 *Example of a strict screening test meant to activate potential failure mechanisms in multi-chip, hybrid integrated circuits for space travel. The failed components are removed.*
N.B.: *the order may not be changed, otherwise a subsequent test may cause a failure mechanism which will not be screened out any more.*

Problems

2.1. One wants to determine the MTTF (Mean Time To Failure) of a new monolithic digital-to-analog converter at 25 °C / 77 °F / 298 K. For that purpose 60 converters are operated for 1000 hours at 100 °C / 212 °F / 373 K and 60 converters for 1000 hours at 85 °C / 185 °F / 358 K.
At 100 °C the MTTF turned out to be 6.5×10^3 hours. At 85 °C this was 2.4×10^4 hours. Assume that the failure process behaves as a chemical process with a reaction rate:

$$Q = Q_0 \cdot e^{(-E_A/kT)} \quad \text{(Arrhenius)}.$$

What is the MTTF of this converter at 25 °C?
N.B.: *T is the absolute temperature in Kelvin, k is Boltzman's constant.*

2.2. Would electromigration also occur in heavy current equipment such as generators and transformers? Why (not)?

2.3. Do you know how galvanic corrosion of a ship's propeller is prevented (or reduced)?

2.4. What would the dominant failure mechanism of a car be if the oil was never changed? What would the failure mode be?

2.5. What is 'screening' and what what does one try to achieve with it?

2.6. 'Early failures' (infant mortalities) occur rather quickly directly after components have been taken into operation. Explain this phenomenon. Give a few examples.

3
Statistical Reliability

Statistical reliability engineering is too broad a subject to be completely covered in one book. In the choice we had to make, it was, for example, decided not to treat the measurement of reliability and its related topics such as: Statistical sampling methods and strategies, estimation theory, decision theory, and statistical data analysis. In this book only those subjects that lead to a direct insight into stochastic failure processes, reliable system configurations, calculation methods, reliability models, and maintenance strategies will be treated.

3.1 Nomenclature

The useful life of a system is assumed to end when the first failure occurs. We shall therefore define *failure* as the end of the ability of a system to realise the function required from that system.

- *Implicit restrictions.* It is assumed, in accordance with the reliability definition given in Section 1.1, that the system failure is not (also) caused by *misuse*. So the system is always used within the specified environment. It is further assumed that the failure is not *intermittent*, i.e. a failure does not correct itself without human intervention. Once failed, the system remains broken until repaired. In addition, it is assumed that the system *either functions correctly or is broken*; a state in between the two is not possible. Lastly, it is assumed that the *life variable* is the time t after delivery of the system by the manufacturer to the user. In Figure 3.1 it is indicated that a system with continuously variable parameters (for example, an analog electronic system) can show failures in two ways. The respective parameter x (the amplification) has a nominal value of x_0 and an allowed tolerance interval of $\pm 1\%$. A *degradation failure* is a failure where this tolerance interval is gradually exceeded, whereas a *catastrophic failure* is a failure where the function (amplification) suddenly falls away completely. In this book we shall not treat calculation methods specifically suitable for degeneration failures. The parameter drift in time viewed as a statistical process can often not be described analytically and one has to resort to 'Monte Carlo' simulations. These are statistical model experiments performed on a computer. Here we shall not digress on such special approaches to degeneration failures. So we hold the viewpoint: outside the tolerances means failed, within the tolerances means functioning correctly. Further, we assume that a system with degeneration failures breaks down at the first moment the tolerances are exceeded and remains broken, even if the parameter drifts back to within the tolerances.

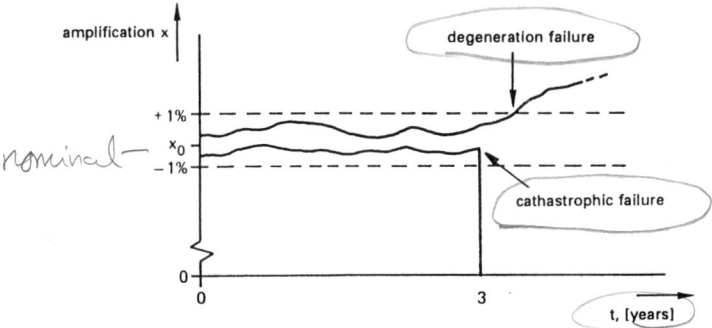

Figure 3.1 *Two different failures in an analog electronic amplifier; a total or catastrophic failure, for example caused by a short circuit which terminates the amplification function completely and a gradual and partial failure, for example caused by drift of the amplifier gain.*

■ *Distinct time intervals*

The useful life of a system is the time that elapses after delivery to the customer and before the system is finally discarded. This useful life span can be divided into a number of time intervals that are meaningful with regard to the operational quantities to be defined in the next section. For the most general system, that is a system which is also maintained, the time intervals are shown in Figure 3.2.

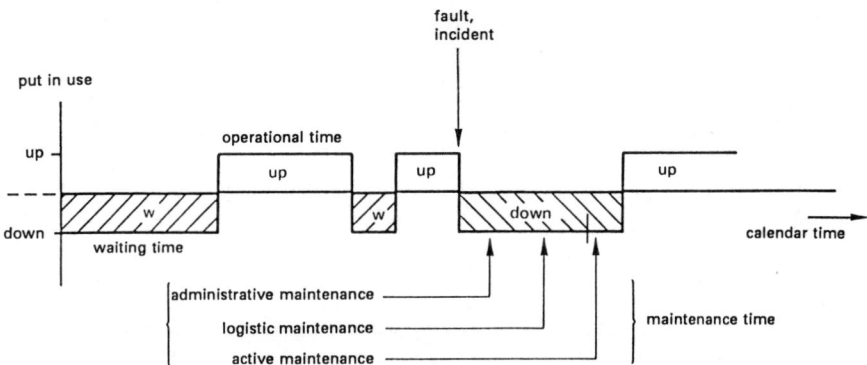

Figure 3.2 *Main subdivision of the calendar time of a maintained system.*

Two cases are to be distinguished here.

• The system is in use or 'up'.

During the so-called *up-time* the system functions correctly. During this time, the owner will utilise the system to the full (example: The activity of actually driving a car) or he will keep it ready for direct use (the car is idling while waiting at a traffic light).

• The system is not in use.

This time can be divided into two classes, depending on whether a system is down on purpose (car parked in front of the house) or unintentionally (car not available because of required maintenance). The first time is called *stand-by* or *waiting time*, the second time *maintenance* or *down time*. The maintenance time can be used for preventive maintenance (changing oil, lubrication, checking the pressure in the tyres), but also for corrective maintenance (fixing a failed ignition). The maintenance time can usually be divided into three classes.

The *active* maintenance time: The time during which a mechanic actually fixes the failure. The *logistic* time: The time needed to obtain the parts required for maintenance. The *administrative* time: The time elapsing while the cause of failure is established, parts are ordered, a mechanic is travelling to make a house call and such things.

Especially in large maintainable systems such as transport systems (airplanes, ships, etc.) a good definition of the various time intervals is of great importance for the operational quantities defined below. Compare: The time needed to man a moored vessel, supply it with fuel and food, start up the engines, etc. before it can sail (the total preparation time) is usually considered to be part of the active use time or 'up' time. The time intervals mentioned in Figure 3.2 mutually exclude one another; a repair performed during the waiting time interrupts this waiting time; the system is down.

3.2 Operational reliability quantities

For the practical use of a system, but also for the management of an organisation, service, etc., a number of operational quantities are important from a reliability viewpoint:

- *System effectiveness*
  This is the probability that one can, within a given time, complete a given task successfully with a system, if this system is used within a predefined environment. So, effectiveness is a broader concept than reliability. It is *task oriented*. It emphasises the purpose for which the system was acquired. If this task can be completed within a specified time, the system is effective for its intended purpose. This specified time also includes warm-up time, preparation time, time needed for small repairs, in addition to the time necessary for the actual task.
- *Reliability*
  The definition of reliability has already been given in Section 1.1. In short it is the probability $R(t)$ that a system functions correctly until and including the time t, provided it is not misused.
  This definition does not mention a task or a purpose, but requires correct functioning, i.e. in accordance with the specification. A system is only reliable if it functions correctly during the entire time interval $[0,t]$; reliability is the survival probability of this interval.
- *Mission reliability*
  This is an even more restricted concept. It is defined as: The probability that a system functions reliably during a certain task, if it is given that it functioned reliably at the beginning of the task. Example: On a transatlantic flight (task) a high mission reliability is of the greatest importance. The correct functioning of the airplane can (to a large extent) be tested before take-off. Once airborne, especially over the ocean, interim repair is virtually impossible.

N.B.: The above-mentioned operational quantities are nuances of the reliability concept. For convenience we shall set aside these nuances and only use the concept reliability $R(t)$.

Beside *non-maintained* systems, such as components, ICs, but also pocket calculators and similar things, one distinguishes *maintained* systems. These are systems in which the correct operation can be restored by human intervention. Human intervention is essential; if a system 'repairs itself' by built-in redundancy, for instance, it is not regarded as a maintained system. The (human) maintenance may be precautionary or preventive, but also restorative or corrective. Preventive maintenance is conducted according to a predetermined plan, for example after a certain interval of the life variable (every month, every 7500 km/4500 mi, every 1000 operational hours, etc.). It can also be conducted based on the system's condition. For example, changing the bearings or balancing a flywheel when a certain vibration level is exceeded.

Corrective maintenance, also called repair, only takes place after a system has failed.

For maintained systems there are two operational quantities which indicate the pace at which the maintenance progresses.

- *Maintainability*
  The maintainability of a system is defined as the probability that the system is restored to a correctly functioning state in a specified down-time interval.
  The maintainability concerns the total down time, so the time needed for administrative and logistic operations is relevant as well as the time needed for the actual corrective intervention.
- *Repairability*
  This is a concept equal to that above, on the understanding that it only pertains to the time needed for the actual corrective intervention (the so-called active maintenance time).

N.B.: We shall again neglect this nuance in the following and only use the concept maintainability $M(t)$.

Three other quantities define the continuity in time with which a maintained system is at the user's disposal.

- *Operational readiness*
  This is the probability, at any time, that a system either functions correctly or is suitable to function correctly, provided it is not misused. When defining 'suitable to function correctly', one may include a certain warning or warm-up time.
- *Availability*
  This is the probability that the system functions correctly or can function correctly at the moment t, provided it is not misused. The time considered here comprises only the total maintenance time (administrative, logistic, and active maintenance time).
- *Intrinsic availability*
  This definition is the same as that of availability with the exception that the considered time now only consists of the operational time (when the system functions correctly) and the active repair time. The other time elements are left out in the definition of this quantity.

N.B.: We shall again ignore these details and only use the concept availability $A(t)$.

The last important operational system quantity is one which concerns the consequences of a system failure for human beings, the environment or other (especially valuable) systems. For that reason one distinguishes failures of which the consequences are dangerous or harmful (*unsafe failures*) and failures for which this is not the case (*safe failures*).

- **Risk and safety**
  Risk is the probability that a system fails in such a way that humans, the environment, and/or other systems suffer dangerous or harmful consequences.
  N.B.: It does not follow from this definition that a small risk means that only a few failures occur. The system could fail often but mostly in a safe mode.
  A system is *safe* if the risk associated with the use of that system is considered 'acceptable'. The concept 'acceptable' is a subjective one, and therefore the concept 'safety' is also subjective. It will be treated in more detail in Section 8.4.

3.2.1. Derived quantities

In statistical reliability engineering the failure time t is a stochastic variable $\underline{t}$, whose distribution, the so-called *failure distribution* or *life distribution*, is given by:

$$F(t) = P(\underline{t} \leq t).$$

So $F(t)$ is the probability that the system perishes before or at the moment t. This probability of failure is also called *unreliability*. Since any system that can be realised physically will eventually break down; $F(\infty) = 1$. It is usually assumed that $F(0) = 0$, which means that all systems function correctly upon delivery to the customer, at the beginning of their useful life. However, for most calculations this assumption is not necessary. If $F(0) = \gamma$, $0 \leq \gamma \leq 1$, then $1 - \gamma$ is called the *yield* of the production or delivery process.

The probability of survival for the time interval $[0,t]$ is:

$$R(t) = P(\underline{t} > t) = 1 - F(t).$$

This relation is clear after the restrictions we made in Section 3.1; a system has either failed or it still functions correctly: The sum of the two probabilities is one. So the probability of survival is identical to the *reliability*.

The *failure probability density function* $f(t)$ associated with the failure distribution $F(t)$ is:

$$f(t) = \frac{dF(t)}{dt} = -\frac{dR(t)}{dt},$$

if $F(t)$ is differentiable. So the inverse also holds:

$$F(t) = \int_0^t f(t) \, dt,$$

and:

$$R(t) = \int_t^\infty f(t) \, dt,$$

Derived quantities 37

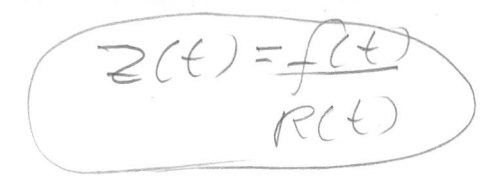

since:

$$\int_0^\infty f(t) = 1.$$

The conditional probability of the system failing in the time interval $(t, t + \Delta t]$ on the condition that it still functioned correctly at time t is the *hazard rate* $z(t)$. Heuristically the hazard rate can be determined as follows:

$$z(t) \cdot \Delta t = P(t < \tau \leq t + \Delta t \mid \tau > t) = \frac{P(t < \tau \leq t + \Delta t)}{P(\tau > t)}$$

$$= \frac{f(t)}{R(t)} \Delta t.$$

Therefore, the hazard rate has been defined as:

$$z(t) = \lim_{\Delta t \to 0} \frac{F(t + \Delta t) - F(t)}{\Delta t} \cdot \frac{1}{R(t)}$$

$$= \frac{dF(t)}{dt} \cdot \frac{1}{R(t)} = \frac{f(t)}{R(t)}$$

The relation between $R(t)$ and $z(t)$ is obtained by integrating the above expression:

$$\int_0^t z(t) \, dt = \int_0^t \frac{f(t)}{R(t)} \, dt = -\int_{R(0)}^{R(t)} \frac{1}{R(t)} \, dR(t) = -\ln \frac{R(t)}{R(0)}.$$

We can therefore also write:

$$R(t) = R(0) \exp\left[-\int_0^t z(t) \, dt\right]$$

The above-mentioned quantities $R(t)$, $F(t)$, $f(t)$ and $z(t)$ can be converted into each other. They therefore contain all information about the failure process of the systems under consideration.

N.B.: If $z(t)$ is independent of t, the resulting constant is denoted as $z(t) = \lambda$. This constant is called the *failure rate*, whereas the time-dependent $z(t)$ is called the *hazard rate*.

The mean life θ of a system is equal to the mathematical expectation of the stochastic variable t, so:

$$\theta = \int_0^\infty t f(t) \, dt.$$

The mean life is a global quantity that does not contain the time dependence information any more that $R(t)$, $F(t)$, $f(t)$ and $z(t)$ contain.

In reliability engineering, the (mean) time it takes for the system to fail is expressed in three ways. In non-maintained systems one speaks of MTTF (Mean Time To Failure). This is the same quantity as derived above, so MTTF $\equiv \theta$. In maintained systems two different mean times are to be distinguished; the mean time between two successive failures: MTBF (Mean Time Between Failures), and the mean time that passes until the first failure occurs: MTTFF (Mean Time To First Failure). The latter two will be discussed in Section 7.3.

The mean life of a non-maintained system, i.e. the MTTF, can be expressed using the reliability function $R(t)$ as follows:

$$\text{MTTF} \equiv \theta = \int_0^\infty t f(t)\, dt = -\int_0^\infty t \frac{dR(t)}{dt}\, dt = -\int_{R(0)}^{R(\infty)} t\, dR(t).$$

Integration by parts results in:

$$\text{MTTF} = \lim_{T \to \infty} \left[(-t R(t)) \Big|_0^T + \int_0^T R(t)\, dt \right]$$

If it is assumed that:

$$\lim_{T \to \infty} T R(t) = 0,$$

(which holds for virtually all practical $R(t)$ functions) the first term in the above expression disappears and it becomes simply:

$$\text{MTTF} = \int_0^\infty R(t)\, dt.$$

So it turns out that the mean life is equal to the area under the reliability function $R(t)$.

In summary, we may state the following about the derived quantities mentioned above:
- $F(t)$ is a failure distribution (a cumulative failure distribution).
- $R(t)$, on the other hand, is no distribution but a (mathematical) function: The reliability function.
- $f(t)$ is the failure probability density function associated with the distribution $F(t)$.
- The fact that $f(t)$ goes to zero when t goes to infinity might suggest that a system, if it is very old, does not fail any more. A more relevant quantity is therefore arrived at by making it a prerequisite that the system still has to function at t and then ask for the probability of the system perishing in the next time interval Δt. It is thus that one arrives at the hazard rate $z(t)$.
- $z(t)$ is far more sensitive to failures than $R(t)$. When $R(t)$ is still practically equal to one, the curve for $z(t)$ shows very clearly what will happen to $R(t)$ when t increases. Since:

$$z(t) = -\frac{dR(t)}{dt}\frac{1}{R(t)} = -\frac{dR(t)}{R(t)}\frac{1}{dt},$$

we see that $z(t)$ may also be regarded as the relative decrease in $R(t)$ per unit of time. The expression above may also be written as:

$$z(t) = -\frac{d\{\ln R(t)\}}{dt}.$$

So the hazard rate is also equal to the decrease in the logarithm of $R(t)$ per unit of time.

Problems

3.1. What is the definition of the *hazard rate*?

3.2. How is *maintainability* defined?

3.3. What is *availability*?

3.4. Prove that:

$$\lim_{t \to \infty} \int_0^t z(t)\, dt \to \infty.$$

3.5. The mean life of a non-repairable system is equal to the area under the $R(t)$ function, provided that .... Which condition must $R(t)$ meet to this end?

3.6. Prove that the hazard rate $z(t)$ of a system is equal to the relative decrease of the reliability $R(t)$ per unit of time.

3.7. Find an equation for the failure probability density function $f(t)$ expressed in terms of the hazard rate $z(t)$.

3.8. (a) Draw an arbitrary $R(t)$ function and name all the conditions which a valid reliability function has to meet.
(b) If it is further given that:

$$R(t) = \exp\left[-\int_0^t z(t)\, dt\right],$$

which conditions must a valid hazard rate function meet?

3.9. The hazard rate of a system is given by $z(t) = A + Bt$ for $t \geq 0$. Which conditions (dimensions, range, sign) must the constants A and B meet in order for this function $z(t)$ really to describe a valid hazard rate? Determine the failure probability density function $f(t)$ of this system.

3.10. Denote the sensitivity coefficient S of the reliability function $R(t)$ for the time t as S_t^R. This sensitivity coefficient is defined as:

$$S_t^R = \lim_{\Delta t \to 0} \frac{\Delta R(t)}{R(t)} \cdot \frac{1}{\Delta t}.$$

To which derived quantity: $F(t), f(t), z(t),$ or MTTF, is $-S_t^R$ equal (note the minus sign!)?

3.11. A piece of equipment has a constant hazard rate of $\lambda = 10^{-6}$/hour.
(a) What is the reliability for $t = 1000$ hours?
(b) If there are 10,000 of these instruments in use, what is the expected number that will fail within these 1000 hours?
(c) What is the reliability after a period t equal to the mean life (MTTF)?
(d) What is the probability of survival for another 1000 hours, if it is given that the instrument has survived the first 1000 hours?

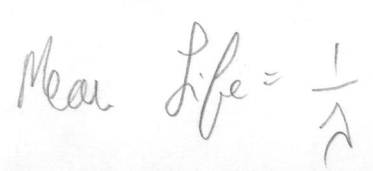

40 Statistical Reliability

3.12. When the hazard rate $z(t)$ of a system is constant and equal to λ in the interval $[t_1, t]$, for which $t \geq t_1$, prove that:

$$R(t) = R(t_1) \exp\{-\lambda(t-t_1)\}.$$

N.B.: $z(t)$ for $t < t_1$ is not known!

3.13. Somebody wants to take a 1000 km trip by car. The car has a constant failure rate of $\lambda = 10^{-4}$ per kilometre travelled. (So, here the life variable is the distance travelled.) What is the probability that the destination is reached without the car breaking down?

3.14. Of a system the following failure probability density function is given:

$$f(t) = at \exp\left(-\frac{a}{2} t^2\right).$$

(a) Determine $R(t)$ and $z(t)$ of this system.

At time $t = 0$ one has 5000 well-functioning systems, each with the above failure probability density function. Of these 5000 systems 4700 turn out to still function correctly after 10 hours.

(b) What is the approximate number of failures to be expected in the interval from 10 to 20 hours?

4
Statistical Failure of Components

A system can be regarded as consisting of components or subsystems which realise the required system functions by means of the mutual interactions between these components. A system component does not necessarily have to be an electronic component such as an IC, transistor, diode or resistor; it may, for example, also be a mechanical component such as a connector, a bolt, a lever, a ball-bearing, etc. Nor does a component have to be at the lowest complexity level in a system. A component itself may again consist of subsystems. Compare system components such as printed-circuit boards with electronic components mounted on them, plug-in units for larger equipment, generator units in a power station, trains in a transport system, pumps in a chemical installation, etc.

The way in which the considered subsystem is partitioned depends, among other things, on the reliability data available. If one does have data on a unit (or a module), but not on the components of which this unit consists, it is, of course, not useful to go into detail further than the level of this unit. If one has reliability data at any complexity level, or if one is prepared to collect this data, the detail is limited by the usefulness. In general, it is not useful to divide the system into parts which do not form a functional physical entity. If that were done one would have taken the reliability study unnecessarily far. In maintained systems it is only useful to go to the level of components that can be replaced completely (light bulbs, ICs in sockets, PC cards in connectors, etc.) In general, a more minute account than to the level of groups of related components which perform a complete (sub)task, is hardly ever necessary or useful (functional physical entity).

In the following we shall study in more detail the statistical failure behaviour of such distinguishable system entities (units, modules or components). The statistical failure behaviour of such system units is often determined from case histories (*ad hoc*) or life tests (*a priori*).

4.1 Failure distributions

Depending on the kind of failure mechanism that causes a system component or unit to fail, many different failure distributions are possible. These failure distributions are sometimes referred to as life distributions. We see a failure distribution as the distribution of a random variable (the failure time) which is arrived at in the following manner. Suppose we have a (theoretically) unlimited collection of components at our disposal (the total population) of which the failure pattern in time is recorded. From this time-dependent failure pattern, the failure distribution follows. In reality, one can only have the disposal of a limited number of components (a sample) which can only be observed for a limited period of time (truncated observation). This results in an *estimated* distribution. We will not go further into these statistical aspects.

In reliability engineering a certain failure distribution for a component is chosen for one or more of the following reasons.
- The dominant failure mechanism satisfies to most or all assumptions which underlie a certain statistical distribution.
- Reliability data is available. The choice is limited to the failure distributions which best fit these data (curve fitting).
- A simple distribution, well suitable for analytical computation, is chosen. A rough estimate of the reliability may be arrived at in this way. The estimate is more accurate as the two above-mentioned criteria for chosing a distribution are better fulfilled.

Besides characterising the distribution of component failures in time some of the same distributions may also serve as *repair time distributions*. They do not then describe the time-to-failure, but rather the time elapsing until the repair is successfully completed.

Once a failure distribution has been established, one can use specially printed probability paper on which the vertical and horizontal scales are arranged so that the distribution concerned shows up as a straight line. In this way, it is possible to discover relatively minor deviations from the expected failure pattern. One sees these deviations immediately when plotting the measurement results gathered in the field or in the laboratory.

A failure pattern often encountered in practice is that in which relatively many components fail in the beginning, subsequently relatively few in the period thereafter and, eventually, after a long time, relatively many again. Such a failure pattern is associated with a hazard rate function which is shaped like a bath tub (see Figure 4.1). This can be seen as follows: The hazard rate is indicative of the percentage of the surviving components which expire in the next unit of time. The infancy failure period, also called the *early failure* period, is the period in which the hazard rate is decreasing monotonically. These early failures are caused by weak components. The weak components are often the result of irregularities in the production process, craftmanship, etc. After the production process they can, in general, be (partially) removed by the application of screening (see Section 2.3).

The intermediate time interval where the hazard rate is approximately constant and during which relatively few failures occur, is called the *useful life* or *normal life period*. The failures are occurring randomly here. The end of the normal life period, where the hazard increases again, is determined by the beginning of *wear out*. This end-of-life period of a component is accompanied by ageing phenomena which will eventually result in failure.

There are no analytical expressions for the 'bath tub distribution'; however, by using the distributions which will be examined below its shape can be fairly well approximated.

N.B.: Many components (especially electronic ones) have such a long life that it has not (yet) been proven that they have a wear out area. On the other hand, mechanical components in particular sometimes have failure distributions in which the early failure period directly changes over into the ageing period without any intermediate 'useful life'. So it may certainly not be assumed that the 'bath tub distribution' is a form of higher wisdom governing the failure of all technical systems; there are many components that do not fail in accordance with this distribution. For that reason, the bath tub curve should only be used to discern three periods: The guarantee period (early failures), the

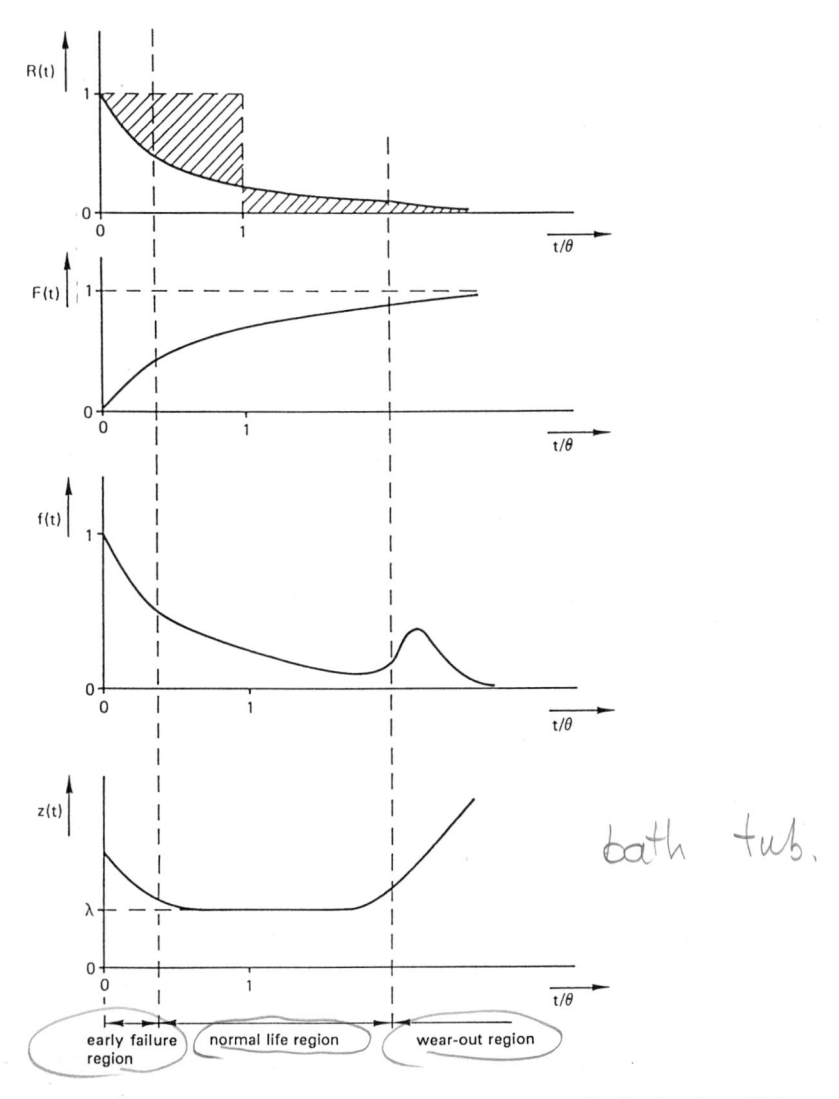

Figure 4.1 *Example of the so-called 'bath tub distribution'. This is a distribution for which no analytical expression exists. For clearness' sake the three separate areas of this distribution (early failures, normal life, and wear out) are shown exaggerated. The parameter θ is the mean life. The time axis scale has been normalised with respect to the mean life θ.*

operational period, and the replacement period (wear out failures). In these periods the hazard rate is decreasing, constant, and rising in time respectively.

We shall now give an example of the bath tub distribution from everyday life: The reliability of a human being. For convenience it is assumed that human beings function reliably as long as they are alive. Below, in Table 4.1, is listed the number of people $N(t)$ that are alive at age t in a certain country. It is usually assumed that these numbers do not change if they are taken over the life of the people (in time) rather than over the ensemble

44 Statistical Failure of Components

of all people. In other words, if the process of death as a stochastic process is ergodic, the reliability function $R(t)$, the failure distribution $F(t)$, the failure probability density function $f(t)$, and the hazard rate $z(t)$ can be derived from $N(t)$.

t	N(t)	t	N(t)	t	N(t)	t	N(t)
0	1 025 601	15	965 251	50	801751	85	75 122
1	1 120 532	20	957 832	55	760 251	90	20 102
2	985 250	25	938 275	60	675 325	95	3 100
3	981 125	30	930 010	65	574 288	99	122
4	980 526	35	900 210	70	451 989		
5	979 852	40	880 201	75	320 289		
10	975 815	45	851 157	80	180 326		

Table 4.1 Population composition of a small, fictitious country.

For $R(t)$ the following equation holds:

$$R(t) = N(t)/N(0).$$

For $F(t)$ the expression is:

$$F(t) = 1 - R(t) = \{N(0) - N(t)\}/N(0).$$

$f(t)$ is determined by:

$$f(t) = \frac{F(t+\Delta t) - F(t)}{\Delta t},$$

in which the table interval of 5 years is taken for Δt. Finally, $z(t)$ follows from:

$$z(t) = \frac{f(t)}{1 - F(t)}.$$

The results have been plotted in Figure 4.2. These results correlate very well with the 'maintenance' that has to be conducted on human beings to keep them alive (and living comfortably).

To illustrate this, part of the costs for this 'maintenance' are shown in Figure 4.3: The hospital costs. These costs are closely correlated to the hazard rate in Figure 4.2.

In Figure 4.2 one notices a period of a decreasing hazard rate, a period of a (practically) constant hazard rate and a period of an increasing hazard rate. These respectively define an early failure period (infant mortality) and an old age period (end-of-life period). In this regard the hazard rate is often called the 'force of mortality'.

We shall now discuss a number of important failure distributions. These distributions are also used to describe other reliability parameters, e.g. for repair distributions, waiting time distributions, and so on.

4.1.1 Negative-exponential distribution

The negative-exponential distribution is by far the most used failure and repair distribution in reliability engineering. It may be regarded as having been derived as follows. It is observed in a failure pattern that components fail independently of one another, at random moments, and also that the average number failing in a certain time interval $(t, t+\Delta t)$ does not depend on the time t but increases proportionally with Δt. Such a failure process constitutes a so-called *Poisson process*. The probability $P_n(\Delta t)$ that exactly n systems fail in a time interval Δt, if on average $\lambda \Delta t$ systems fail in that interval, is equal to:

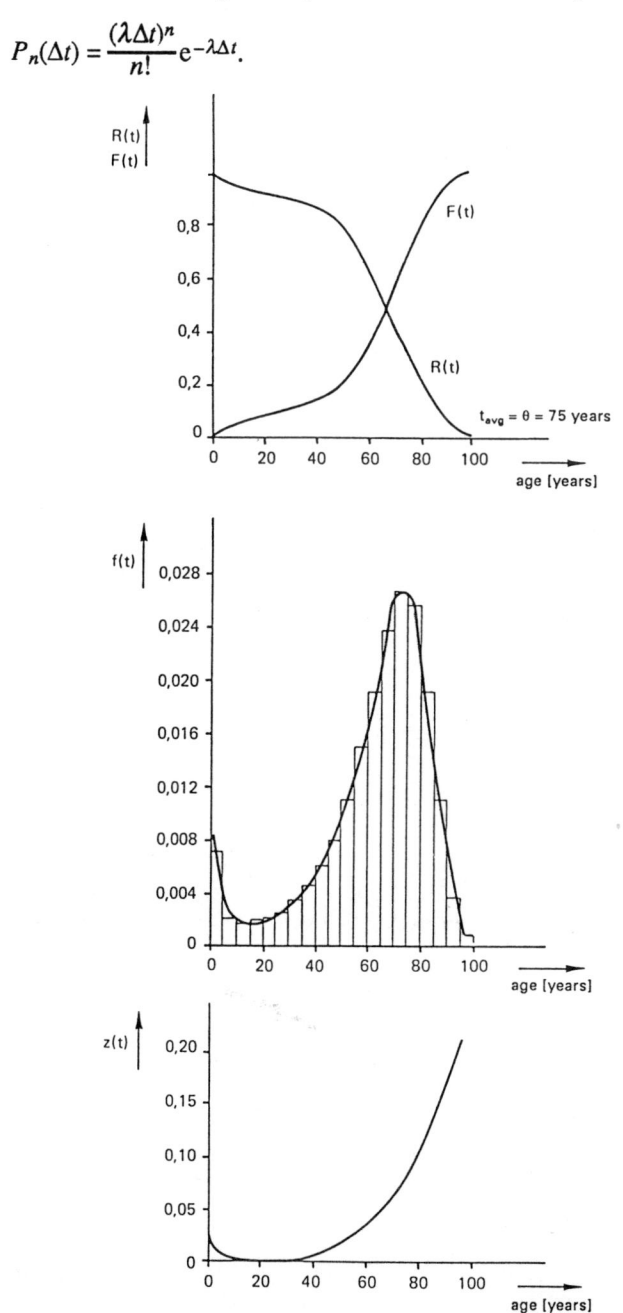

$$P_n(\Delta t) = \frac{(\lambda \Delta t)^n}{n!} e^{-\lambda \Delta t}.$$

Figure 4.2 Life distribution of the population of the country of Table 4.1.

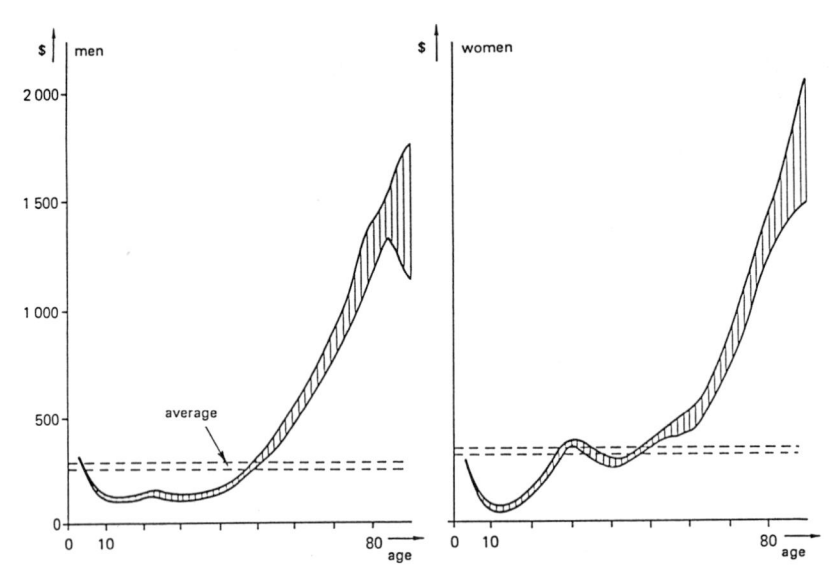

Figure 4.3 Hospital costs per year as a function of age of the Dutch population in 1981. The shaded area indicates the 75% confidence level.

The reliability definition requires that there be no failures in the time interval [0,t]. With $\Delta t = t$ and $n = 0$ this results in the following reliability function:

$$R(t) = P_0(t) = e^{-\lambda t}.$$

So the failure distribution is:

$$F(t) = 1 - e^{-\lambda t},$$

the failure probability density function:

$$f(t) = \lambda e^{-\lambda t},$$

the hazard rate:

$$z(t) = \lambda,$$

the mean life:

$$\theta = 1/\lambda,$$

the standard deviation of the mean life:

$$\sigma = 1/\lambda,$$

and, finally, the median life:

$$\theta_m = \frac{\ln 2}{\lambda}.$$

The negative-exponential distribution is not symmetric; i.e. it does not have a symmetric probability density function. With respect to the point $F(t) = 0.5$ the distribution is 'leaning to the right'. It holds for a distribution leaning to the right (as that of $F(t)$ in

Figure 4.4) that the *most frequently occurring value* (the top of the probability density function $f(t)$) is the most to the left. Then the *median value* follows with 50% of the area of the probability density function $f(t)$ on both sides of the median. The value most to the right (the largest) is the *mean value*. The opposite is true for a distribution that leans to the left.

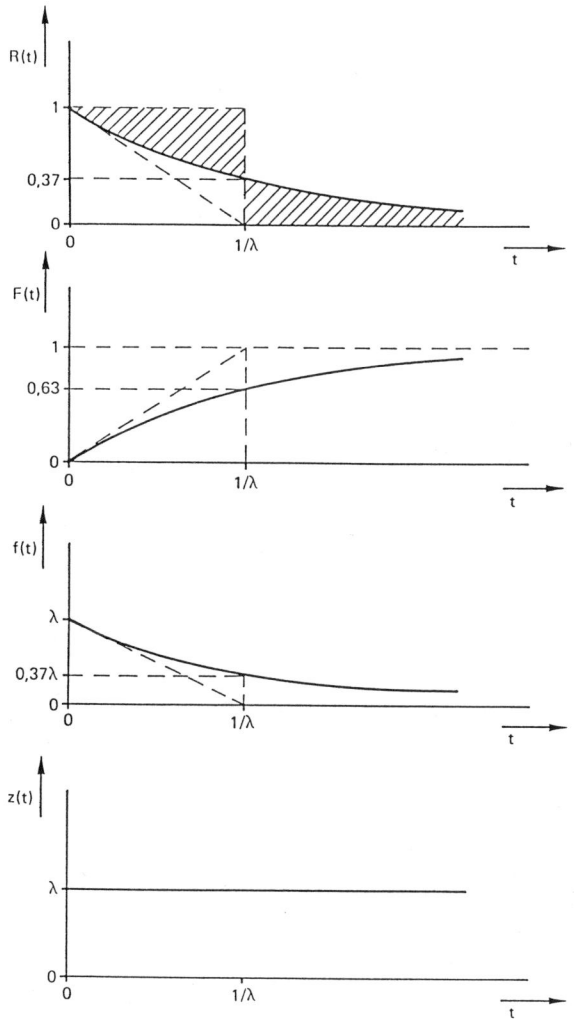

Figure 4.4 *The negative-exponential distribution. The mean life θ is equal to $1/\lambda$. Both cross-hatched areas are equally large.*

The Poisson failure and repair process is *without memory* for the time passed since the inception of the component's use, or the outset of the repair. In other words: If a system with such a failure process still functions correctly at a random time τ, then statistically speaking it will behave as a new system after time τ. This can be seen as follows: The remaining life Δt after the instant $t = \tau$ is determined by:

$$R(\tau + \Delta t \mid \underline{t} > \tau) = R(\tau + \Delta t)/R(\tau).$$

48 Statistical Failure of Components

For a negative-exponential distribution:

$$R(\tau + \Delta t) = R(\tau)R(\Delta t).$$

So the remaining life becomes:

$$R(\tau + \Delta t \mid \underline{t} > \tau) = R(\Delta t).$$

We see that the remaining life is independent of τ.

N.B.: It is often assumed in reliability engineering that components have a constant failure rate. This assumption should be regarded as a first-order approach to a time-dependent failure rate (resulting from lack of data) and not as a proven fact.

In Figure 4.4 the various quantities of a negative-exponential distribution are shown. Examples of a negative-exponentially distributed failure mechanism are: Punctures in tyres caused by nails or other sharp, loose objects versus the number of kilometres/miles travelled and the distribution of the impact of radioactive particles in semiconductor memories, in which the released charge causes so-called soft errors.

Summarising we may say that the negative-exponential distribution describes the time between the independent failures occurring randomly at a constant pace. Therefore, one can prove that the life distribution of complex systems with a wide variety of components with different life distributions, which do not necessarily have to be negative-exponential, is distributed negative-exponentially. This applies only if the system is not redundant.

4.1.2 Normal distribution

The normal distribution is one of the best known distributions from the theory of De Moivre probability. It is also called the Gauss distribution. This is not accurate, however, since Moivre was the first to use this distribution in 1733. Laplace, too, used it as early as 1774. Nevertheless, by an historic mistake, the distribution was named after Gauss who only introduced it in 1809.

A normal distribution describes components which only fail as a result of a wear process. For that reason the hazard rate (initially) increases monotonically. The component's life is distributed normally $N(\theta, \sigma)$ with a certain mean life θ and a standard deviation σ (see Figure 4.5).

The failure distribution is:

$$F(t) = \frac{1}{\sigma\sqrt{2\pi}} \int_{-\infty}^{t} e^{-\frac{(t-\theta)^2}{2\sigma^2}} dt$$

so:

$$f(t) = \frac{1}{\sigma\sqrt{2\pi}} e^{-\frac{(t-\theta)^2}{2\sigma^2}},$$

and:

$$z(t) = \frac{e^{-\frac{(t-\theta)^2}{2\sigma^2}}}{\int_t^\infty e^{-\frac{(t-\theta)^2}{2\sigma^2}} dt}.$$

This failure distribution may be regarded as resulting from the *sum* of a large number of stochastically independent failure causes, of which the individual failure distributions may

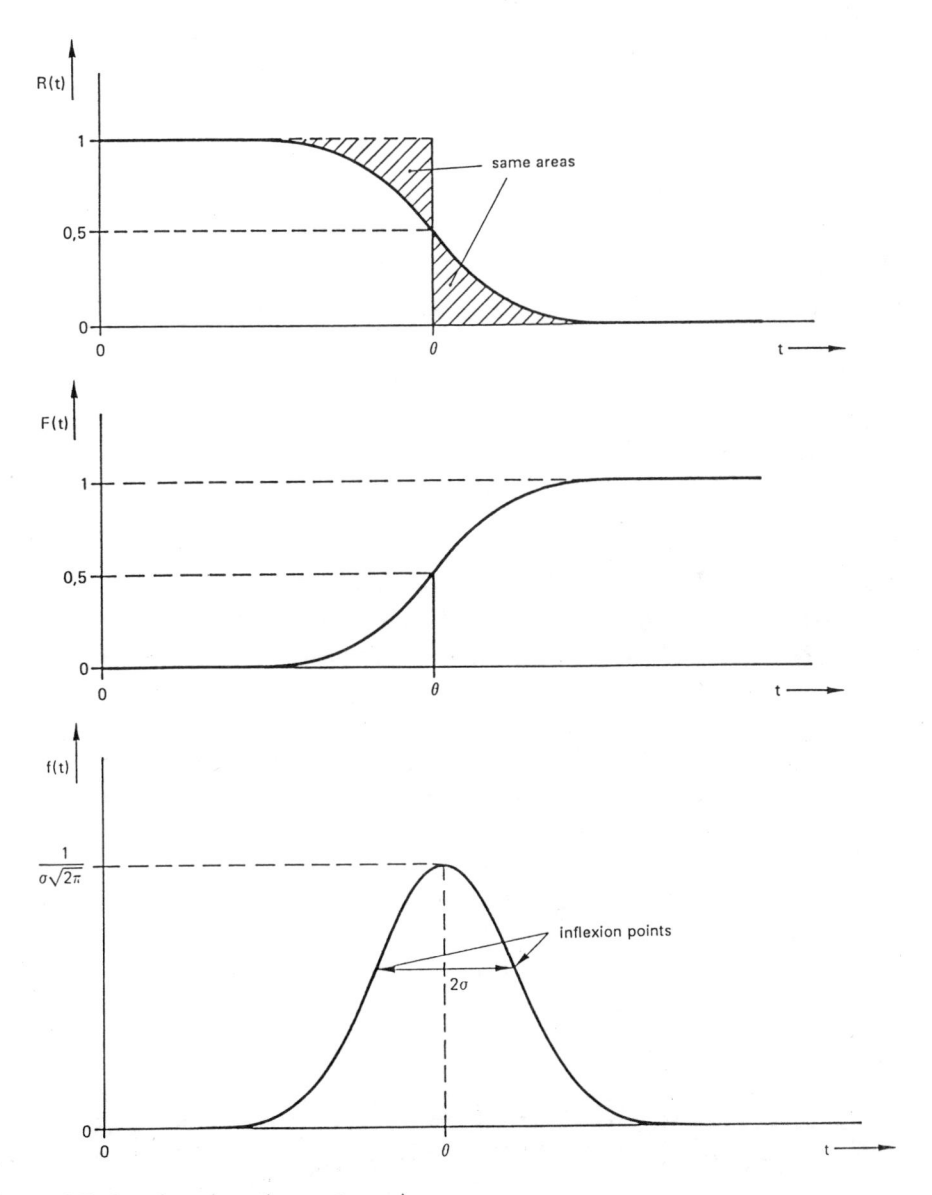

Figure 4.5 (continued on the next page)

Figure 4.5 *(continued)*

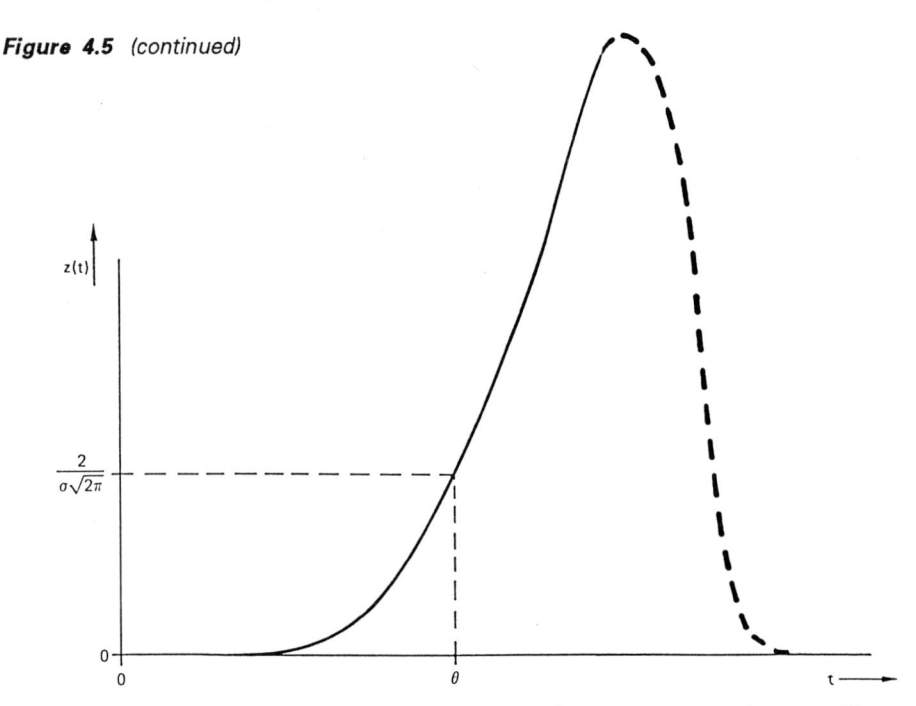

Figure 4.5 *The normal distribution for a standard deviation σ = 0.2 θ and a mean life θ.*

have an arbitrary, and thus also non-normal, shape (*central limit theorem*). A disadvantage of this distribution is that the distribution for negative time it is non-zero. When the variation coefficient θ/σ is large this is usually no problem, since the distribution is then very narrow: One can simply truncate the distribution at $t = 0$.

Examples of the use of the normal distribution in reliability engineering are the failure distribution of light bulbs, the distribution of parameter values of unselected analog electronic components, the distribution of the size of mechanical components, and so on.

4.1.3 Lognormal distribution

From the normal distribution one can derive other distribution families, which are more important in reliability engineering than the normal distribution itself. To that end it is assumed that, instead of the random variable $\underline{t}$, the random variable $\underline{g(t)}$ is distributed normally. $g(t)$ is here a single-valued explicit function of t. One of these transformed distributions is the logarithmic normal distribution which only exists for $t > 0$. This *lognormal failure distribution* can be simply computed with the transformation:

$$g(t) = \ln kt \qquad (t > 0),$$

resulting in the following two-parameter lognormal failure distribution:

$$F(t) = \frac{1}{\sigma'\sqrt{2\pi}} \int_0^t \frac{1}{t} \exp\left\{-\frac{(\ln kt - \theta')^2}{2(\sigma')^2}\right\} dt.$$

Here θ' and τ' are dimensionless and the constant k has the dimension T^{-1}. Further, $e^{\theta'}$ is the location parameter and σ' the shape parameter. The failure probability density is:

$$f(t) = \frac{1}{\sigma' t\sqrt{2\pi}} \exp\{-\frac{(\ln kt - \theta')^2}{2(\sigma')^2}\}.$$

The hazard rate can be calculated with the expressions from Section 3.2.1:

$$z(t) = \frac{f(t)}{1-F(t)}.$$

The various derived quantities are shown in Figure 4.6. A number of important values of the lognormal distribution are:

median value	$k^{-1} e^{\theta'}$,
mean value	$k^{-1} e^{(\theta' + (\sigma')^2/2)}$,
most frequent value	$k^{-1} e^{(\theta' - (\sigma')^2)}$,
variance	$k^{-2} e^{(2\theta' + (\sigma')^2)}(e^{(\sigma')^2} - 1)$.

In Figure 4.6 a number of lognormal distributions for different values of σ' have been plotted. In this figure we see that for small σ' the lognormal distribution turns into the normal distribution. The hazard rate $z(t)$ begins at zero, then rises to a peak, and subsequently falls off asymptotically to zero (for all values of σ'!).

N.B.: Just as the *sum* of a number of independent, random variables produces a normal distribution, the *product* of a number of independent random variables gives rise to a lognormal distribution.

The lognormal distribution represents stochastic variables of which the logarithm is normally distributed. This occurs, for example, if the considered stochastic variable is the product of a large number of independent stochastic variables.

The lognormal distribution represents the life distribution of many semiconductor components very well. It is also used fruitfully to describe the repair time distribution of technical systems. Also failure as a result of cracks, fissures, etc. caused by material fatigue can be described with this distribution.

4.1.4 Weibull distribution

The Weibull distribution may be considered as transformed version of the exponential distribution described in Section 4.1.1. The transformation is given by:

$$g(t) = t^c, \qquad (t > 0, c > 0).$$

Thus if $g(t)$ is distributed negative-exponentially, $\underline{t}$ has a Weibull distribution. c produces a change in the shape of the distribution.

The reliability function R(t) and the failure distribution now become:

$$R(t) = e^{-\alpha t^c}, \qquad (\alpha \geq 0),$$

$$F(t) = 1 - e^{-\alpha t^c}.$$

52 *Statistical Failure of Components*

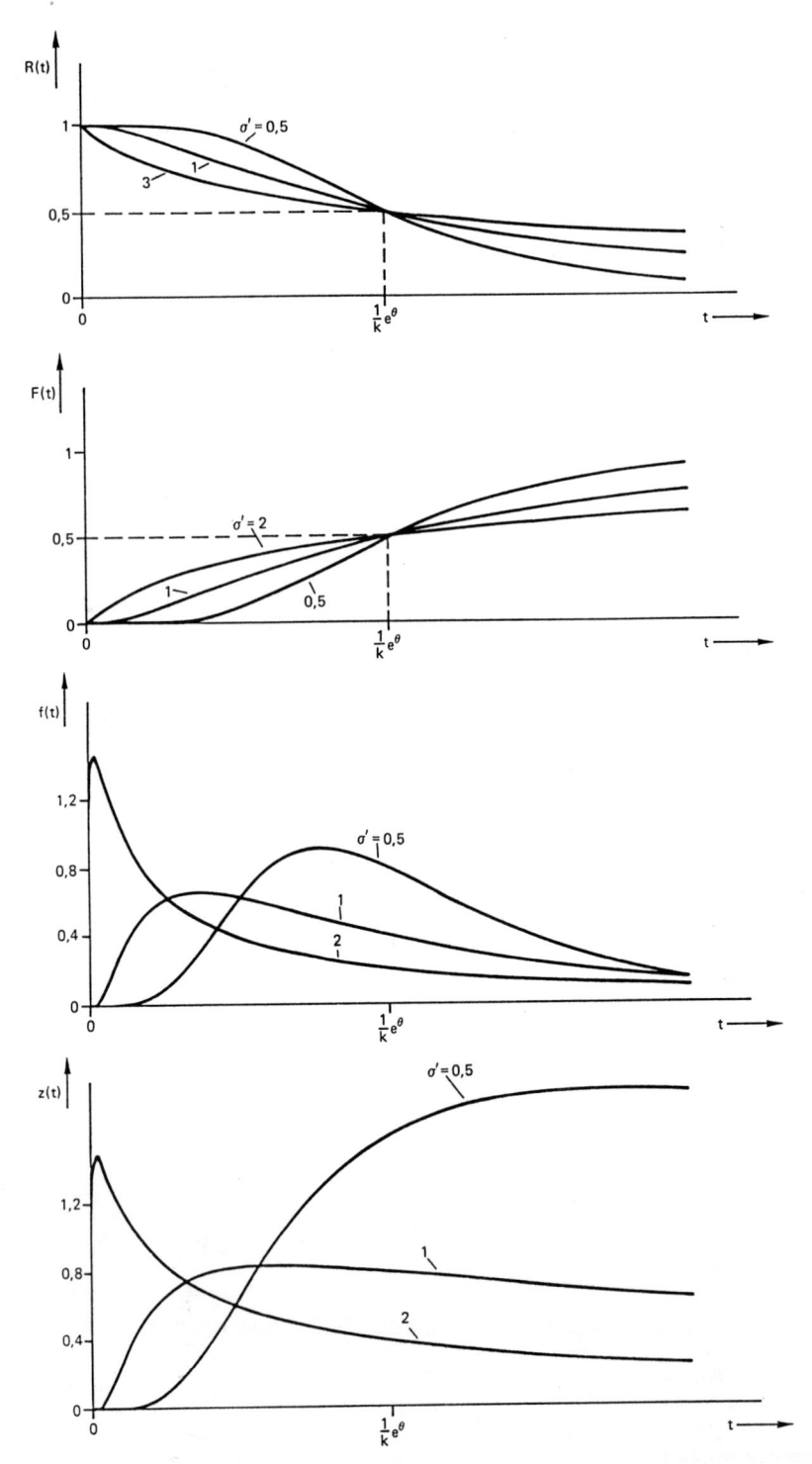

Figure 4.6 *The lognormal distribution. The median life is $k^{-1} e^{\theta}$, the location parameter is e^{θ}. The shape parameter is σ'*

The failure density becomes:

$$f(t) = \alpha c(t)^{c-1} e^{-\alpha t^c},$$

and the hazard rate becomes:

$$z(t) = \alpha c t^{c-1}.$$

So c is the shape parameter and α the location parameter.

In Figure 4.7 the Weibull distribution is shown for $\alpha = 1$. Note that for $t = 1$, the reliability $R(t) = e^{-1} \approx 0.37$, so becomes independent of c.

The Weibull distribution is named after the Swedish physicist Waloddi Weibull who used this distribution in 1939 to describe the strength of various materials.

The power of the Weibull distribution lies not so much in a certain theoretical failure model which might give rise to this distribution, but in its flexibility as a feasible approximation for empirically determined distributions. This can be seen from Figure 4.7 where we see that for $c = 1$ the negative-exponential distribution arises (in which α is equal to the hazard rate λ) and for $c > 1$ a failure probability density with a peak at:

$$t = \left(\frac{c-1}{\alpha c}\right)^{1/c}.$$

For $c > 1$ the hazard rate rises monotonically, corresponding to systems exhibiting wear-out (the right-hand part of the 'bath tub distribution' in Figure 4.1). For $0 < c < 1$ and $t \to \infty$ the hazard rate $z(t)$ asymptotically approaches zero, a characteristic of systems with early failures only (the far left part in the 'bath tub distribution' of Figure 4.1). This very flexibility makes the Weibull distribution suitable for curve fitting to experimental values.

The median life is given by:

$$\theta_m = \left(\frac{1}{\alpha} \ln 2\right)^{c-1}.$$

The mean life by:

$$\theta = \left(\frac{1}{\alpha}\right)^{c-1} \Gamma\left(\frac{1}{c} + 1\right),$$

in which the function $\Gamma(x)$ is the so-called gamma function (see Section 4.1.5). The standard deviation σ of the life variable is:

$$\sigma = \frac{1}{\alpha} \sqrt{\Gamma\left(\frac{2}{c} + 1\right) - \Gamma^2\left(\frac{1}{c} + 1\right)}.$$

4.1.5 Gamma distribution

The *gamma distribution* is determined by:

$$F(t) = \int_0^t \frac{\lambda}{\Gamma(c)} (\lambda t)^{c-1} e^{-\lambda t} dt, \qquad (c > 0, \lambda > 0, t \geq 0).$$

Here c is the shape parameter and λ the time scale parameter with the dimension T^{-1}. The *gamma function* $\Gamma(c)$ is further given by:

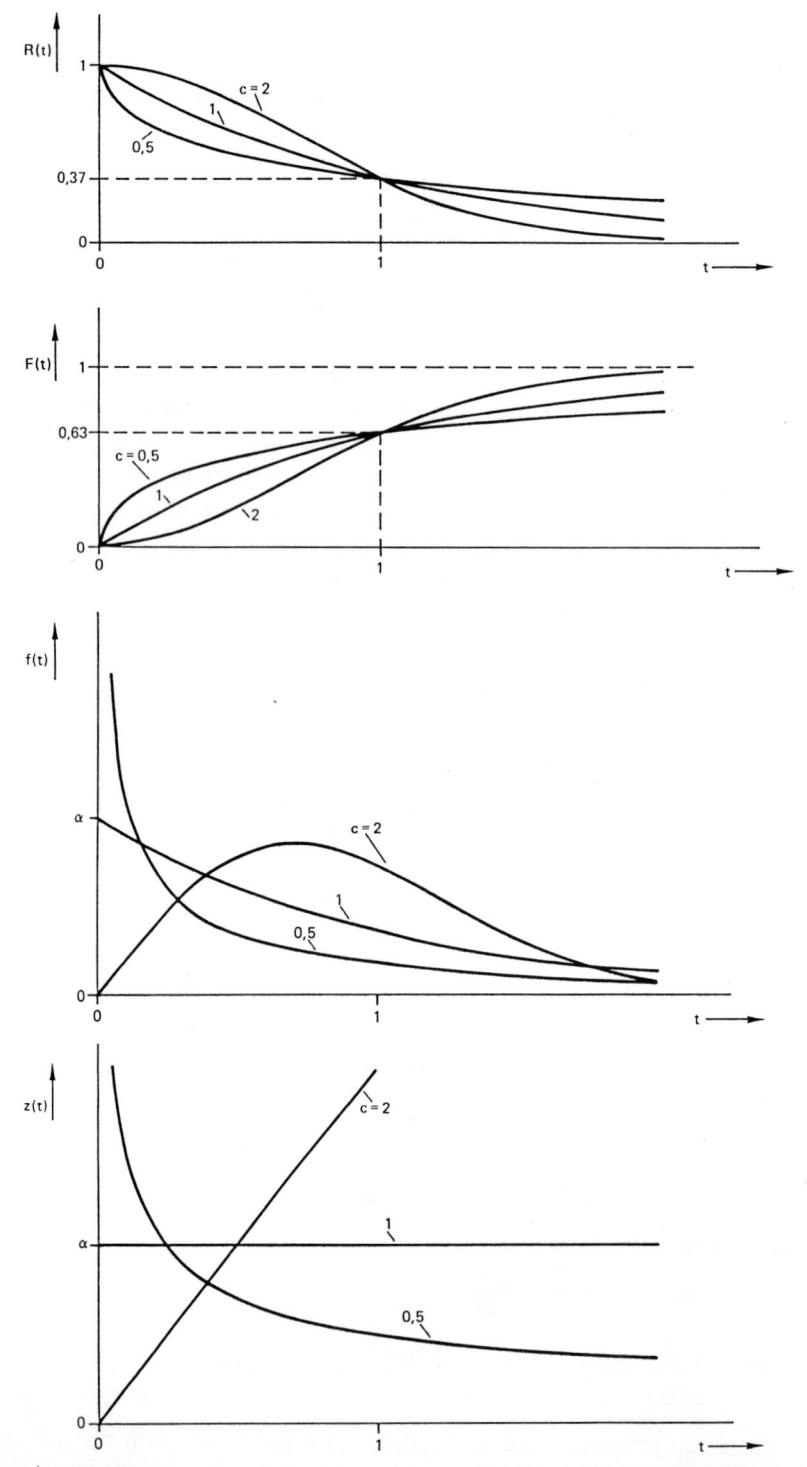

Figure 4.7 The Weibull distribution. If c = 1 a negative-exponential distribution results, in which α is equal to the hazard rate λ. If c = 2 the Raleigh distribution arises.

$$\Gamma(c) = \int_0^\infty \lambda(\lambda t)^{c-1} e^{-\lambda t} dt.$$

The gamma function has the following characteristics:

$$\Gamma(c) = (c-1)\, \Gamma(c-1),$$

$$\Gamma(c) = (c-1)!\ \text{for positive integers},$$

$$\Gamma(\tfrac{1}{2}) = \sqrt{\pi}.$$

So with:

$$f(t) = \frac{dF(t)}{dt},$$

the failure probability density is:

$$f(t) = \frac{\lambda}{\Gamma(c)} (\lambda t)^{c-1} e^{-\lambda t}.$$

The mean life is $\theta = c/\lambda$ and the standard deviation of the component's life is $\sigma = \sqrt{c/\lambda}$.

Notes
- If $c = 1$ the gamma function degenerates into a negative-exponential distribution.
- If c is a positive integer a distribution arises which is called Erlang's distribution.
- If c is an integer the integral expression for $F(t)$ can be solved by repeated integration by parts.
This results in:

$$F(t) = \sum_{i=c}^{\infty} \frac{(\lambda t)^i e^{-\lambda t}}{i!},$$

and hence:

$$R(t) = 1 - F(t) = \sum_{i=0}^{c-1} \frac{(\lambda t)^i e^{-\lambda t}}{i!}.$$

Figure 4.8 shows the gamma distribution for a number of different values of the shape parameter c.

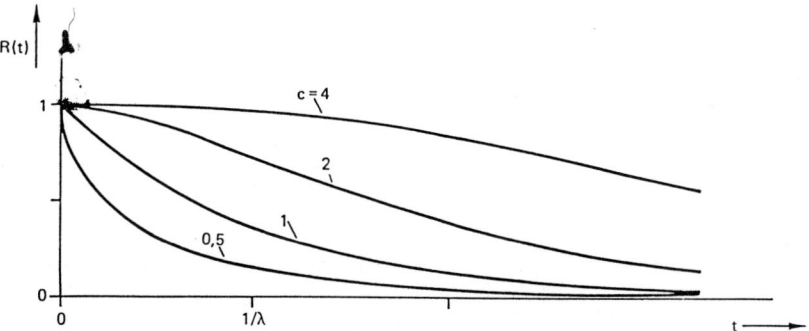

Figure 4.8 (continued on the next page)

Figure 4.8 *(continued)*

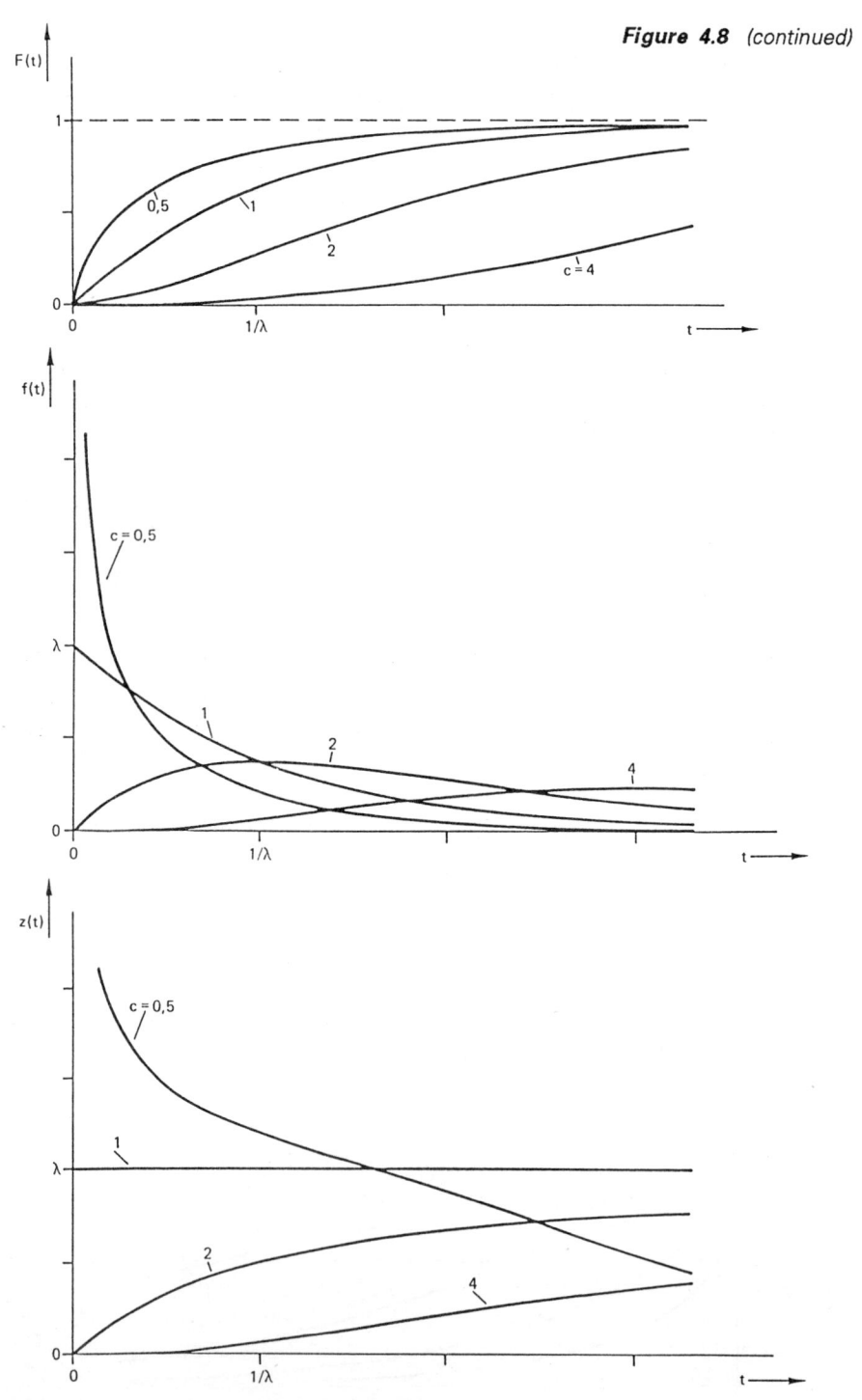

Figure 4.8 The gamma distribution and its different shapes caused by varying the shape parameter c.

The gamma distribution takes on shapes that much resemble those which the Weibull distribution can assume. For curve fitting, however, the Weibull distribution is more suitable. Certain shapes of the gamma distribution, particularly Erlang's distribution, are better suited for computation in an analytical form. Erlang's distribution is known from queueing theory: If the time to failure t_i is a stochastic variable with a negative-exponential distribution with failure rate λ, the distribution of the time t_m which passes until m units have failed is given by Erlang's distribution with parameters λ and $c = m$.

So we see that a certain class of gamma distributions (Erlang's distributions) occur in systems with (passive) redundancy in which there are m identical units in parallel. So, unlike the Weibull distribution, the gamma distribution has a theoretical background.

An example of an application of gamma distributions from maintenance engineering is the use of this distribution for the time elapsing between two calibrations of an instrument that has to be recalibrated after being used m times.

4.2 Life distribution measurements

Measuring failure distributions is a simple process, at least in principle. As we shall see below, however, there are a number of practical factors that throw a spanner in the works. In principle, measuring a life distribution is done as follows: A life test is performed and the mortality recorded (or the mortality occurring during the practical use of the product is recorded). From these data one can construct a failure distribution and determine the characteristic parameters of this distribution.

There is, however, a practical consideration. It is often known in advance which distribution one can expect. It is useful then to plot the results of the life test on special probability paper on which the axes are printed in such a way that the relevant distribution results in a straight line. If the cumulative failure frequency is plotted on this paper, the advantage is that large random deviations can be avoided (after all, all other reliability quantities contain the derivative of the failure distribution and are therefore very sensitive to deviations from the anticipated distribution). In addition, deviations from the anticipated distribution show up during the test.

In Section 4.2.1 we shall further address the availability of *a priori* knowledge about the failure distribution which is to be expected from the life test.

A fundamental problem occurring in life tests is that one does not want to sacrifice all available components. On the one hand, it is advantageous from an economic point of view to use as few components as possible in a life test, on the other hand, the accuracy (or rather confidence) of the measurement requires a sample size that is as large as possible. We shall discuss this dilemma further in Section 4.2.2.

The reliability of semiconductor components is relatively high. Consequently one should design very long life tests. For that reason artificially accelerated life tests are usually conducted. The question now arises: How large is the acceleration factor exactly. We shall discuss this problem in Section 4.2.3.

4.2.1 Failure distribution from life tests

It is to be recommended that some restraint be shown when determining the nature of a failure distribution exclusively based on the fact that this distribution fits the generated measurement data well. Since, in practice, life tests are performed on finite (and for economical reasons as small as possible) samples taken from the total production population, and since a life test has to last as short as possible (truncated life test), one has to live with non-zero confidence intervals around the measured life values. The smaller the sample and the shorter the duration of the test, the wider the confidence interval around the measured value. So the greater is the likelihood of drawing the wrong conclusions from a specific (coincidental) outcome of the measurement.

The approach is less arbitrary if a rational correlation can be proven between the failure distribution on the one hand and the physicochemical processes which cause failures on the other hand.

For example, one has been able to determine that most life tests of semiconductor components can be best represented by the lognormal distribution. This is explained by the fact that most failure processes in semiconductors are chemical or physicochemical, and are often caused by a number of random variables having a multiplicative effect on the shortening of the component's useful life.

As we have seen, the distribution of the product of a (large) number of (statistically independent) variables is lognormal, despite the distribution of the individual variables.

4.2.2 Confidence level of life tests

In the discussion of the failure distributions in Section 4.1 an infinite sample size was assumed. From an infinite sample size, so from the total population, we can determine the failure distribution exactly; we do not need to make estimates. With a sample of finite size, we can only estimate the distribution (or certain stochastic parameters thereof). The smaller the sample, the greater the inaccuracy of the estimate.

As a measure for the accuracy of statistical experimental results the concept *confidence level* was introduced. The confidence levels (lower limit and upper limit) mark off an interval around the estimated value of the considered statistical parameter.

If the probability of exceeding this interval is taken to be α, one may expect that for $100 \cdot \alpha$ % of the components (randomly drawn from the same population as the life test was conducted on) the actual life is outside the confidence interval around the estimated (measured) value. For the estimated value $\widehat{F}(t_1)$ of the (cumulative) failure distribution $F(t_1)$ at an arbitrary moment t_1, determined on the basis of a sample of size n we take:

$$\widehat{F}(t_1) = m(t_1)/n,$$

in which $m(t_1)$ is the number of failed specimens in the sample through the time t_1. The actual value is $F(t_1)$.

It can be simply seen that $\widehat{F}(t_1)$ is distributed binominally (the likelihood of $m(t_1)$ failures out of n with an average failure probability $F(t_1)$). So the lower limit F_l and the upper limit F_u for $F(t_1)$ with a confidence $1 - \alpha$ can be calculated from respectively:

$$\sum_{i=m(t_1)}^{n} \binom{n}{i} F_l^i (1-F_l)^{n-i} = \frac{1}{2}\alpha,$$

and:

$$\sum_{i=0}^{m(t_1)} \binom{n}{i} F_u^i (1-F_u)^{n-i} = \frac{1}{2}\alpha.$$

In Figure 4.9 the confidence limits are shown for $\alpha = 0.1$, 90% confidence level and different values of the sample size n.

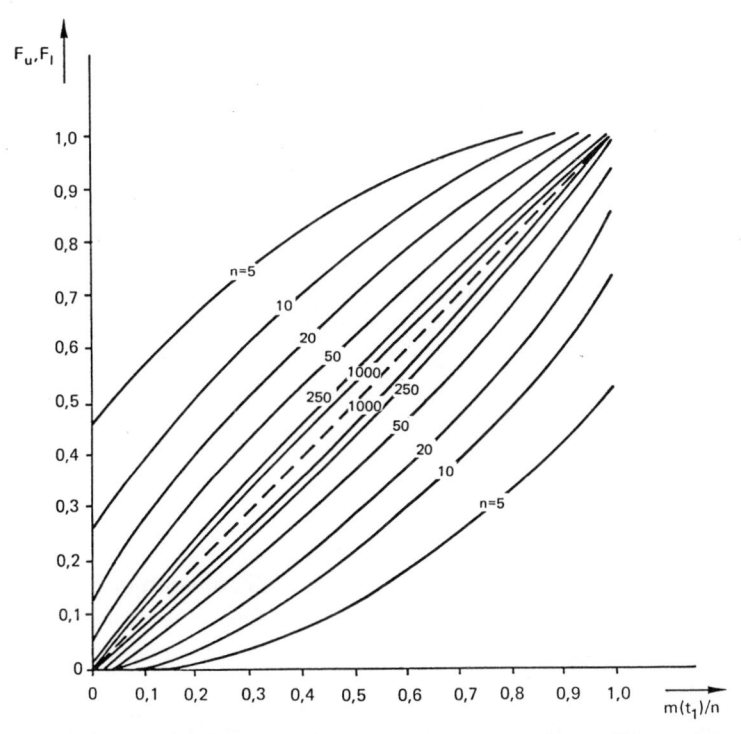

Figure 4.9 Upper limit F_u and lower limit F_l of the 90% confidence interval ($\alpha = 0.1$) around the estimated (measured) value $m(t_1)/n$ for a sample size n and a measured mortality $m(t_1)$.

In this way one can indicate a confidence interval around the measured failure distribution $F(t)$ (and, of course, also around the measured reliability $R(t)$). The actual distribution will lie within this interval with a probability $1 - \alpha$. It will be clear that there are infinitely many distributions one can draw through this confidence 'band'. Therefore one has to have extra information in order to decide for a certain distribution (with regard to this see also the thesis posed in Section 4.2.1).

If it is known in advance which distribution to expect, one may calculate estimators for the parameters of that distribution. Since more information is available now than in the above example (where we assumed the distribution to be unknown), one may expect that the 1−α confidence intervals around these estimated (or measured) values are narrower than without the distribution information. Below we shall give a number of these estimators and their confidence intervals for reference purposes.

By far the most important parameter of a failure distribution is the mean value (or mathematical expectation) of the life τ_g. We shall therefore give an estimator for the mean life with the corresponding confidence interval for several well-known distributions.

We assume that the sample size is n and all specimens are tested completely until all n have failed. The moments of failure are t_i ($i = 1,2,...,n$).

For the normal distribution we take as an estimator:

$$\hat{\tau}_g = \frac{1}{n} \sum_{i=1}^{n} t_i.$$

(This is a so-called *unbiased maximum likelihood estimator*). If the standard deviation σ of the normal distribution is known ahead of the game, the 1−α confidence interval around this estimator is:

$$\hat{\tau}_g \pm N_{1-\alpha/2} \, \sigma/\sqrt{\pi}.$$

Here $N_{1-\alpha/2}$ is the upper α/2 percentile of the standard normal distribution N.

If the standard deviation σ also has to be estimated we use as an estimator:

$$\hat{\sigma} = \sqrt{\frac{1}{n-1} \sum_{i=1}^{n} (t_i - \hat{\tau}_g)^2}.$$

We then get the confidence interval:

$$\tau_g \pm T_{1-\alpha/2,\, n-1} \, \hat{\sigma}/\sqrt{n}.$$

Here $T_{1-\alpha/2,\, n-1}$ is the upper α/2 percentile of the T distribution of Student with $n - 1$ degrees of freedom. This distribution can be found in tables in most statistical handbooks. It is named after the English statistician W.S. Gosset who published under the pseudonym Student.

The above-mentioned estimators and confidence intervals can also be used for the lognormal distribution if one realises that with $g(t) = \ln t$ the lognormally distributed variable t is transferred into the normally distributed variable $g(t)$. The estimator:

$$\hat{g} = \frac{1}{n} \sum_{i=1}^{n} \ln t_i,$$

for the mean value of the natural logarithm of the failure time t_i then has the same 1−α confidence interval as indicated above for $\hat{\tau}_g$. The estimator for the standard deviation of the resulting normal distribution is:

$$\sqrt{\frac{1}{n-1} \sum_{i=1}^{n} (\ln t_i - \hat{g})^2}.$$

The estimator for the mean life of a negative-exponential distribution is:

$$\hat{\tau}_g = \frac{1}{n}\sum_{i=1}^{n} t_i.$$

We may generalise the life test by no longer assuming that all n components fail but only m ($m \leq n$, truncated test) and that after a component has failed it is replaced by a new one. This results in the estimator:

$$\hat{\tau}_g = \frac{T}{m}.$$

Here T is the total accumulated test time of all components. It is the sum of all measured life spans plus the sum of the time that the components which have not yet failed have spent under the test conditions. The distribution of $\hat{\tau}_g$ is $\tau_g/2m$ times the chi-square distribution with $2m$ degrees of freedom. So the limits:

$$2m\,\hat{\tau}_g/\chi^2_{\alpha/2,2m}, \quad 2m\,\hat{\tau}_g/\chi^2_{1-\alpha/2,2m}$$

then determine the $1-\alpha$ confidence interval for $\hat{\tau}_g$.

For more complex distributions and for interrupted (truncated) life tests one often can no longer give an analytical expression for the confidence intervals. They have to be determined experimentally by means of Monte Carlo simulations.

4.2.3 Accelerated life tests

The useful life of most electronic components is so high that one cannot conduct feasible life tests without resorting to an *increased stress*. With such a stress increase, one also introduces a certain degree of arbitrariness; namely the choice of the stress parameter, the magnitude of the stress increase, the associated acceleration factor and the maximum allowable acceleration before another failure mechanism is triggered.

In general one may say that the stress parameters to be used (such as temperature, humidity, vibration level, and so on) have to reflect as well as possible the practical operational environment to which the component will later be subjected. This means that failure mechanisms that do not occur in practice are not allowed to occur during the increased stress experiment, or the other way around; no failure mechanisms that do occur in practice may be masked by the experiment. Further, the experiment has to be done in such a way that the acceleration factor associated with the increased stress is known. Since the acceleration factor varies for different stress parameters and since it also depends on the nature of the (dominant) failure mechanism that is accelerated by it, the accelerated life tests are usually designed with only one stressed parameter.

Concerning the magnitude of the acceleration factor, the smaller the acceleration factor (so the closer the stress in the experiment approaches the operational stress) the more reliable the results of the experiment. The stress increase may on no account be so high that it triggers other dominant failure mechanisms.

As a rule, the allowable range of stress during an accelerated life test is determined by a so-called *step-stress test*. This is an experiment in which the stress quantity is not kept constant, as in an accelerated life test, but in which the stress level is increased in steps.

62 Statistical Failure of Components

This increase continues until a new failure mechanism is activated that is highly unlikely to occur in normal use (for example, the melting or distortion of a plastic IC-casing from high temperatures).

Once the 'safe area' of a stress quantity is known one may determine the magnitude of the acceleration factor as a function of the intensity of the stress quantity by physical observation of the failure mechanism, or by a number of accelerated life tests with different stress levels. The most reliable method (but also the most expensive and time-consuming) is to do both.

In an accelerated life test (with a constant, increased stress level) one must not only ensure that the stress is not chosen so high that it triggers another (dominant) failure mechanism but one must also take care (in theory at least) that the shape of the failure distribution does not change with these applied higher stress levels. After all, if that were the case, one could no longer use the mean life as statistical parameter for the characterisation of the failure distribution. In general this requirement has been met if the stress level in the test is kept much lower than the level at which other failure mechanisms become dominant.

In Section 2.2 we have already discussed a number of life-shortening stressors which follow Arrhenius' model. In principle any stress that the product will also experience in its later operational environment qualifies for accelerated stress testing, e.g. load, speed, larger periods between maintenance (lubrication), shocks, vibration, electric discharge (lightning stroke in vehicles, ships, and airplanes), and so on.

Problems

4.1. A failure distribution of a (passive) redundant system consisting of two units with a constant failure rate λ is described by a gamma distribution with a shape factor c which equals 2.
  (a) Determine the failure rate $z(t)$ of this system.
  (b) What is the maximum failure rate of this system and explain the value found.

4.2. In a truncated life test 100 power transistors, which are biased at twice the nominal power dissipation, are tested for 100 hours. After these 100 hours it turns out that none of the 100 transistors has failed yet. If we may assume that the components have a constant failure rate, how large is this failure rate with a 90 % confidence interval?

4.3. What can be concluded from a decreasing hazard rate?

4.4. Prove that for the hazard rate of a component with a normally distributed life holds:
$$\lim_{t \to \infty} z(t) = 0.$$

4.5. Wat are accelerated life experiments and under which conditions do they yield useful results?

5
Reliability Models

This chapter will discuss a number of frequently used reliability models. A reliability model is determined by a number of premises about the failure of system components. Taken together, these premises form the model on which the reliability computation is based.

In Section 3.1 we have already mentioned a number of premises. This was necessary, because otherwise concepts such as $R(t)$, $F(t)$, $f(t)$ and $z(t)$ could not be defined without ambiguity. These premises are:

1. A component is correct or has failed; for degeneration failures the component is faulty from the first time it exceeds the tolerances.

2. Once failed, a component remains defective (until maintenance is completed); no intermittent failures occur.

3. The life variable is the calendar time (see Section 1.1).

The reliability models that will be discussed in this chapter also include other assumptions, still to be introduced, in addition to those above.

N.B.: One can also do without the above-mentioned assumptions, but the theory will become complex. Just think of omitting (2) which has the result that the $R(t)$ function will no longer be a single-valued function. The reliability function would increase if the likelihood increases that more components will function again.

In addition to reliability models we also come across other models in reliability engineering. Examples are *schematic models* (wiring diagrams, component layouts, and the like) and *functional models*. A schematic model contains the highest useful degree of detail with the component parts, parameter values and other useful data (see Figure 5.1). A functional model is, in fact, a drawing indicating how a system is built up from subsystems. In such a functional model the information and energy flows are clearly reflected (see Figure 5.2).

5.1 Catastrophic failure model

The catastrophic failure model is the simplest failure model, in which, besides the above-mentioned premises, it is also assumed that if a component fails it does not matter how it fails. So it is, in fact, assumed that the component has only one single distinguishable failure mode (*single mode failure*). Representing failures by means of a model now becomes very simple. As indicated in Figure 5.3, the component (or the relevant part of

64 *Reliability Models*

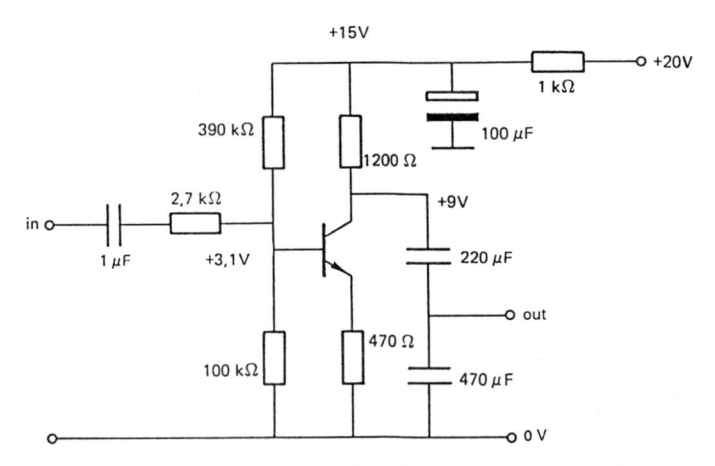

Figure 5.1 Example of a schematic model of part of an electronic circuit, usually called 'schematic' for short.

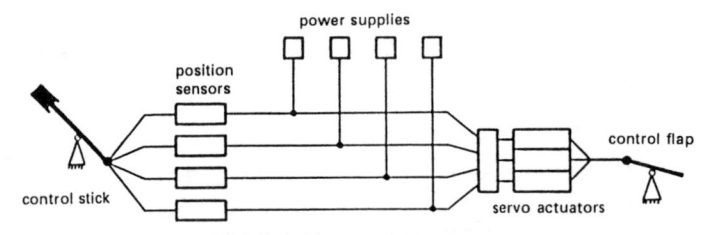

Figure 5.2 Functional model of a 'fly-by-wire' system as used in modern airplanes (fourfold redundant power supplies, sensors and cabling).

the system) may be replaced by a black box in which there is a hypothetical switch.

If the component is functioning correctly this switch is closed and the *state* of the component is defined as '1'. The state is '0' if the component has failed; the switch is then open. In the corresponding reliability model there must always be at least one path of closed switches in oder for the system to function. Compare Figures 5.3b and 5.3c. The state of a system is therefore fully determined by the state of its component parts. One can now simply draw up a *truth table* with the state of the components as binary input variables and the state of the system as the binary output variable. The structure of the reliability model determines the relationship between these variables, which can be described by Boolean algebra. Based on this is a particular calculation method for the reliability of systems, which will discussed in Section 6.9.2.

To demonstrate the limitations of catastrophic failure models we use Figure 5.3c for reference. For the RC-member depicted a certain nominal value for R and C will be specified, together with certain tolerance limits if it is to function correctly as a filter. It is assumed that the resistor exhibits the failure 'open' if it exceeds the upper tolerance limit, while it shows the failure 'short-circuited' if the lower tolerance limit is exceeded. Subsequently, we assume that both failure modes, as far as the reliability is concerned, do not need to be distinguished. We therefore label them jointly as: 'Broken resistor' with a

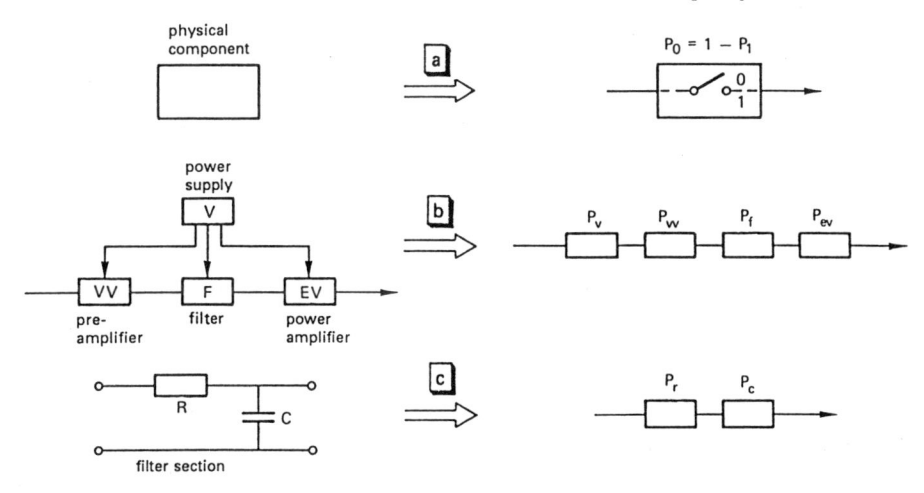

Figure 5.3 Catastrophic failure model. (a) The state of the component is 1 if it is in good working order and 0 if it is defective. (b) The illustrated functional model of an electronic system has a series or chain connection for a reliability model. (c) The illustrated RC-member does not function any more if the resistor is open or short-circuited (probability P_r) or if the capacitor is open or short-circuited (likelihood P_c).

failure probability P_r. For the capacitor the same assumptions are made. We see that a catastrophic failure model contains no information as to how a component fails. It is not always possible to lump together the transgression of a component parameter beyond an upper and a lower tolerance limit. This can be easily seen from Figure 5.4. The two amplifier stages have to be insulated for DC voltage (this is to keep the biasing of the transistors correct). For that reason, two AC signal-coupling capacitors are used in series. One of the capacitors may exhibit an electrical breakdown of its dielectric (short circuit) without influencing the operation of the circuit (redundancy). If only this failure mode is taken into consideration the reliability model consists of two branches in parallel, each with a failure probability P_k. If only the 'open capacitor' failure mode is considered the reliability model consists of two branches in series, both with failure probability P_o.

N.B.: A series circuit or a parallel circuit of components in a schematic diagram or functional model therefore does not necessarily result in a series or parallel circuit in the associated reliability model!

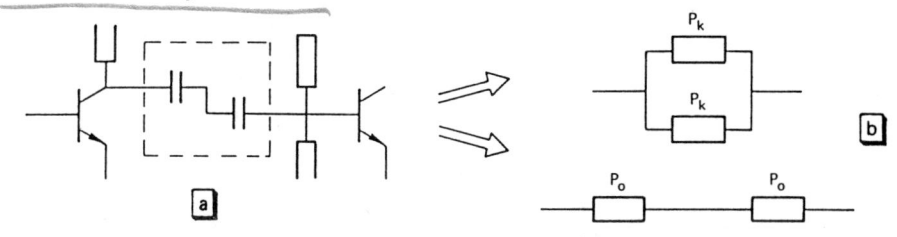

Figure 5.4 (a) Redundant coupling capacitors between two amplifier stages with a different DC voltage level. (b) Failure model for, respectively, short-circuits or open failures only. P_k is the probability of a short-circuit in a capacitor, P_o the probability of an open capacitor.

66   Reliability Models

In reality, of course, both failure modes occur. Components like these exhibiting *multi-mode failures* cannot be described with the simple catastrophic failure model. It is tempting to do this anyway, by putting both reliability models from Figure 5.4b in series. However, this is not a correct representation of the reality because the failures which the various blocks now reflect are stochastically dependent (a short-circuit in a capacitor precludes an open failure). The simplicity of the modelling for components with only one failure mode is lost in the case of multi-mode failures. In the latter case, by listing all combinations of acceptable failures (i.e. those failures that do not cause a system breakdown) and their probability, one can arrive at the reliability of the total system. We shall now discuss an example of such a system with multi-mode failures.

Example

In addition to the state 'correct operation' (state x) a semiconductor diode has two failure modes, namely open (state x_o) and short-circuited (state x_s). It is assumed that these states are mutually exclusive (disjunct). In formula form:

$$P(x \cup x_o \cup x_s) = P(x) + P(x_o) + P(x_s) = 1.$$

The reliability of a diode is:

$$R = P(x) = 1 - P(x_o \cup x_s) = 1 - P(x_o) - P(x_s).$$

If two diodes are connected in series, the combination will fail if one of the diodes is open or if both are short-circuited. These events are given by:

$$x_1 x_{2o} + x_{1o} x_2 + x_{1o} x_{2s} + x_{1o} x_{2o} + x_{1s} x_{2o} + x_{1s} x_{2s}.$$

As a matter of convenience the symbol $\cup$ is replaced by an arithmetical plus sign and the symbol $\cap$ is dropped, so that a product results. Each of the combinations of events in the above expression leads to a separation of the output from the input in the corresponding (multi-mode) reliability model. Each combination is therefore called a *cut*. The complete set of all cuts forms the *cut set* (see also Section 6.9.2). In terms of this cut set the reliability of the two diodes in series is:

$$R = 1 - P(x_1 x_{2o} + x_{1o} x_2 + x_{1o} x_{2s} + x_{1o} x_{2o} + x_{1s} x_{2o} + x_{1s} x_{2s}).$$

The set of all correct diode state combinations, the so-called *tie set*, is:

$$x_1 x_2 + x_1 x_{2s} + x_2 x_{1s}.$$

Consequently, in terms of the tie set, the reliability of two diodes in series is:

$$R = P(x_1 x_2 + x_1 x_{2s} + x_2 x_{1s}).$$

If the failures in both diodes are stochastically independent, and if it is assumed that: $P(x) = p$ and $P(x_o) = q_o$ then $P(x_s) = 1 - p - q_o = q_s$ and:

$$R = p^2 + 2pq_s.$$

In the case where two diodes are connected in parallel, the shortest expression can be obtained by forming the tie set:

$$x_1 x_2 + x_1 x_{2o} + x_{1o} x_2.$$

With the above assumptions the reliability is:

$$R = P(x_1x_2 + x_1x_{2o} + x_{1o}x_2) = p^2 + 2pq_o.$$

N.B.: The terms 'cut', 'tie' and 'set' are from graph theory. These terms are convenient because the reliability models of Figures 5.3 and 5.4 can also be represented as graphs. We shall return to this issue in Section 6.9.2.

N.B.: The terms 'disjunct' and 'stochastically independent' must be kept well apart. Two events are disjunct if they exclude one another and therefore cannot occur simultaneously. An example is electric components with two connection wires which cannot be open and short-circuited at the same time. (This is not the case with components with three or more connection wires, such as transistors, ICs and the like!) For these disjunct events it holds that the probability of the occurrence of the union is equal to the sum of the probabilities of each of the individual events:

$$P(x_o \cup x_s) = P(x_o) + P(x_s).$$

This can be seen quite easily because for two events x_o and x_s it generally holds that:

$$P(x_o \cup x_s) = P(x_o) + P(x_s) - P(x_o \cap x_s).$$

For disjunct events the intersection $x_o \cap x_s$ is empty, so $P(x_o \cap x_s) = 0$.

Two events are independent in the stochastical sense if the occurrence (or non-occurrence) of one event does not influence the probability of occurrence (or non-occurrence) of the other event. Such an influence does exist if, for example, the failures in separate components would have a common cause. Compare the development of condensation on the components of an airplane transceiver that takes off in the tropics (from hot and humid, to the intense cold of high altitudes). If there is additional chemical contamination (solder resin residue) electrochemical corrosion of the components will result. This common cause may eventually lead to the failure of many components. Another type of dependent failures is created when one failure brings about the other or accelerates the failure process in the other component. Compare a broken component (ball bearing) running hot resulting in the melting of electric insulation near by and thus in a short-circuit in adjacent wiring.

For independent events it holds that the probability of the intersection of events is equal to the product of the probabilities of the individual events:

$$P(x_1 \cap x_2) = P(x_1)P(x_2).$$

Since it generally holds that:

$$P(x_1 \cap x_2) = P(x_1 \mid x_2) P(x_2),$$

and

$$P(x_1 \cap x_2) = P(x_2 \cap x_1),$$

it can easily be seen that the following equation also holds:

$$P(x_1 \cap x_2) = P(x_2 \mid x_1) P(x_1).$$

68   Reliability Models

If the events x_1 and x_2 are independent, the conditional probability $P(x_2 | x_1)$ is equal to the probability $P(x_2)$. Similarly $P(x_1 | x_2) = P(x_1)$. Thus for the independent events x_1 and x_2 it holds that:

$$P(x_1 \cap x_2) = P(x_1) P(x_2).$$

5.2 Stress-strength model

In the *stress-strength model* used in reliability engineering it is assumed that a component fails only if the magnitude of the *stress* $\underline{x}$ becomes larger than the component's *strength* $\underline{y}$. The stress can be mechanical, electrical, or of some other kind; for example, thermal stress. As we have already seen in Chapter 2, the stress is the sum of the *internal stress* of the component created by its operational use and the *external stress* imposed by the environment in which the component is placed.

Here the stress $\underline{x}$ is assumed to be a stochastic variable. The reliability of a component is now given by:

$$R = P(\underline{x} \leq y) = 1 - P(\underline{x} > y).$$

If the probability density function of the stochastic variable $\underline{x}$ is known and is given by $g(x)$, the reliability becomes:

$$R = \int_{\infty}^{y} g(x)dx = 1 - \int_{y}^{\infty} g(x)dx.$$

So the shaded part of $g(x)$ in Figure 5.5a indicates the failure region. It will be clear that the strength $\underline{y}$, and thus the 'resistance' of a component to breakdown, varies from component to component; i.e. $\underline{y}$, too, has its own probability density function $f(y)$. Since:

$$f(y) = \lim_{\Delta y \to 0} \frac{F(y + \Delta y) - F(y)}{\Delta y},$$

in which:

$$F(y) = P(\underline{y} \leq y),$$

we may write for the probability that $\underline{y}$ lies in the interval $(y, y + dy]$:

$$P(y < \underline{y} \leq y + dy) = f(y)dy.$$

It is further assumed that the stress $\underline{x}$ and the strength $\underline{y}$ are stochastically independent. The probability that the strength $\underline{y}$ is in the interval $(y, y + dy]$ and that the stress $\underline{x}$ does not exceed this strength becomes:

$$f(y)dy \int_{-\infty}^{y} g(x)d(x).$$

The reliability of the considered components is the probability that the stress $\underline{x}$ does not exceed strength $\underline{y}$ for all possible values $\underline{y}$ can assume, so:

$$R = \int_{-\infty}^{\infty} f(y) \left[\int_{-\infty}^{y} g(x)dx \right] dy.$$

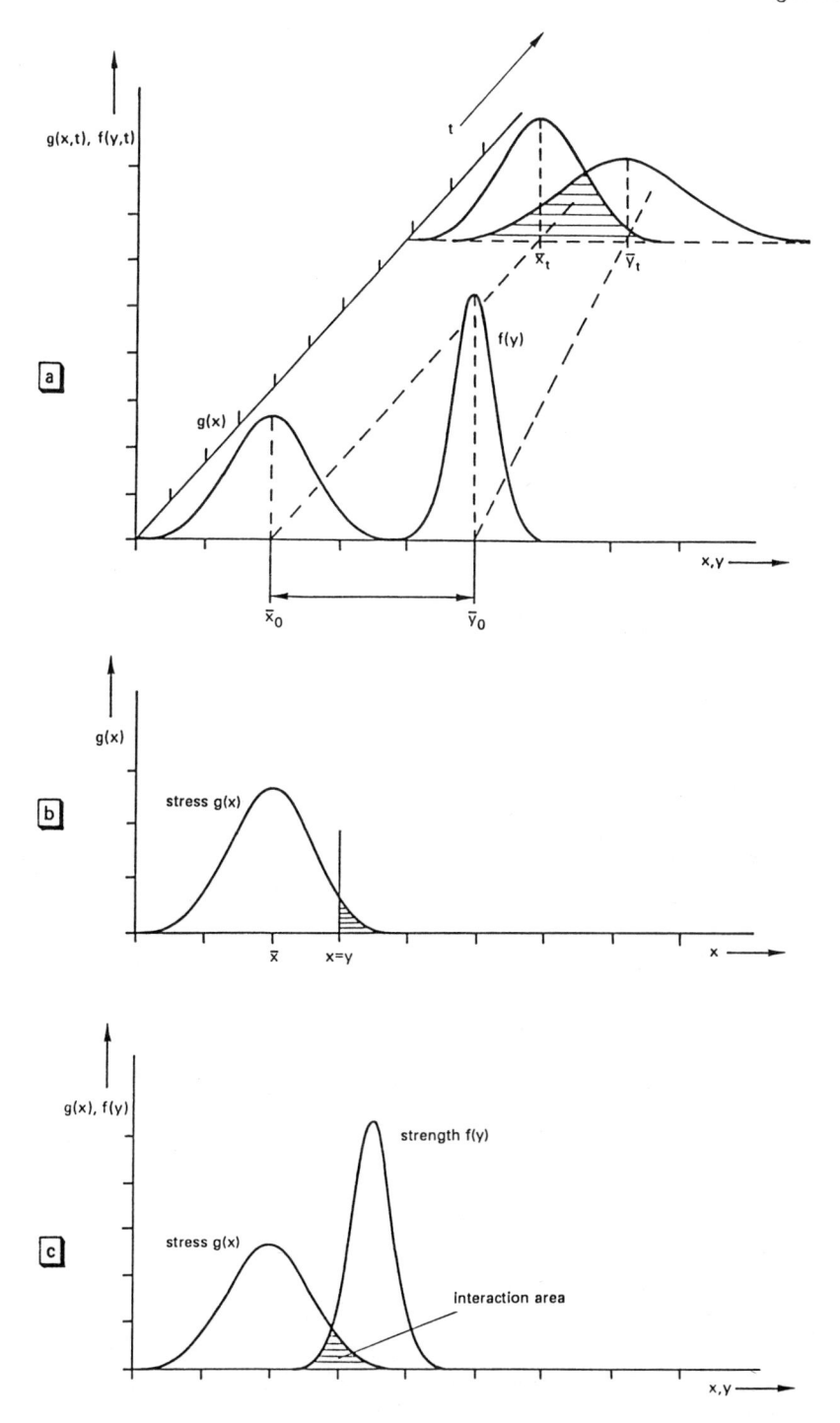

Figure 5.5 Stress probability density g(x) and strength probability density f(y) as a function of stress x, strength y and time t.

If the problem is put the other way, namely: The reliability is the probability that the strength $\underline{y}$ is not smaller than the stress $\underline{x}$ for all values $\underline{x}$ can assume, the result is:

$$R = \int_{-\infty}^{\infty} g(x) \left[\int_{x}^{\infty} f(y) dy \right] dx.$$

Of course, as one might expect, both expressions are identical.

This situation of distributed stress *and* strength is illustrated in Figure 5.5b. The overlapping area of both probability density functions indicates the interval in which failure may occur (and will occur eventually). If the average values of $\underline{x}$ and $\underline{y}$ are denoted by, respectively, $\bar{x}$ and $\bar{y}$, so if:

$$\bar{x} = \int_{-\infty}^{\infty} x g(x) dx \text{ and } \bar{y} = \int_{-\infty}^{\infty} y f(y) dy,$$

the *safety factor* of the considered component, given the respective stress distribution, is defined as:

$$\eta = \frac{\bar{y}}{\bar{x}}.$$

This safety factor indicates to what degree the strength exceeds the stress. Since this factor pertains to averages only, it does not contain information about the distribution of $\underline{x}$ and $\underline{y}$ which would be necessary for the determination of the reliability R. For example, in a system with a certain safety factor η_o the reliability will decrease as the standard deviations σ_x and σ_y of $\underline{x}$ and $\underline{y}$ become larger. So for a high reliability one should aim at both a large safety factor and a low standard deviation of the strength (and if possible also the stress) distribution. A low standard deviation of the strength distribution of a component is obtained by a good *quality control* during the production process. This will keep the stochastic fluctuations in the production process, which are the cause of strength variations, as small as possible.

A low standard deviation of the stress is obtained by properly defining the environment in which the component has to operate. Just compare the vibration level, temperature, and humidity range of components used in airplanes with the environment to which components are exposed, for example, in a telephone exchange. By means of insulation and other conditioning measures this stress can often be reduced and/or the range of stress variation be decreased.

From the above it will be clear that the same components may fail more often or earlier in different environments depending on the hostility of such an environment. To what extent this is the case is expressed by the *environmental factor*. This factor is a coarse measure of the ratio of the failure rate in the respective environment compared to that in a standard environment. Some examples of these environment-dependent failure rate multiplication factors are given below:

Environment	Factor
standard, conditioned (laboratory)	1
non-conditioned, immobile	5 – 10
vessels (marine)	10 – 25
trains (land, mobile)	25 – 50
airplanes	100 – 200
rockets/missiles	500 – 1500

Such environmental factors can, of course, be no more than a rough indication of the 'hostility' of the respective environment.

Another way to realise a higher reliability is to increase the safety factor η. To that end the average strength $\bar{y}$ of a component has to be made much larger than the average stress $\bar{x}$. To achieve this, the components are *overdimensioned* (increase of $\bar{y}$) for instance by choosing other, stronger subcomponents, or the stress $\bar{x}$ to which the components are subjected is decreased. The latter is called (stress-)*derating*. This can, for example, be done by distributing the stress over several components instead of letting only one component bear the brunt. Compare distributing the dissipated power in an amplifier output stage over several parallel transistors. This results in a lower junction temperature and, consequently, in a lower failure rate. Other examples of stresses that can relatively easily be derated by distributing them over more components are: Breakdown voltage of diodes and capacitors, the maximum wheel load of vehicles and the landing weight of airplanes. Derating and overdimensioning result in a factor (*derating factor*) by which the failure rate improves. The derating factor decreases the failure rate: It has an effect opposite to the environmental factor.

Let us illustrate this with an example.

If it is assumed that the fluctuations in stress and strength are created by a great number of stochastically independent causes, both x and y will be distributed normally and will be independent. This facilitates the introduction of the difference $z = y - x$ as a new stochastic variable. This new variable will also have a normal distribution: The distribution of the sum and the difference of independent stochastic variables with a normal distribution is again a normal distribution. The average value of z is $\bar{y} - \bar{x}$ and the standard deviation of z is:

$$\sigma_z = \sqrt{\sigma_x^2 + \sigma_y^2},$$

The probability density function $f(z)$ is given by:

$$f(z) = \frac{1}{\sqrt{2\pi(\sigma_x^2 + \sigma_y^2)}} \exp\left(-\frac{(z + \bar{x} - \bar{y})^2}{2(\sigma_x^2 + \sigma_x^2)}\right).$$

The reliability then is:

$$R = P(z \geq 0) = 1 - P(z < 0) = \int_0^\infty f(z)\, dz.$$

In Figure 5.6 the reliability of this system has been plotted for the normalised value z' of

72 Reliability Models

z. We see clearly that it is the difference $\bar{y} - \bar{x}$ that increases the components' reliability. The environmental factor works to increase $\bar{x}$, whereas the derating works to increase the effective value of $\bar{y}$.

As shown in Figure 5.5c the distribution of the stress x and the strength y will be a function of the time. This causes the calculated reliability to be a function of time as well. This will particularly happen because the strength probability density function $f(y)$ will not only exhibit a lower average strength $\bar{y}$ with the passing of time, but also an increasing standard deviation σ_y with time. This is brought about by various sorts of ageing processes occurring in the components of such a system (for instance fatigue phenomena).

Figure 5.5c shows that for $t = 0$ the safety factor η and the standard deviations σ_x and σ_y are such that even at this early a time there already exists an overlapping area, which leads to the so-called 'early failures'. This means that in this early-failure-period the weakest components perish by (relatively rare) peaks in the stress. So at the beginning of the system's life, immediately after $t = 0$, the overlapping tails of both distributions result in the 'infant mortality'.

The time-dependency of the strength and the stress will not be discussed here. We will confine ourselves to noting that this time-dependency is usually modelled in a greatly simplified way. In practice this is often done by measuring the time-dependency at a certain value of the stress, for example in the form of the system's hazard rate. Subsequently, a correction factor is applied for the influence of the stress. This can be simply explained as follows.

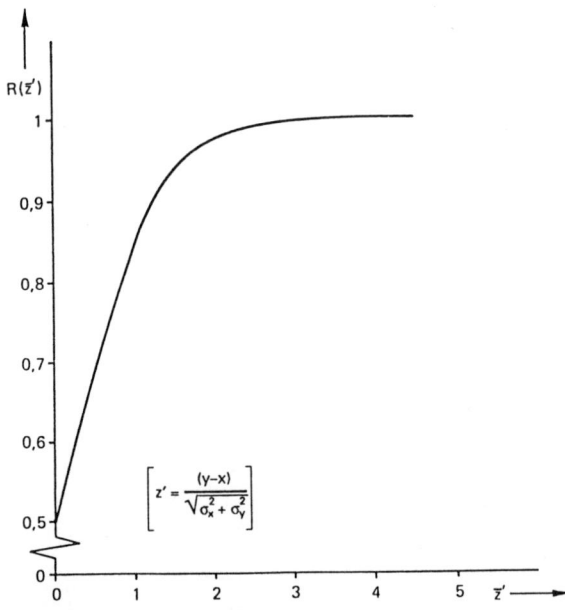

Figure 5.6 The reliability of a component of which both the strength $\bar{y}$ and the stress $\bar{x}$ have a normal distribution, plotted as a function of the normalised difference $\bar{z}'$ between strength and stress.

Suppose that the stress is caused by the temperature. Further suppose that the hazard rate is measured at a certain temperature T_0, so $z(t, T_0)$ is known. It is now assumed that the failure accelerating influence of the temperature can be indicated by an *acceleration factor* $\beta(T)$ which is derived from the equation of Arrhenius in Section 2.1:

$$\frac{z(t,T)}{z(t,T_0)} = \exp\left(-\frac{E_A}{nk}\left(\frac{1}{T} - \frac{1}{T_0}\right)\right) = \beta(T).$$

N.B.: Since we are concerned here with hazard rates instead of life expectancies the right-hand side is the inverse of the one in the corresponding expression in Section 2.1.

For $t \leq t_0$ (early failure region) the hazard rate at temperature T can be written as:

$$z(t,T) = \beta(T)\, z(t,T_0),$$

and for $t > t_0$ (operational region):

$$z(t,T) = \beta(T)\, z(t_0,T_0).$$

The next example clarifies this.

Example

For analogue ICs it turns out that the hazard rate as a function of time can be approximated relatively accurately by a Weibull distribution. With $T = T_0$ and $t \leq t_0$ we can write:

$$z(t,T_0) = \alpha c t^{c-1}.$$

At $T_0 = 373$ K, measurements show that $c = 0.5$ and $\alpha^2 = 4.5 \times 10^{-9}$ per hour. The acceleration factor $\beta(T)$ as a result of an increased temperature is equal to the above expression. With the following data: Effective activation energy $E_A/n = 0.7$ eV (1 eV = 1.6×10^{-19} J), $t_0 = 2000$ hours, Boltzmann's constant $k = 1.38 \times 10^{-23}$ J/K, this results in the curves depicted in Figure 5.7. This example clearly shows the effect of derating. If such an analogue IC is used at a lower environmental temperature and/or dissipates less power internally to the IC, the hazard rate decreases considerably and, as a result, the reliability will be higher.

5.3 Markov model

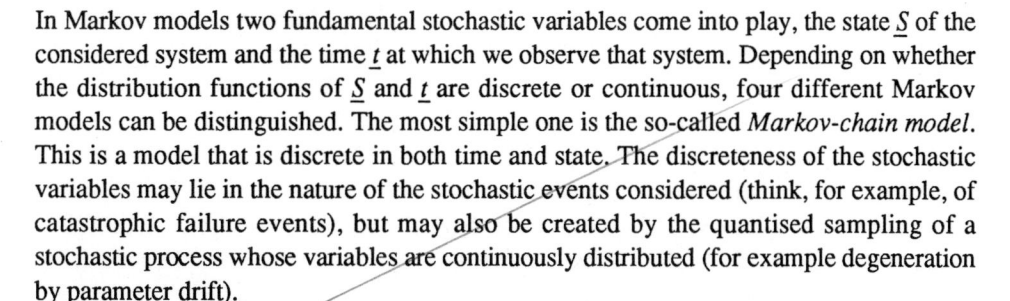

In Markov models two fundamental stochastic variables come into play, the state $\underline{S}$ of the considered system and the time $\underline{t}$ at which we observe that system. Depending on whether the distribution functions of $\underline{S}$ and $\underline{t}$ are discrete or continuous, four different Markov models can be distinguished. The most simple one is the so-called *Markov-chain model*. This is a model that is discrete in both time and state. The discreteness of the stochastic variables may lie in the nature of the stochastic events considered (think, for example, of catastrophic failure events), but may also be created by the quantised sampling of a stochastic process whose variables are continuously distributed (for example degeneration by parameter drift).

For reliability engineering the most important combination of the four models above is the one with continuous time $\underline{t}$ and discrete state $\underline{S}$. This continuous time, discrete state

74 Reliability Models

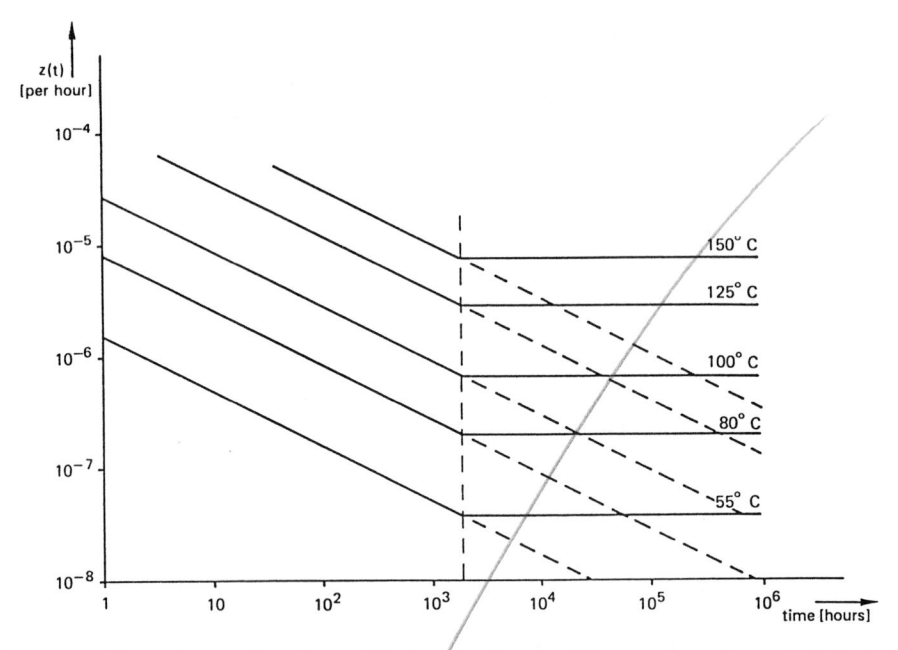

Figure 5.7 Hazard rate of analogue ICs versus time and temperature (temperature scale: ambient temperature plus internal temperature rise due to dissipation).

stochastic failure model is called a *Markov process*. In the following, we will confine ourselves to these models only.

In a Markov model a number of assumptions have been made:

- *States*. The state $\underline{S}$ of the system under observation is one element from the complete set of all discrete, mutually exclusive states in which the system can arrive due to all probable physical failure modes. If the associated reliability model of the system consists of more than one element (component), the state of the system is an ordered row (array) of the states of these elements.

Example

If we assume that the four elements of Figure 5.3b only can assume two states (good = 1, defective = 0), then every state of the system is an array of four binaries representing the states of the elements. With the (arbitrarily ordered) listing below, the states S_i become:

Elements				System	
V	W	F	EV		
1	1	1	1	S_0	1
0	1	1	1	S_1	0
0	0	1	1	S_2	0
1	0	1	1	S_3	0
–	–	–	–	–	–

Table 5.1 The states S_i in which the system of Figure 5.3b can be.

Since we are dealing here with a pure series system, it only functions correctly in state S_0, i.e. in all other states the system is down. Note: The number of states that each of the elements can assume is not necessarily limited to two states; more states are allowed, they only complicate the listing of all possible system states S_i. This would create a multi (component) mode system.

- *Transition probabilities* . The dynamics, i.e. the time dependencies, of a Markov model are governed by the transition probabilities between the various states of the model. It is not necessarily so that all transitions have a transition probability different from zero. In the above example the transition probabilities from S_i ($i \neq 0$) to S_0 are all equal to zero because the system is not maintained. An important property of Markov models is that the transition probability p_{ij} from state S_i to state S_j only depends on S_i and S_j and not on previous or later states. If this transition probability depended on a (finite) number of past states, new states S_i' and S_j' can be introduced for which the above-mentioned property holds. Because of this property, p_{ij} can be written as a conditional probability $P(j|i)$; the probability that the state of the system switches to state S_j if it is in state S_i. Thus, in the above example, the probability p_{01} only depends on S_0 and S_1 and is equal to the probability of a breakdown of the power supply V on the condition that all other elements remain functioning correctly. The transition from S_1 to S_2 depends on these two states only ; p_{12} is the probability that the preamplifier VV breaks down on the condition that the power supply is down already and that the other elements remain functioning. The transition from S_0 to S_2 is the probability that both V and VV break down on the condition that F and EV remain operative, and so on. Check for yourself what comprises the transition from state S_2 to S_3.

In Figure 5.8 this is illustrated for the first four states.

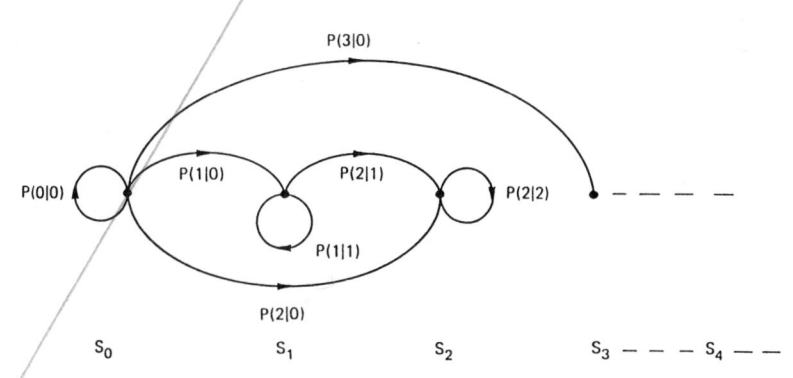

Figure 5.8 (Partial) State model for the system of Figure 5.3b.

- *Time continuity*. In the above, the time element did not play a role. The system was considered as being static; i.e. the moment of a state transition (i.e. a break down) was not involved in the considerations. This ever so important time element in reliability engineering can be introduced by sampling the (actually time-continuous) state of the system in time at intervals Δt. We take, as it were, a snap-shot of the system after every Δt and record in which state the system is at that moment; which elements are still functioning and which have failed.

The conditional transition probabilities $p_{ij} = P(j|i)$ mentioned above will then change into conditional transition probability densities. This can be understood as follows. The transition probability from state S_i to state S_j in the time interval $(t, t + \Delta t]$ with the transition moment (failure time) $\underline{t}$ as a continuous stochastic variable is:

$$p_{ij}(t, t + \Delta t) = P(t < \underline{t} \le t + \Delta t \mid \underline{t} > t) = \frac{P(t < \underline{t} \le t + \Delta t)}{P(\underline{t} > t)}.$$

In terms of failure, failure distribution and reliability this becomes:

$$p_{ij}(t, t + \Delta t) = \frac{f(t)}{R(t)} \Delta t = z(t) \Delta t.$$

In the time-continuous case the hazard rate $z(t)$ is apparently equal to the rate at which the transitions occur with respect to time: $z(t)$ may therefore also be referred to as a conditional failure probability *density* (see Section 3.2.1).

If, in a Markov model, the transition rates do not depend on time (so if we are dealing with constant failure rates) that model is called a (time-)*homogeneous* Markov model. If time-dependence is involved the model is *inhomogeneous*.

We shall now demonstrate how simple it is to perform reliability calculations with these Markov models. For clarity we will use a very simple, although basic, system, namely a component that cannot be repaired and that has only one failure mode.

First let us list all possible disjunct states. For this simple component there are only two states, viz. S_0 and S_1, in which the component is operative or broken respectively. So, in fact, we consider only one transition in a (more complex) Markov model (see Figure 5.9). The hazard rate of the component is given to be $z(t)$ and we consider the state of the component at the moments t and $t + \Delta t$.

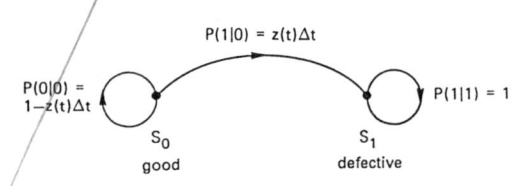

Figure 5.9 *State model of a single component that can only fail catastrophically.*

The probability that the component is in state S_0 at the moment $t + \Delta t$ shall be denoted $P_{S_0}(t + \Delta t)$. This probability is equal to the probability $P_{S_0}(t)$ that the component is in S_0 at t, times the probability that there are no failures in the interval Δt given by $1 - z(t)\Delta t$, added to the probability that the component is in S_1 at t, times the probability that its repair will be completed in Δt. We are dealing with a non-repaired component so:

$$P_{S_0}(t + \Delta t) = [1 - z(t)\Delta t] P_{S_0}(t) + 0 P_{S_1}(t).$$

Similarly, the probability that the component is in S_1 at $t + \Delta t$ is:

$$P_{S_1}(t + \Delta t) = [z(t) \Delta t] P_{S_0}(t) + 1 P_{S_1}(t).$$

Here, we have used the knowledge that S_0 and S_1 were *all* states in which the component could be and that these states are *disjunct*.

We have assumed that multiple events do not occur in Δt (e.g. by an intermittent failure); the probability of more than one transition in Δt is an infinitesimal of a higher order and will therefore become zero for $\Delta t \to 0$ (which, as we will later see, is necessary to return to the time-continuous case).

The above equations can be rewritten as follows:

$$\frac{P_{S_0}(t+\Delta t) - P_{S_0}(t)}{\Delta t} + z(t) P_{S_0}(t) = 0$$

and

$$\frac{P_{S_1}(t+\Delta t) - P_{S_1}(t)}{\Delta t} + z(t) P_{S_0}(t) = 0$$

These difference equations will have to be converted to differential equations by taking the limit for $\Delta t \to 0$, in order to obtain the time-continuous case. This results in:

$$\frac{dP_{S_0}(t)}{dt} + z(t) P_{S_0}(t) = 0,$$

and:

$$\frac{dP_{S_1}(t)}{dt} - z(t) P_{S_0}(t) = 0,$$

This system of differential equations can be solved by separating the variables:

$$\frac{dP_{S_0}(t)}{dP_{S_0}(t)} = -z(t)dt, \quad \text{so} \quad \ln P_{S_0}(t) = -\int_0^t z(t)dt + C.$$

With the initial condition $P_{S_0}(0) = \alpha$ ($0 \le \alpha \le 1$) this results in the well-known expression (see Section 3.2.1):

$$R(t) = P_{S_0}(t) = \alpha \exp[-\int_0^t z(t)dt].$$

In a similar way for $P_{S_1}(t)$ it is found that:

$$P_{S_1}(t) = 1 - \alpha \exp[-\int_y^t z(t)dt].$$

One does not have to solve $P_{S_1}(t)$ from the differential equations here, since:

$$P_{S_0}(t) + P_{S_1}(t) = 1.$$

As was already known, the solution will be very simple for $z(t) = \lambda$, i.e. a time-independent transition rate.

N.B.: The above-mentioned expressions are more general than those from Section 3.2.1, since they do not assume that $R(0) = 1$ but $R(0) = \alpha$. Apparently the Markov model allows an $R(0)$ different from 1. It also allows multi-mode failures in the system components, i.e. a component may have more than two possible states it can be in. The introduction of repair does not cause problems either in this model. It can also be used to

78　Reliability Models

model and calculate dependent failures. Finally, this method gives very simple equations for a constant $z(t)$, so for constant transition rates.

For more complex Markov models with three or more states there often is only an analytical solution for the reliability if the corresponding system of differential equations only contains constant coefficients. This is the case when all transition rates are time-independent (so for constant failure and repair rates). In that case the simplest solution is to use the Laplace transformation for conversion of the differential equations with constant coefficients (for a Markov model with n states there are n such equations) into n linear equations with n unknowns.

We may also derive the differential equations directly from the state model. If we look more closely at the state model of Figure 5.9, we see that a transition is indicated by a branch with a corresponding transition probability. This probability can be seen as the 'transfer' of the relevant branch. The states S_i are represented by nodes. In each node we assume a signal source with a signal strength $P_{S_i}(t)$. We may now state:

- The sum of the transition probabilities of all branches leaving a node is one; after all, the system always has to transfer to one of the states S_i ($i = 0,1,\ldots, n$).
- The probability that the system is in state S_i at the moment $t + \Delta t$ can be directly derived form the state model. It is equal to the sum of all signals entering at the node corresponding to S_i. All nodes are regarded as signal sources with a strength equal to the probability of the system being in the state corresponding to that node at the moment t. An incoming signal is now equal to the strength of the node source times the transfer of the branch between this node and the one corresponding to S_i.

The differential equations can also be derived directly from the state model. Here the time derivative of the probability that the system is in a certain state S_i is equal to the sum of the 'incoming signals' from the other nodes S_j ($j \neq i$), minus the sum of the 'signals departing' to other nodes.
This results in:

$$\frac{dP_{S_i}(t)}{dt} = \sum_{k=0}^{n} P_{S_k}(t)\, z_{ki}(t) - \sum_{k=0}^{n} P_{S_i}(t)\, z_{ik}(t) =$$

$$= \sum_{k \neq 0}^{n} P_{S_k}(t)\, z_{ki}(t) - P_{S_i}(t) \sum_{k=0}^{n} z_{ik}(t) \qquad (i = 0,1,\ldots,n).$$

The 'signal' of those branches that return to the same node S_i, which represents the probability that the system remains in the same state S_i during the infinitesimally small time interval dt, is:

$$P_{S_i}(t)\, z_{ii}(t).$$

These contributions would be counted in both the incoming branches and the outgoing branches and therefore do not contribute to the differential equations. So, when forming differential equations, these contributions have to be included consistently or they have to be omitted consistently. The latter has been done in the equations below:

$$\frac{dP_{S_i}(t)}{dt} = \sum_{\substack{k=0 \\ k \neq i}}^{n} P_{S_k}(t)\, z_{ki}(t) - P_{S_i}(t) \sum_{\substack{k=0 \\ k \neq i}}^{n} z_{ik}(t) \qquad (i = 0,1,\ldots,n).$$

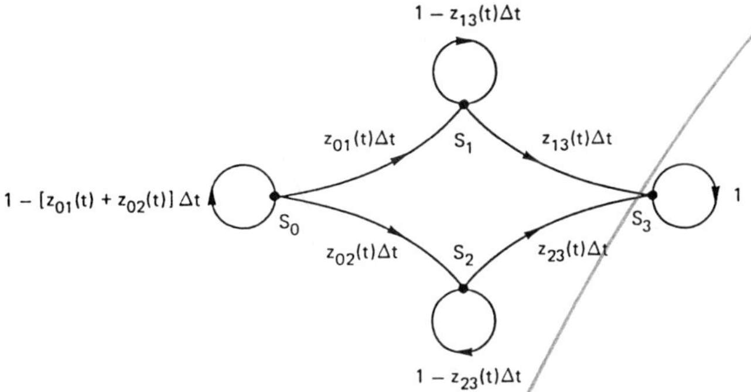

Figure 5.10 State model for a system consisting of two elements. See Table 5.2 for the state identification.

Example

We shall finish with an example that is more challenging than that of Figure 5.9. In Figure 5.10 the associated Markov model is given. The system consists of two components A and B. It does not matter in the calculation (at least for the time being) whether these components are in series or in parallel, or whether the second component is used to repair the first one. It is assumed that A and B can only be functioning or defective. As indicated in Table 5.2 the system can then only be in one of four states S_i. Now try for yourself to form the differential equations directly from the Markov state model. You will find:

$$\frac{dP_{S_0}(t)}{dt} = -[z_{01}(t) + z_{02}(t)]\, P_{S_0}(t),$$

$$\frac{dP_{S_1}(t)}{dt} = z_{01}(t)\, P_{S_0}(t) - z_{13}(t)\, P_{S_1}(t),$$

$$\frac{dP_{S_2}(t)}{dt} = z_{02}(t)\, P_{S_0}(t) - z_{23}(t)\, P_{S_2}(t),$$

$$\frac{dP_{S_3}(t)}{dt} = z_{13}(t)\, P_{S_1}(t) - z_{23}(t)\, P_{S_2}(t),$$

A	B	S
1	1	S_0
0	1	S_1
1	0	S_2
0	0	S_3

Table 5.2 States S_i in which a system with two single-mode components A and B can be. It is not indicated yet in which of the four states the system is functioning or defective.

These differential equations cannot be solved for a general hazard rate $z(t)$. For simplicity's sake we assume that:
- $z_{01}(t) = \lambda_1$, $z_{02}(t) = \lambda_2$, $z_{13}(t) = \lambda_3$, $z_{23}(t) = \lambda_4$, and
- initial conditions: $P_{S_0}(0) = 1$, $P_{S_1}(0) = P_{S_2}(0) = P_{S_3}(0) = 0$.

The solution is now:

$$P_{S_0}(t) = e^{-(\lambda_1 + \lambda_2)t},$$

$$P_{S_1}(t) = \frac{\lambda_1}{\lambda_1 + \lambda_2 - \lambda_3} [e^{-\lambda_3 t} - e^{-(\lambda_1 + \lambda_2)t}],$$

$$P_{S_2}(t) = \frac{\lambda_1}{\lambda_1 + \lambda_2 - \lambda_4} [e^{-\lambda_4 t} - e^{-(\lambda_1 + \lambda_2)t}],$$

$$P_{S_3}(t) = 1 - \sum_{i=0}^{2} P_{S_i}(t).$$

We now have to introduce the configuration of the two components A and B in the system:

(a) Let us first assume that we are dealing with two components in series. Neither is allowed to fail. So only S_0 is a correctly functioning state, therefore:

$$R(t) \equiv P_{S_0}(t) = e^{-(\lambda_1 + \lambda_2)t}.$$

(b) Let us now assume that the two components are in parallel. The system functions as long as no more than one component has failed, so:

$$R(t) \equiv P_{S_0}(t) + P_{S_1}(t) + P_{S_2}(t) = 1 - P_{S_3}(t).$$

(c) Let us now suppose that we have a parallel system of two similar components which together share the stress and because of this derating these components have a lower failure rate λ_0. If one component breaks down, the other component is exposed to the full stress and therefore exhibits a higher failure rate λ_h. With the above expression and the substitution $\lambda_0 = \lambda_1 = \lambda_2$ and $\lambda_h = \lambda_3 = \lambda_4$ we get the following expression for the reliability of that system:

$$R(t) = \frac{2\lambda_0 e^{-\lambda_h t} - \lambda_h e^{-2\lambda_0 t}}{2\lambda_0 - \lambda_h}.$$

What would this system result in for $\lambda_0 = \lambda_h = \lambda$, and for $\lambda_0 = \frac{1}{2}\lambda_h$?

(d) The general solution for the reliability of a parallel system found in (c) can also be interpreted in the following manner. A component with failure rate λ_1 is switched on. The second component is not activated and has a failure rate of zero in this mode; so $\lambda_2 = 0$ (at least $\lambda_2 \ll \lambda_1$). If the second component, which is different from the first one, is activated it has a failure rate λ_3. We then find:

$$R(t) = \frac{\lambda_1}{\lambda_1 - \lambda_3} e^{-\lambda_3 t} - \frac{\lambda_3}{\lambda_1 - \lambda_3} e^{-\lambda_1 t}.$$

N.B.: $P_{S_2} = 0$ for any t since $P_{S_2}(0) = 0$ and $\lambda_2 = 0$. So the value of λ_4 does not matter. Since in practice for two identical components it should hold that $\lambda_1 \approx \lambda_3$, the reliability found by differentiating the numerator and the denominator of the above expression with respect to λ_3, using De l'Hospital's rule, is given by:

$$R(t) = e^{-\lambda t} + \lambda t e^{-\lambda t}.$$

(After all, a direct substitution would lead to an undetermined expression.)

As the number of states and the number of allowable transitions between them increases, solving the corresponding system of differential equations becomes more and more complex. For systems with four or more states it is therefore advisable to use a solution method in which the clarity is maintained throughout and which keeps the possibility of mistakes small. For a Markov model with $n + 1$ states the following set of differential equations can be written:

$$\frac{dP_{S_0}(t)}{dt} + \lambda_{u_0} P_{S_0}(t) - \lambda_{10} P_{S_1}(t) - \lambda_{20} P_{S_2}(t) - \ldots - \lambda_{n0} P_{S_n}(t) = 0$$

$$\frac{dP_{S_1}(t)}{dt} - \lambda_{01} P_{S_0}(t) + \lambda_{u_1} P_{S_1}(t) - \lambda_{21} P_{S_2}(t) - \ldots - \lambda_{n1} P_{S_n}(t) = 0$$

$$\vdots$$

$$\frac{dP_{S_n}(t)}{dt} - \lambda_{0n} P_{S_0}(t) - \lambda_{1n} P_{S_1}(t) - \lambda_{2n} P_{S_2}(t) - \ldots + \lambda_{u_n} P_{S_n}(t) = 0.$$

Here λ_{u_i} is the sum of the failure rates corresponding to all branches leaving node S_i. In matrix notation we get the following equation:

$$\begin{bmatrix} \frac{dP_{S_0}(t)}{dt} \\ \frac{dP_{S_1}(t)}{dt} \\ \vdots \\ \frac{dP_{S_n}(t)}{dt} \end{bmatrix} + \begin{bmatrix} +\lambda_{u_0} & -\lambda_{10} & -\lambda_{20} & \ldots & -\lambda_{n0} \\ -\lambda_{01} & +\lambda_{u_1} & -\lambda_{21} & \ldots & -\lambda_{n1} \\ \vdots & & & & \vdots \\ -\lambda_{0n} & -\lambda_{1n} & -\lambda_{2n} & \ldots & +\lambda_{u_n} \end{bmatrix} \begin{bmatrix} P_{S_0}(t) \\ P_{S_1}(t) \\ \vdots \\ P_{S_n}(t) \end{bmatrix} = \begin{bmatrix} 0 \\ 0 \\ \vdots \\ 0 \end{bmatrix}$$

which can be rewritten as:

$$\left[\frac{dP_{S_i}(t)}{dt} \right] + U \cdot [P_{S_i}(t)] = [0],$$

in which U is called the transition matrix.

Reliability Models

After the Laplace transformation of all time functions we get the following equation:

$$s \cdot [P_{S_i}(s)] - [P_{S_i}(0)] + U \cdot [P_{S_i}(s)] = [0].$$

Here s is the Laplace variable. By bringing the vector with the initial conditions $P_{S_i}(0)$ to the right-hand side of the equation we get:

$$[s \cdot I + U] \cdot [P_{S_i}(t)] = A \cdot [P_{S_i}(s)] = [P_{S_i}(0)].$$

Consequently matrix A looks as follows:

$$A = \begin{bmatrix} s + \lambda_{u_0} & -\lambda_{10} & -\lambda_{20} & \cdots & -\lambda_{n0} \\ -\lambda_{01} & s + \lambda_{u_1} & -\lambda_{21} & \cdots & -\lambda_{n1} \\ \cdot & & & & \cdot \\ \cdot & & & & \cdot \\ \cdot & & & & \cdot \\ -\lambda_{0n} & -\lambda_{1n} & -\lambda_{2n} & \cdots & s + \lambda_{u_n} \end{bmatrix}$$

Since the following holds:

$$A^{-1} A [P_{S_i}(s)] = A^{-1} [P_{S_i}(0)],$$

it also holds that:

$$[P_{S_i}(s)] = A^{-1} [P_{S_i}(0)].$$

For the initial conditions it is usually assumed that:

$$P_{S_0}(0) = 1 \text{ and } P_{S_i}(0) = 0 \qquad (i = 1,2,\ldots,n).$$

Only the first column of the inverse matrix A^{-1} is important. The elements a_{ij} of this matrix can be found with:

$$a_{ij} = \frac{(\text{cofactor})_{ij}}{|A|}.$$

Example

It is easy to check that the following matrix A results from the Markov state model of Figure 5.10:

$$A = \begin{bmatrix} s + \lambda_{01} + \lambda_{02} & 0 & 0 & 0 \\ -\lambda_{01} & s + \lambda_{13} & 0 & 0 \\ -\lambda_{02} & 0 & s + \lambda_{23} & 0 \\ 0 & -\lambda_{13} & -\lambda_{23} & s \end{bmatrix}.$$

Here λ_{ij} are the failure rates corresponding to the transitions from S_i to S_j.

Once A is known $P_{S_i}(s)$ can be written out and transformed back to arrive at $P_{S_i}(t)$.

Concludingly, we may now state:

The Markov model is easily applied to systems with a finite number of states, with time-independent failure processes exhibiting constant failure and/or repair rates. This leads to the restriction that the failure distributions of the elements of the system considered should be negative-exponential, or should have a failure distribution which results from a series/parallel combination of such elements (in that case the introduction of dummy states leads again to a Markov model with only time-independent failure processes).

In Section 6.9.4 we shall use Markov models to calculate the reliability of more complex systems. Later on, the Markov model will also be used for the evaluation of maintained systems.

Problems

5.1. In order to tolerate short- and open-circuit failures, a redundant configuration of four diodes might be used for every single diode operation: Two branches with two diodes in series, connected in parallel. If q_o is the probability of a diode failing open-circuit mode and q_s is the probability of a diode failing short circuit, determine for which range of q_o and q_s it is useful to make a connection between the centres of the two parallel branches.

5.2. Can two disjunct events, each with a probability of occurrence different from zero, be independent?

5.3. Given is the following uniform distribution:

$$f(t) = \frac{1}{L} \quad \text{for } 0 \leq t < L \text{ and } f(t) = 0 \text{ for } t \geq L.$$

Determine $R(t)$, $F(t)$ and $z(t)$.

5.4. The weather radar system of an airliner has an MTTF of 1140 hours. Assuming that the failure rate is constant solve the following problems:
   (a) What is the probability of failure during a 4-hour flight?
   (b) What is the maximum duration of a flight when the reliability may not drop below 0.99? (During the flight the system is in continuous operation.)

5.5. A failure density function for a system is given:

$$f(t) = at \, e^{-\frac{1}{2}at^2}.$$

   (a) Determine $R(t)$ and $z(t)$ for this system.
   At the moment $t = 0$ there are 5000 well-functioning systems, each with the failure density function given above. Of these 5000, 4700 turn out to be still functioning correctly after 10 hours.
   (b) What is (approximately) the expected number of failures in the time interval from 10 to 20 hours?

5.6. Demonstrate in a sketch of a stress probability density function $g(x)$ and a strength probability density function $f(y)$ what is meant by the safety factor η. (In one figure!) Also explain from this sketch how quality control affects reliability.

5.7. Hospitals always have emergency provisions to take over the supply of electricity if the public power distribution lines fail. Suppose the failure rate of the lines is λ_n and that of the emergency generator is λ_{g_1} in the non-used state, λ_{g_2} while operational. Draw a Markov diagram for this electricity supply system, in which repair has to be left out of consideration.

6
Non-Maintained Systems

We shall distinguish between *maintainable systems* and *maintained systems*. The latter concept is a more restricted one than the first: A car is a maintainable system, yet whether it is actually maintained depends on the owner. From a reliability point of view there is little to be gained if a system is made maintainable (easily accessible to maintenance etc.) and the user neglects maintenance for instance because of misplaced thrift. As far as reliability is concerned, such a system is a non-maintained system.

6.1 Introduction

A *non-maintained system* either is a non-maintainable system or a maintainable system that is not actually maintained. A *maintained system* has to be both; maintainable and maintained. Examples of non-maintainable systems are inexpensive systems for which maintenance is not an economically feasible proposition (pocket calculators, and so on), systems for which maintenance is not possible (earth satellite), or systems which are used only once (solid fuel rocket).

A *maintainable system*, as we have already seen in Section 3.2, is a system that can be restored to an *operational state* after a failure has occurred by means of *human intervention*. Human intervention is essential in this definition; a system that 'maintains itself' by switching on a built-in redundant subsystem need not be maintainable. This distinction is made because human intervention only takes effect after a certain time delay referred to as *maintenance time*. This maintenance time includes: Alerting the maintenance crew, actually locating the failure, and ordering the necessary components and tools. The maintenance time depends on the experience and the training of the maintenance people and on the components and tools which they have locally available. All in all, this delay time can only be described by a stochastic process: The maintenance-time distribution. It also turns out that human intervention is not always 100% effective. After a maintenance operation, a system is generally not '*as new*'; it is merely restored to an '*operational state*'. This is caused by matters such as damage done elsewhere to the system which was caused by careless repairs, use of the wrong components or of tools that were not designed for the respective purpose, or the following of incorrect repair procedures (nuts tightened with the wrong torque, and so on). These effects, too, can only be described by stochastic processes.

A very powerful maintenance strategy is the external human maintenance of a system which is equipped with internal, self-actuating redundancy. A strategy that is very effective from a reliability point of view is the *preventive maintenance of a redundant*

86 Non-Maintained Systems

system. The maintenance here replaces the failed redundant subsystems by new ones before the system can fail from an exhausted redundancy. This type of maintenance can often be conducted without closing down the system. This shall be discussed in more detail in Section 7.3.4.

For convenience two things are assumed in this chapter:
- The system components have a *constant failure rate*.
- The system components fail *stochastically independent*.

The assumption about the constancy of the failure rate implies that the components do not have a life memory; if they are still operating they are 'as new'. This assumption is not necessary, it merely simplifies our calculations.

About the latter assumption: The failures in the various components may, for example, not have a common cause, nor may failures in one component increase the probability of failures in other components. This means that common-cause failures as well as secondary failures are excluded.

N.B.: Secondary failures are failures resulting from primary failures. Generally: Failures of which the likelihood is increased by primary failures are secondary failures. If a secondary failure does not occur in the same component as the primary failure but in a redundant component, the redundancy is deceptive. It is exhausted far earlier than expected in first instance (i.e. without this statistical dependency).

N.B.: In practical situations these common-cause failures are greatly feared. One should duly keep in mind the distinct possibility that these failures may occur in any design of a redundant system and account for these dependent failures in the modelling of such a system. Dependent failures will be discussed in more detail in Section 6.4.1.

An example of a common-cause failure is an earthquake which may cause failures at several places in a nuclear reactor system and, what is worse, if one unit fails because of this sudden earth movement, identical redundant units will also fail with a high likelihood. To avoid this, redundant subsystems are often non-identical (i.e. made by different manufacturers, based on different operating principles, etc.). Avoiding common-cause failures due to flooding, collapsing structures, and the like are usually avoided by putting the redundant subsystem in different locations.

We shall now address the influence of the *structure of a system* on its reliability.

6.2 Series systems

In reliability engineering, a series system is understood to be a system whose reliability model has a *series* or *chain* structure. This means that *all* components of that system have to function properly for the system to function properly. The breakdown of any one element ends an operational period of the system. Since it was assumed that failures in components occur independently, the reliability $R(t)$ of such a series system with n components can simply be written as:

$$R(t) = \prod_{i=1}^{n} R_i(t).$$

Here $R_i(t)$ is the reliability of the i-th component. From the above so-called product rule, which applies regardless of the particular failure distributions of the system components, it is easily seen that the reliability decreases as more components are added to a series system; the greater the (numerical) complexity, the lower the reliability. Further, the reliability of a series system turns out to be lower than that of the weakest component:

$$R(t) < \min_i \{R_i(t)\}, \qquad (i = 1,2,\ldots,n).$$

The above expression can therefore be used as a simple, albeit coarse, upper limit of the reliability.

N.B.: As was illustrated in Figure 5.4, the prefix 'series' has *nothing* to do with the physical structure of the system. Thus, even though it mee seem odd, when driving a car at night, the lights are in series with the steering system, the drive train, the brakes and other necessary functions in the car.

With the aid of:

$$R(t) = R(0) \exp \left[-\int_0^t z(t)\, dt\right],$$

from Section 3.2.1, a simple sum rule can be arrived at for the hazard rate $z(t)$ of a series system:

$$z(t) = \sum_{i=1}^n z_i(t),$$

in which $z_i(t)$ is the hazard rate of the i-th component.

For a negative-exponential failure distribution the expressions do become very simple. If it is assumed that $z_i(t) = \lambda_i$, we find for a series system with n independently failing components:

$$R_i(t) = e^{-\lambda_i t},$$

$$z(t) \equiv \lambda = \sum_{i=1}^n \lambda_i,$$

so:

$$R(t) = e^{-\lambda t} = \exp\left[-\left(\sum_{i=1}^n \lambda_i\right) t\right].$$

We conclude that the failure distribution of a series structure of components with negative-exponential failure distributions is again a negative-exponential distribution. Therefore, we can write the mean useful life of a series system as:

$$\theta = \frac{1}{\lambda} = \frac{1}{\sum_{i=1}^n \lambda_i}.$$

88 Non-Maintained Systems

This shows that the weakest component in such a system results in the, relatively, largest shortening of the system's life.

On the above is based a simple but primitive method to estimate the reliability of a system. One simply assumes that all components are necessary for a proper functioning of the system; one adds the failure rates of all components. In the case that the system is not a true series system, the above procedure gives a pessimistic estimate for the system reliability i.e. it estimates the reliability too low.

Example

For an environment (train) with a known failure accelerating environmental factor (35) the failure rates apply given in Table 6.1. It concerns a certain electronic (filter) circuit consisting of components on a printed-circuit board. The mean useful life is 380 years, i.e. after 380 years 63 % of the initial number of circuits would have failed. If for a certain critical application, such as the central automatic traffic control system for these trains, it is required that the reliability may drop to 0.999 at the most, the corresponding lifetime is only 139 days. The application of redundancy is therefore essential for such systems.

Remarks

- As already mentioned, the above method for estimating a system's reliability yields a lower limit on the reliability of the system (so a pessimistic estimate) if it is not a true series system. The estimate is more and more pessimistic as more components can fail without causing the system to break down.

Number	Component	Failure rate	Sub-total
10	metal film resistance	$2\times10^{-9}/h$ =	$20\times10^{-9}/h$
4	tantalum capacitor	$10\times10^{-9}/h$ =	$40\times10^{-9}/h$
5	low-power transistor	$5\times10^{-9}/h$ =	$25\times10^{-9}/h$
1	electrolytic capacitor	$100\times10^{-9}/h$ =	$100\times10^{-9}/h$
2	analogue IC	$45\times10^{-9}/h$ =	$90\times10^{-9}/h$
2	digital IC	$7\times10^{-9}/h$ =	$14\times10^{-9}/h$
75	soldering joints	$0.1\times10^{-9}/h$ =	$7.5\times10^{-9}/h$
20	printed circuit board traces	$0.01\times10^{-9}/h$ =	$0.2\times10^{-9}/h$
		l_{total} ≈	$300\times10^{-9}/h$

Table 6.1 *Failure rates of the components of a certain electronic module. If the components fail independently and if all components are necessary for the correct operation of the module, the failure rates must be added to obtain the failure rate of the module.*

- Also, if dependent failures would occur, the above-mentioned estimate is a pessimistic one.
- In the example illustrated in Table 6.1 the electrolytic capacitor is responsible for 1/3 of the total failure rate. If the capacitor could be avoided by a redesign, a considerable improvement in inherent reliability would result. Therefore, preliminary designs

always have to be followed by reliability calculations and by one or more redesigns to arrive at a design that is as good as possible given the cost and time constraints.
- Preventive maintenance to series systems with components that fail negative-exponentially is meaningless. After all, preventive maintenance can only be performed as long as the system has not failed yet. In a series system this is only the case if none of the components have failed. However, replacing components with a negative-exponential failure distribution as a preventive measure makes no sense as we have already seen in Section 4.1.1. This is the case because still functioning components are 'as new'.
- If a component of a system when not in use has a failure rate (λ_0) different from that when the system is in use (λ_u), the introduction of a *duty cycle d* yields an effective failure rate:

$$\lambda_{\text{eff}} = (1-d)\lambda_o + d\lambda_u,$$

in which:

$$d = \frac{\sum t_{ui}}{\sum t_{ui} + \sum t_{oi}},$$

where $\sum t_{ui}$ is the accumulated operation time, and $\sum t_{oi}$ the accumulated not-in-use time.

6.3 Redundancy

With bad or inferior components one can still make good systems, provided that one is allowed to use larger numbers of components than strictly necessary for realising the system function. So a 'trade in' is possible between the number of components on the one hand and the required reliability of those components on the other. The *excess* or *redundancy* in components is utilised here to continue the undisturbed functioning of the system till after the first components have failed.

N.B.: The system designer will, of course, never deliberately choose components that are bad or inferior. Redundancy is more a safeguard against incidentally bad components. In addition to this, as we will see later, the better the components, the higher the reliability improvement attained by applying redundancy.

Remark

Intermittent failures in a system can be counteracted by other means than the *hardware redundancy* mentioned above. It is often sufficient to just repeat the information in the system signals (a few times), or to encode the system signals redundantly. This form of redundancy is called *information* or *signal redundancy*. It does not affect the reliability of a system with non-intermittent failures.

In the case of *hardware* or *structural redundancy* the redundant units, modules or components in the system may be kept at three different levels of activity: Fully active, only partially active, or fully passive. In the case of *active* or *hot redundancy*, all the extra components are in full operation. This hot redundancy does not differentiate between primary engaged components and components held in reserve; all components are equally

90 Non-Maintained Systems

active. In the case of *passive* or *dormant redundancy,* the reserve components are switched off completely. In general, this results in the longest life of the system. Dormant redundancy is often difficult to realise in operational practice, since it requires a certain warming-up time before the passive, reserve components are able to take over the task of the operational system components. A solution, in between this dormant or cold redundancy and the previously mentioned hot redundancy, is the so-called *stand-by redundancy*. In this case the reserve is already half switched on in order to take over the task of the operational components without any appreciable lead time. An example is the stand-by redundant output stage of a broadcasting transmitter. The reserve transmission tube is waiting here to be fully engaged at a certain (lowered) filament current to allow a quick take-over in approximately 1 second.

In Figure 6.1 an example is given of the various kinds of redundancy that may be utilised in a system.

N.B.: Passive redundant components do not always necessarily have a longer life in the reserve state than components that are held in a stand-by mode. Electrolytic capacitors, for instance, have a longer life with a certain small voltage applied than without such a voltage. In many electronic systems the thermal dissipation occurring during operation prevents the formation of a water film due to condensation of water vapour from the air. Such condensation may occur in passive units. Together with air pollution contaminants the condensed water gives ionised salts, bases and acids that corrode the metalisation traces of such components as printed-circuit boards, thereby shortening their life.

Also, keeping living creatures such as human beings and animals in reserve for a certain task in full passivity is not beneficial for a reliable functioning of the system of which this task is a part. Think, for example, of firemen, security personnel and emergency surgery crews in hospitals. Boredom and loss of skill due to a low active-passive time ratio will eventually lead to errors.

In the reliability model of a redundant system, stand-by and passive components are characterised by their own separate failure distributions that differ from those of an operational component. When a component switches from passive or stand-by to active, the failure rate becomes different.

In a redundant system it is not always the case that one operational component will suffice to carry out a certain task correctly. Often a number of such components (performing mutually identical functions) are needed to handle power dissipation and provide the load carrying capability. Thus, in the general case m (identical) components will be required for the operational task. Out of the total number of n components there are $n - m$ reserve components. The degree of redundancy η is then defined as:

$$\eta = \frac{n - m}{n - 1}.$$

It will be clear that if $m = 1$ all n units can be used fully ($\eta = 1$) and that if $m = n$ *de facto* there is no redundancy left ($\eta = 0$). In the first case ($\eta = 1$) the system is a *pure parallel system*, in the second case ($\eta = 0$) it is a *pure series system*. In the intermediate case $0 < \eta < 1$ the system is called an *m-out-of-n system*. A special case of *m-out-of-n*

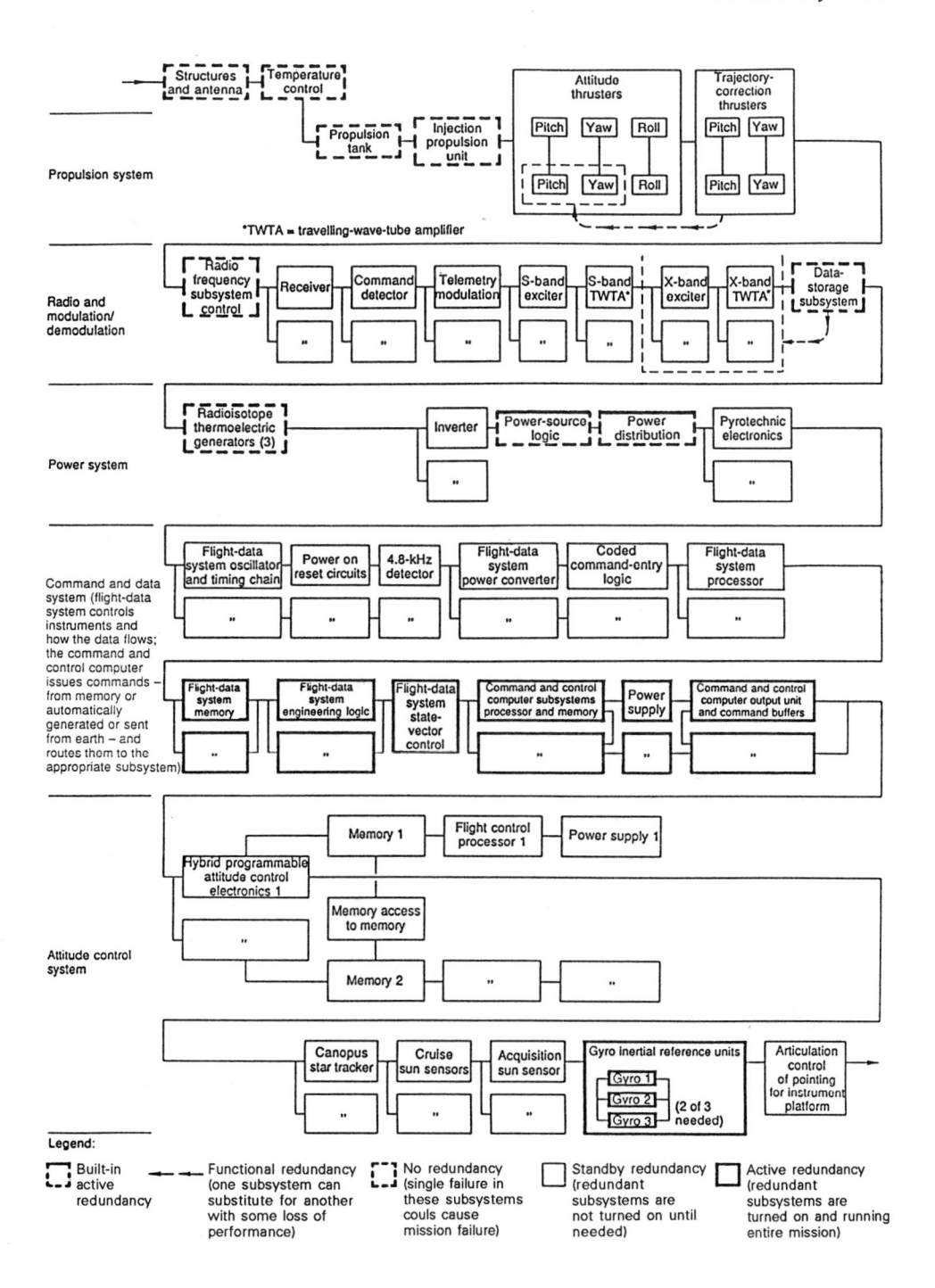

Figure 6.1 Catastrophic failure model of the space probe Voyager in which various kinds of redundancy are applied (see legend). Courtesy W.C. Williams, 'Lessons from NASA', IEEE Spectrum, October 1981.

systems is the so-called *majority voting system*, for which $\eta \approx 0.5$. Most systems have a mixed structure, i.e. a structure that is neither purely parallel nor purely series in nature. In the following we shall therefore also discuss mixed systems, in addition to parallel, *m*-out-of-*n*, and majority voting systems.

6.4 Parallel systems

A system with a pure parallel structure (i.e. with a degree of redundancy $\eta = 1$) needs only one component to be operational to perform its function. If a component fails, any other component from the system may take over this function. The system fails if and only if all *n* components of the system have failed. Expressed in probabilities we find:

$$F(t) = \prod_{i=1}^{n} F_i(t).$$

This is the *product rule* for the *unreliability* $F(t)$ of a pure parallel system with independent failures. The reliability and the hazard rate of such systems can be determined with:

$$R(t) = 1 - F(t)$$

and

$$z(t) = \frac{1}{R(t)} \frac{dF(t)}{dt}.$$

If a negative-exponential distribution is assumed for $F_i(t)$ we find:

$$F(t) = \prod_{i=1}^{n} (1 - e^{-\lambda_i t}),$$

$$R(t) = 1 - \prod_{i=1}^{n} (1 - e^{-\lambda_i t}).$$

Integration of $R(t)$ with respect to time gives for the mean life θ or MTTF:

$$\theta = \left(\frac{1}{\lambda_1} + \frac{1}{\lambda_2} + \ldots + \frac{1}{\lambda_n}\right) +$$

$$- \left(\frac{1}{\lambda_1 + \lambda_2} + \frac{1}{\lambda_1 + \lambda_3} + \ldots + \frac{1}{\lambda_{n-1} + \lambda_n}\right) +$$

$$+ \left(\frac{1}{\lambda_1 + \lambda_2 + \lambda_3} + \frac{1}{\lambda_1 + \lambda_2 + \lambda_4} + \ldots + \frac{1}{\lambda_{n-2} + \lambda_{n-1} + \lambda_n}\right) +$$

$$- \ldots + \frac{(-1)^{n+1}}{\sum_{i=1}^{n} \lambda_i}.$$

For a parallel system consisting of *n* identical components with failure rate λ this becomes:

$$\theta = \frac{1}{\lambda} \sum_{i=1}^{n} \frac{1}{i}.$$

The system hazard rate $z(t)$ can then also be calculated easily:

$$z(t) = n\lambda(1 - \frac{1 - (1 - e^{-\lambda t})^{n-1}}{1 - (1 - e^{-\lambda t})^n}).$$

To illustrate the above, the $R(t)$, $F(t)$, $f(t)$ and $z(t)$ of one component and of two such components have been plotted in Figure 6.2. The plots were made for two components in series as well as in parallel. The components are identical and have a constant failure rate λ. The most drastic differences in the three curves are found for the time interval just after $t = 0$. The useful life corresponds with this time interval. This useful life is always much smaller than $t = 1/\lambda$; the mean life of the components.

Conclusions

- A parallel structure generates a time-dependent hazard rate, even for components with a constant failure rate. At the start $z(t)$ is zero (after all, the redundancy has to be used up before the system can break). For very large t, $z(t)$ asymptotically approaches the constant failure rate of the one, longest lasting component, i.e. the one with the lowest λ_i.
- The MTTF (so θ) depends on the failure rate of the components. The better the components are (the lower λ_i is), the larger is the absolute increase of the MTTF.
- Addition of redundant components to an existing system causes the largest increase in system reliability when the components' reliability is high.
- The addition of extra, redundant components is most effective when there is still little redundancy (for small n).
- In order to achieve the largest reliability gain for a given total number of available components the redundancy has to be applied at so elementary a level as possible. For a certain system one should not, for example, simply duplicate the entire system, but parallel every component within that system by another. This takes the same number of components, but is far more effective. So redundancy has to be used at so low a hierarchical level as possible in a system, preferably even within the components.
- From the above expression for θ (the MTTF) it can be observed that if a component with failure rate λ is provided with an identical redundant parallel component ($\lambda, n = 2$) the MTTF goes from $1/\lambda$ to $1.5/\lambda$; an improvement by a factor 1.5. The costs, however, will have (practically) doubled since instead of one component two are needed. If one is prepared to spend this double amount of money on one more expensive component, one can buy far better, so-called 'high-rel' components. The MTTF of these components is usually much more improved than a factor 1.5. Therefore, it can (usually) be said that, to get good system reliability, much attention has to be paid to procuring good components. One subsequently should apply these components in systems having a structure that is failure forgiving, such as a redundant structure.

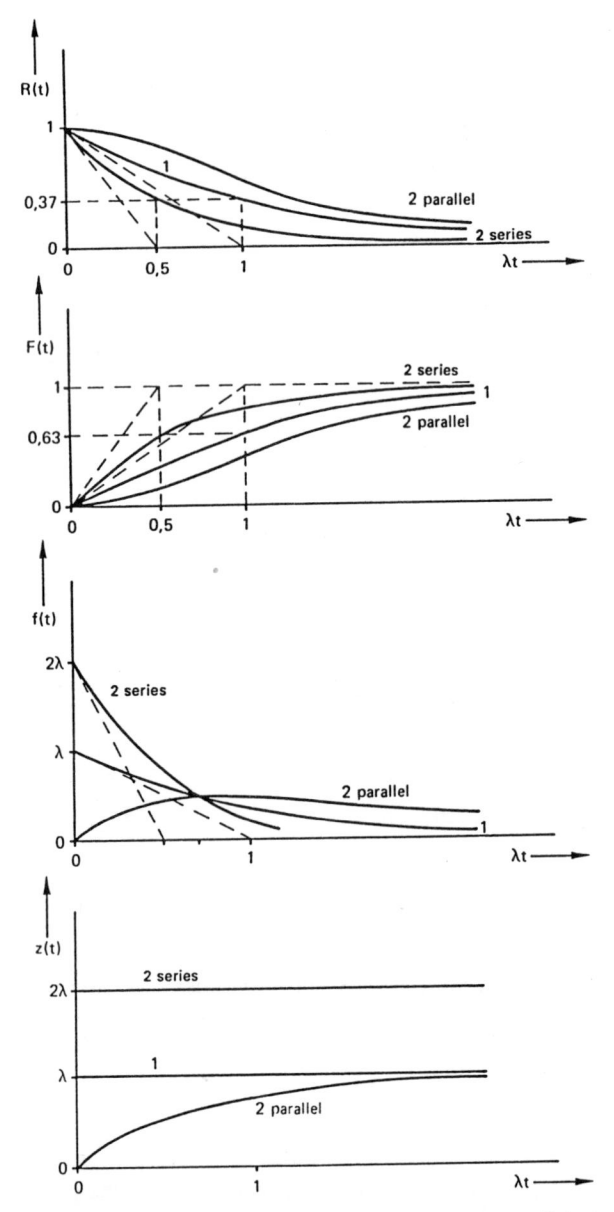

Figure 6.2 Reliability parameters of a pure series and a pure parallel structure consisting of two identical components compared with the parameters of one such a component.

Passive redundancy

So far it has been tacitly assumed that we were concerned with active redundancy. For *passive redundancy* the picture looks completely different. A passive redundant system has been illustrated in Figure 6.3. In the general case, the switches S_1 and S_2 may fail because they switch over without good cause, remain stuck in one position, present a short-circuit or give an open circuit. The components 1 through n may fail in the active

Parallel systems

mode but also in the passive mode. Let us first assume that the switches are ideal, that the components do not fail in the passive mode and, for convenience, that $n = 2$.

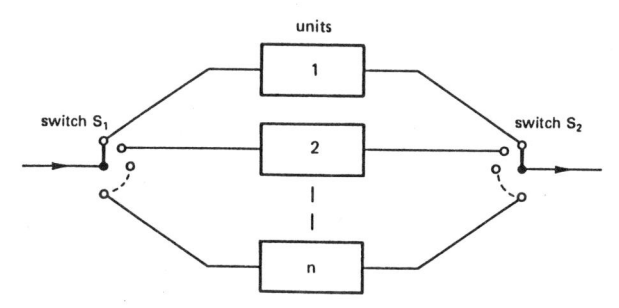

Figure 6.3 *Passive redundancy: $S_1 - S_2$ is a coupled switch, multi-path valve or routing element. C_1 through C_n are the task performing units.*

The failure moment t_1 of component 1 is a stochastic variable $\underline{t}_1$, and that of component 2 is $\underline{t}_2$. The reliability $R_n(t)$ of this system is represented by:

$$R_2(t) = P((\underline{t}_1 > t) \cup (\underline{t}_1 \leq t \cap \underline{t}_2 > t - \underline{t}_1)).$$

The events 'component 1 is still good' and 'component 1 has failed, but 2 is still good' are mutually exclusive:

$$R_2(t) = P(\underline{t}_1 > t) + P(\underline{t}_1 \leq t \cap \underline{t}_2 > t - \underline{t}_1).$$

Consequently:

$$R_2(t) = R_{c_1}(t) + \int_0^t f_{c_1}(t_1) R_{c_2}(t - t_1) \, dt_1.$$

$R_{c_1}(t)$ and $R_{c_2}(t)$ are the reliability of the first and the second component and $f_{c_1}(t)$ is the failure probability density of the first component.

The second term in this expression is the contribution of the second component to the reliability. For three components we can therefore derive:

$$R_3(t) = R_2(t) + \int_0^t f_{c_1}(t_1) \int_0^{t-t_1} f_{c_2}(t_2) R_{c_3}(t - t_1 - t_2) \, dt_2 \, dt_1.$$

Here $R_2(t)$ is the reliability of the dual passive redundant system.

The following approach is somewhat different. The failure probability density of a passive redundant system with n components may be written as:

$$f_n(t) = \int_{t_{n-1}=0}^{t} \int_{t_{n-2}=0}^{t_{n-1}} \ldots \int_{t_1=0}^{t_2} f_{c_1}(t_1) f_{c_2}(t_2 - t_1) \ldots f_{c_n}(t - t_{n-1}) \, dt_1 \, dt_2 \ldots dt_{n-1}.$$

This expression makes use of the fact that the failure probability density of the sum of a number of random variables is equal to the convolution of the individual failure probability densities.

The same expression as for the reliability above is then found with:

$$R_n(t) = 1 - F_n(t) = 1 - \int_0^t f_n(t)dt = \int_t^\infty f_n(t)dt.$$

This can be simply checked if one starts with a system comprising only two components.

Furthermore, it can also be shown that for passive redundancy, in which the components cannot fail in the 'off-state', the mean life of the redundant system is given by:

$$\text{MTTF} = \sum_{i=1}^n \text{MTTF}_i,$$

provided the switches S_1 and S_2 in Figure 6.3 are ideal.

The expressions derived above are independent of the failure distribution. If the system consists of n identical components, all with a negative-exponential failure distribution, it follows that:

$$R_n(t) = e^{-\lambda t} \sum_{i=1}^n \frac{(\lambda t)^{i-1}}{(i-1)!}.$$

If the switches in Figure 6.3 are not ideal, but may, for example, get stuck in one contact position, the calculation is as follows. Suppose that the switch functions with a probability p_s at the moment it has to switch. In that case the expression for $n = 2$ becomes:

$$R_2(t) = R_1(t) + p_s \int_0^t f_{c_1}(t_1) R_{c_2}(t - t_1) \, dt_1.$$

For higher values of n the switching probability p_s can be introduced similarly. For two identical units with failure rate λ and a 'sticking' switch with failure rate λ_s the mean life can be determined to be:

$$\text{MTTF} = \frac{1}{\lambda} + \frac{1}{\lambda + \lambda_s}.$$

So it is, as it were, as if the failure rate of the switch is added to that of a unit. This can also be simply seen directly. After all, if the switch is stuck in the first unit position the second unit can no longer be engaged; a switch stuck on the last unit does not shorten the life of the system.

Spontaneous switching of the switch alone cannot result in a failure. It may disturb the order of engaging the units. In the case where the units are not identical, and the switch also can get stuck on a unit, spontaneous switching can influence the reliability. We shall not analyse this further.

If it is assumed that the switch can fail at any time by an open or a short circuit, the reliability of the switch $R_s(t)$ is in series with the rest of the system $R_n(t)$ and therefore the product rule holds:

$$R(t) = R_s(t) R_n(t).$$

6.4.1 Dependent failures

As we have briefly seen in Section 6.1, the beneficial effect of redundancy in a system can, to a great extent, be undone if the failures in the units of such a system are or become *stochastically dependent*. The usual assumption that failures in a system are mutually independent is not always allowed:

Examples
- The redundant units of a certain system, due to lack of space, are placed quite closely together. The heat resulting from a short circuit in one unit (a primary failure) may cause a failure in another unit (the secondary failure) or it may accelerate the occurrence of a failure in another unit.
- An area where one uses overhead lines to transport electrical energy (high tension grid) is hit by a sluggish winter. Glazed ice and a subsequent winter storm cause some of the lines to break. For a higher reliability the transport system has been given a redundant structure in the form of a ring network. The common failure cause (in this case the winter) makes failures in the two redundant lines that supply any single point with energy stochastically dependent and lessens the effect of the redundancy.
- In an airplane the antenna-grounding has deteriorated by corrosion (primary failure). This causes part of the transmitted energy to enter the airplane during a transmission, which results in a flow of warnings and failure alarms in the cockpit (secondary failures). After the crew makes the correlation with the use of the transmitter, it is decided to continue the flight as planned.
- In the junction box for the distribution of electric energy in a newly built house an electrician has forgotten to fasten all clamping connections properly (primary failure). The energy dissipating in the contact resistance of these bad connections generates so much heat that the plastic insulation of the wiring melts. This fire hazard is discovered in time because of the attendant acrid smell. The electrician renews the wiring and, this time, tightens the connections properly. Some time after this incident, the lady of the house is electrocuted when she touches a defective metal floor lamp with a wet cloth on the tiled floor (primary insulation failure in the lamp). On inspection the ground-fault circuit breaker, in the junction box of the house located right over the terminal board, turns out to have corroded by the corrosive chloric gas that was released from the melting PVC insulation. The corrosion prevented the switching off (secondary failure) of the ground-fault interrupter. This circuit breaker has a caption warning the owner that the test button has to be pushed each month to test its correct functioning. The house owner had neglected to do this (misuse).

From the above examples it will be clear that the cause of dependent failures often escapes detection. This is what makes such failures so dangerous. One thinks one can rely on the installed systems because of their inherent reliability which may have been achieved by using redundancy, whereas in practice one cannot: The dependency of possible failures reduces the effectiveness of the redundancy.

If the failures in the various units of a system are dependent, the product rules for series and parallel systems given in the previous section do not hold. If x_i is the event that the

98 Non-Maintained Systems

i-th unit of a system consisting of n units functions well and $\bar{x}_i$ is the complementary event, then for a *series system*:

$$R_s = P(x_1 \cap x_2 \cap \ldots \cap x_n).$$

Writing this out gives:

$$R_s = P(x_1)\, P(x_2|x_1) \ldots P(x_n|x_1 x_2 \ldots x_{n-1}).$$

Only for independent events may these conditional probabilities be replaced by (unconditional) probabilities. The reliability can then be written as:

$$R_s = P(x_1)\, P(x_2) \ldots P(x_n).$$

Analogous to this, for a *parallel system* the following holds:

$$F_p = P(\bar{x}_1 \bar{x}_2 \ldots \bar{x}_n),$$

and therefore also:

$$F_p = P(\bar{x}_1)\, P((\bar{x}_2|\bar{x}_1) \ldots P(\bar{x}_n|\bar{x}_1 \bar{x}_2 \ldots \bar{x}_{n-1}).$$

For modelling such dependent failures we have to make a number of assumptions. We shall restrict ourselves to a subset of the set of dependent failures, namely to the so-called *common-cause failures; these are failures that occur simultaneously in different units and that are the result of a common failure cause.*

It is assumed that:
- Common-cause failures (CC-failures) only manifest themselves as catastrophic failures that occur simultaneously in two or more different units (CC-failures within one unit are considered to be comprised in the hazard rate of that unit).
- CC-failures and other non-CC-failures are mutually stochastically independent.
- CC-failures have a constant failure rate.
- The units of the considered systems are mutually identical from a reliability viewpoint (homogeneous system).

A system with such CC-failures can be modelled in a simple way, as is shown in Figure 6.4.

It is assumed that the failure rate λ_t of the system units is measured while the unit is part of the system (so in the failure increasing presence of the other units and of the operational environment). It can be written as:

$$R = P(x_i) = e^{-\lambda_t t} \quad (i = 1, 2, \ldots, n).$$

For the series system in the upper left of Figure 6.4a the reliability is:

$$R_s = P(x_1 x_2) = P(x_1)\, P(x_2|x_1),$$

in which:

$$P(x_1) = e^{-\lambda_t t} = e^{-(\lambda + \lambda_{cc})t},$$

and

$$P(x_2|x_1) = e^{-\lambda t}.$$

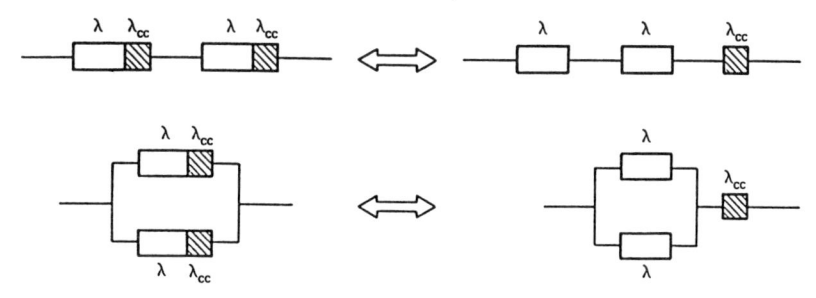

Figure 6.4 (a) Modelling of common-cause failures in series and parallel systems. (b) Reliability of two parallel units with common-cause failures, in which α is the fraction of the unit failure rate resulting from dependent failures.

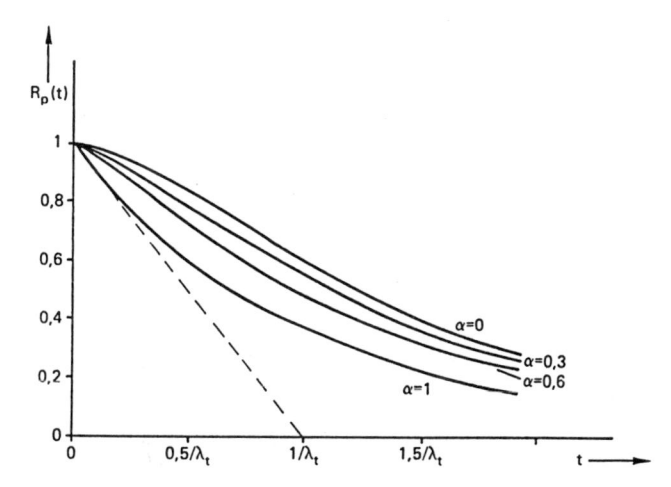

Figure 6.4 (b) Reliability of two parallel units with common-cause failures, in which α is the fraction of the unit failure rate resulting from the dependent failures.

So:

$$R_s = e^{-\lambda_t t} \cdot e^{-\lambda t} = e^{-(2\lambda + \lambda_{cc})t}.$$

Apparently this system can be represented by the model shown in the upper right-hand side of Figure 6.4.a. So in a series system we have to take into account λ_{cc} (the common-cause contribution to the total failure rate λ_t of a unit) only once. We have, as it were, corrected for the fact that the dependent failures were counted twice.

In a similar way we find for a parallel system with:

$$F_p = P(\bar{x}_1 \bar{x}_2) = 1 - P(x_1 + x_2) = 1 - P(x_1) - P(x_2) + P(x_1 x_2),$$

for the failure rate:

$$F_p = 1 - e^{-\lambda_t t} - e^{-\lambda_t t} + e^{-(2\lambda + \lambda_{cc})t} =$$

$$= 1 - 2e^{-\lambda_t t} + e^{-(\lambda + \lambda_t)t}.$$

This turns out to be the failure probability represented by the model in the lower right-hand side of Figure 6.4a. We apparently may move the CC-failures out of the parallel branches and account for them once in the form a single series element.

So, under the conditions described above, we can account for CC-failures as a single series element in the reliability model of a system, independent of whether it is a series or a parallel system. This is allowed provided we correct the failure rate λ_t of a unit as measured in the system into:

$$\lambda = \lambda_t - \lambda_{cc}.$$

If it is assumed that the part λ_{cc} in the total failure rate λ_t of the units is variable, so if:

$$\alpha = \frac{\lambda_{cc}}{\lambda_t}, \qquad 0 \leq \alpha \leq 1,$$

we arrive at the diagram of Figure 6.4b. In this diagram the reliability $R_p(t)$ of a system with two units in an active parallel redundant configuration has been plotted as a function time for various values of α.

From this figure it is clear that for $\alpha = 0$ the CC-failure part is zero. We have then a purely redundant system with only stochastically independent failures. For $\alpha = 1$ the breakdown of both units has become fully correlated. The system will only perform as well as one single unit and the redundancy has been completely eliminated.

N.B.: In safety systems for high-risk processes a high degree of redundancy is usually employed. (For example the safety systems of a nuclear reactor.) A value of α significantly different from zero is therefore very risky for such safety systems; one relies on the built-in redundancy that is merely imaginary due to the dependency of the failures.

So the effectiveness of redundancy is decreased by CC-failures. The principal sources of CC-failures are:

- Secondary failures caused by primary failures occurring in other units. These happen because such failure mechanisms were taken insufficiently into account in the design. The designer may not even have been aware of these primary-secondary failure relationships. For example, overheating of adjacent units in a compactly built redundant system creates dependent failures when one unit becomes too hot because of an internal short-circuit or an overload condition.
  Compare: Some time after an airplane had taken off a warning sounded that one of the engines was on fire. Shortly after that a second engine was reported to be on fire. The flight was cancelled. Back on the ground it became evident that the warning circuits for engine fire, which were located in separate casings, had been placed above one another in a single bin. At the bottom of this bin a control resistor had become overheated by lengthy tests on the ground. This had first triggered the single engine fire alarm and subsequently had spread to the other engine fire alarm located above it. In reality there had been no fire at all, it turned out to be a so-called *nuisance alarm*.
- Electrical, mechanical and thermal dependence between the various units in a system. Examples: A fan breaks down in a rack of cards with many heat producing digital TTL-logic IC's. A short-circuited input of one amplifier out of a redundant amplifier system causes the entire system to fail. Careless operation or poor maintenance may

also be causes. Examples are: The use of wrong components in repair (e.g. replacing bolts by cheap ones that do not have the same mechanical strength), imperfect adjustment because the measuring equipment is out of calibration (e.g. systematically indicates too low).
- Common external causes such as: Dust, dirt, condensation, heat, vibration and mechanical shocks.
- Environmental causes such as: Floods, earthquakes, fire, storm, glazed ice, severe frost, heavy snowfall, etc.
- The use of redundant units of the same make. If one procures the redundant units from the same manufacturer the same failure mechanisms are likely to occur in all the units. One can think of commonality in design, choice of materials and production techniques. An example: Bad soldering joints in the automatic soldering process of printed-circuit boards for surface-mounted IC's.
- Common sources of energy, material and personnel. Examples: A common energy supply for redundant units, common transport pipes, coolant pipes, etc. for redundant processes, as well as general strikes by the personnel or hostage-taking of personnel and installations by terrorists.

CC-failures are the most serious (and therefore also the most feared) form that dependent failures can assume. After all, owing to a common cause all similar units from a system fail simultaneously in a catastrophic way. Many sorts of dependent failures have not so drastic a result. The occurrence of the common failure cause in one unit, for instance, only increases the probability of failure of other units in a system. So, if the system is redundant, it will be possible to replace the subsequently failing units during preventive maintenance and thus keep the system running. All the same, all kinds of, dependent failures in redundant systems degrade the effectiveness of the redundancy and should therefore be avoided as much as possible.

6.5 *m*-out-of-*n* systems

It was explained before that systems with m-out-of-n redundancy are neither pure series systems nor pure parallel systems. In such an m-out-of-n redundant system at least m components have to function well at all times if the system is to function well. It will be clear that the degenerate cases for $m = n$ and $m = 1$ result in a pure series system or a pure parallel system respectively. Thus, m-out-of-n redundancy is a more general form of redundancy than pure parallel redundancy.

We once more use the case of active redundancy and identical components. The probability of k surviving components out of n follows a binomial distribution:

$$P_k = \binom{n}{k} [R_0(t)]^k [1 - R_0(t)]^{n-k}.$$

Here $R_0(t)$ is the reliability of one component. The m-out-of-n system functions as long as at least m components still function. So the system is reliable if m, $m+1$, $m+2$, ... , or n components are functioning correctly. The probability of this subset of events gives the system's reliability:

102 Non-Maintained Systems

$$R(t) = \sum_{k=m}^{n} \binom{n}{k} [R_0(t)]^k [1-R_0(t)]^{n-k}.$$

Example
On board most means of transport one needs a certain amount of electric power for lighting, communication, navigation and other purposes. It appears that for a certain means of transport (an airplane) 15 kW maximum power is needed; 7 kW is needed for reduced but acceptable use, and a minimum of 5 kW of power is required under emergency conditions. Such a means of transport can be equipped with one single 15 kW generator, two 8 kW generators in parallel or three 5 kW generators in parallel. For convenience it is assumed that that the reliability R_g of all three types of generator is equal. We can then compose Table 6.2 for the reliability under full use, (acceptably) reduced use and emergency use.

	1 × 15 kW	2 × 8 kW	3 × 5 kW
Full use	R_g	R_g^2	R_g^3
Acceptable use	R_g	$2R_g - R_g^2$	$3R_g^2 - 2R_g^3$
Emergency use	R_g	$2R_g - R_g^2$	$3R_g - 3R_g^2 + R_g^3$

Table 6.2 *Operational inventory of all modes of an energy generating system.*

The last column is the most illustrative: The system changes from a 3-out-of-3 system, so a pure series system, via a 2-out-of-3 system to a 1-out-of-3 system (a pure parallel system). In Figure 6.5 the system reliability R for these different kinds of use has been plotted against the generator reliability R_g. What is the best design solution: choosing one, two or three generators in this means of transport? The cause of the S-shape of the curve for the reliability of a three-generator system for 'acceptable use' $(3R_g^2 - 2R_g^3)$ shall be discussed in the next section.

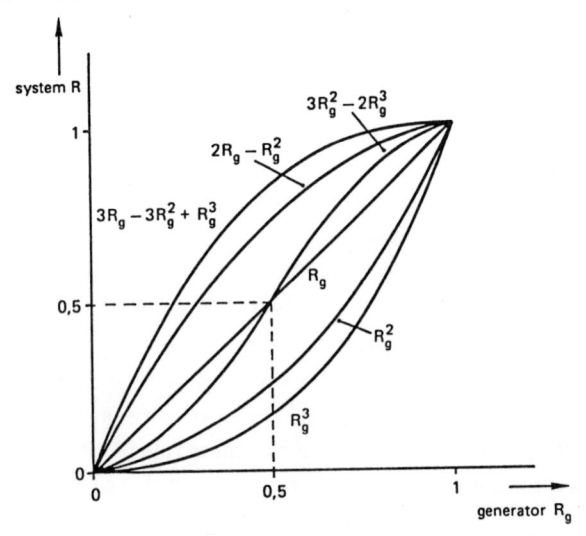

Figure 6.5 *A number of solutions for the energy generating system powering a vehicle. Also see Table 6.2.*

If now an m-out-of-n redundant system is assumed in which all n units are active and exhibit a failure rate λ, it can be derived that such a system obeys:

$$R(t) = \sum_{k=m}^{n} \binom{n}{k} [e^{-\lambda kt}] [1 - e^{-\lambda t}]^{n-k},$$

$$\text{MTTF} \equiv \theta = \frac{1}{\lambda} \sum_{k=m}^{n} \frac{1}{k}.$$

} active m-out-of-n redundancy

For passive m-out-of-n redundancy in which only m units are active with failure rate λ and the other units are passive with zero failure rate one can derive:

$$R(t) = e^{-\lambda mt} \sum_{k=m}^{n} \frac{(m\lambda t)^{k-m}}{(k-m)!},$$

$$\text{MTTF} \equiv \theta = \frac{(n-m+1)}{m\lambda}.$$

} passive m-out-of-n redundancy

The importance of m-out-of-n systems lies in the fact that it is not always feasible to design a system with a redundancy degree $\eta = 1$. For example, it is quite easy to make a redundant amplifying system which allows one failure out of three sub-amplifiers circuited in parallel. This results in a 2-out-of-3 system. Also think of the 'clusters' of parallel wheels of a landing-gear or of a heavy truck. A DC-10 airplane is still able to fly and land normally if at least two of its three engines are functioning.

6.6 Majority voting systems

A majority voting system is a redundant system that functions correctly as long as a majority of the n redundant components or channels is still functioning. In Figure 6.6 an example of such a system is given. The practical use of this form of redundancy is immediately clear here. The signal conditioning and processing in the illustrated transfer channel can be very complex. This makes it extremely difficult to check whether such a channel still functions correctly. In other words: failure detection is not feasible, it would require complex and therefore little reliable subsystems. In Figure 6.6 a voting circuit is used to compare the output signals. A selector subsequently selects an output that concords with the majority of the outputs. In analogue electronic systems the output signals will always differ a little. The selector then usually averages the corresponding signals, i.e. the signals that are the closest together. A certain threshold value built into the selector determines here whether signals should be considered to be similar or unallowably deviating. To that end often the median signal is chosen for a reference. The selector determines then the deviation with regard to this median signal of all but the median signals and deselects that signal which deviates more than a predetermined amount (during more than a predetermined time interval). The selector determines the (weighted) average of all acceptably deviating signals as the best output signal. These kinds of selector or voter circuits are also called *similarity voters* because they actually accept or reject the signals based on their similarity to the other signals.

104 Non-Maintained Systems

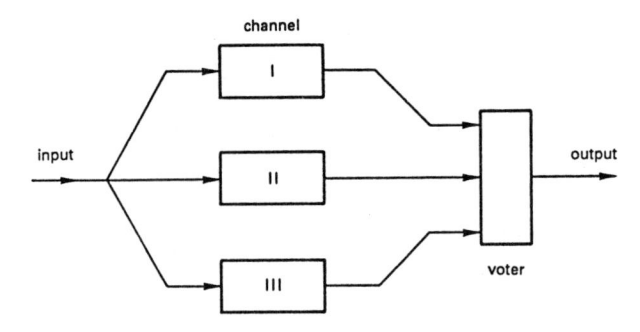

Figure 6.6 Example of a majority voting system applied to information transferring channels.

It is assumed that a voter does not fail, so that it has negligible complexity with regard to the complexity of the redundant channels. If all three channels in Figure 6.6 function, the system functions well. If only two channels function correctly, the system also still functions well and the voter may produce a cautioning signal that a channel has failed (*failure cautioning*). The voter can also report which channel has failed (*failure identification and reporting*). This speeds up maintenance considerably. If two channels fail the system fails. If they would produce the same erroneous output signal, this signal is forwarded to the system output and the voter indicates the correct channel as having failed. If the two faulty channels produce different output signals the voter 'knows' that the system is faulty (three different signals) and will give an alarm (*failure alarm*). The same reasoning can be followed for three faulty channels. Since the channels are usually designed differently it is unlikely that the faulty outputs are the same. So, in most cases the alarm goes off when the system fails.

For majority voting systems with more than three redundant units it can be advantageous to make the voter adaptive (*adaptive majority voting*). This is done as follows: each time the voter detects a failed unit, it is switched off and is therefore no longer part of the set of valid alternatives for the voter. In this way, the system remains functioning correctly as long as at least still two units are good. So redundancy with adaptive majority voting realises a 2-out-of-n system if $n \geq 3$. It can be simply checked what happens after only two units are left.

Adaptive majority voting involves a great risk: if the units show intermittent failures, such as bursts of interference, the system will self-destruct very quickly. This happens especially when many short-lasting failures occur, such as switching transients, glitches, etc. It can be prevented by only switching a unit off for good after it has been 'dissident' for a particular period of time. This means that the system is not adaptive over short time intervals. Such a unit's output is not averaged in the system output but, as it were, put on hold until it mends its ways.

Non-adaptive majority voting requires that $(n/2) + 1$ units out of n are still good if n is even and $(n + 1)/2$ out of n must be good if n is odd. With the redundancy degree:

$$\eta = \frac{n - m}{n - 1},$$

for even n this results in:

$$\eta_{even} = 1 - \frac{n}{2(n-1)},$$

and for odd n:

$$\eta_{odd} = 1/2.$$

Since $\eta_{even} < \eta_{odd}$ it is most efficient to design a majority voring system with an *odd number of* correctly operating redundant units. The difference becomes smaller as n becomes larger. For adaptive majority voting the above does not hold; here the degree of redundancy increases monotonically as n becomes larger: After all, it realises a 2-out-of-n system.

In Figure 6.7 it is illustrated how a non-adaptive majority voting system with a perfect voter behaves as a function of time for units with a negative-exponential failure distribution.

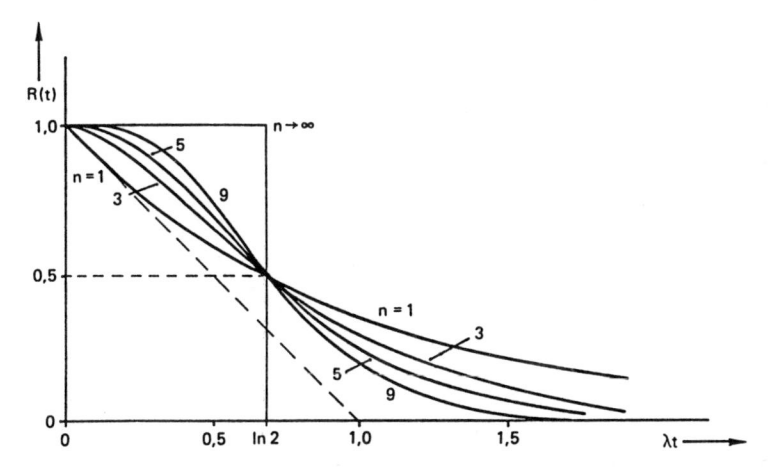

Figure 6.7 *Non-adaptive majority voting applied to a system with n identical redundant units having a failure rate λ. Here $\lambda.t = \ln 2$ is not only the median life of the individual system units, but also that of the majority voting system: The median life does not change!*

This figure shows that as long as the reliability R of the units is still larger than 0.5, the application of majority voting gives an improvement of the system reliability. If $R = 0.5$ majority voting does not make a difference: It is a tie. If $R < 0.5$ the system will become worse by applying majority voting: One votes for the faulty majority. As the number of units increases this behaviour becomes more and more pronounced. Eventually, for $n \to \infty$ the system fails exactly at the moment the units reach their *median life*. The latter statement is true independent of the failure distribution of the units. Why does not the system fail when the units reach their mean life? We see now that the curve labelled $3R_g^2 - 2R_g^3$ in Figure 6.5 is in fact a form of majority voting redundancy. This explains the S-shape of the curve.

6.7 Mixed systems

In the previous sections we have, besides *m*-out-of-*n* redundancy, also discussed special cases of this general form of redundancy. There are systems, however, that cannot be reduced to *m*-out-of-*n* redundancy. These so-called *mixed systems* usually occur in complex equipment with many interconnections such as that used in the process industry. The first thing one can do when determining the reliability of a more complex system is to take all series components together into one component, just as all parallel components. This process of lumping together series elements and parallel elements in the reliability model is illustrated in Figure 6.8a.

There are of course systems with a structure that cannot be taken together into one final element. In Figure 6.8b a bridge structure is given as an example. This structure can be imagined as follows: units A and A' can take over each other's task, just as units B and B'. The nodes 1 and 2, however, cannot be connected just like that. The interconnection requires a buffer operation C. So, the system functions correctly as long as at least one of the following four paths still functions: AB, $A'B'$, ACB' and $A'CB$.

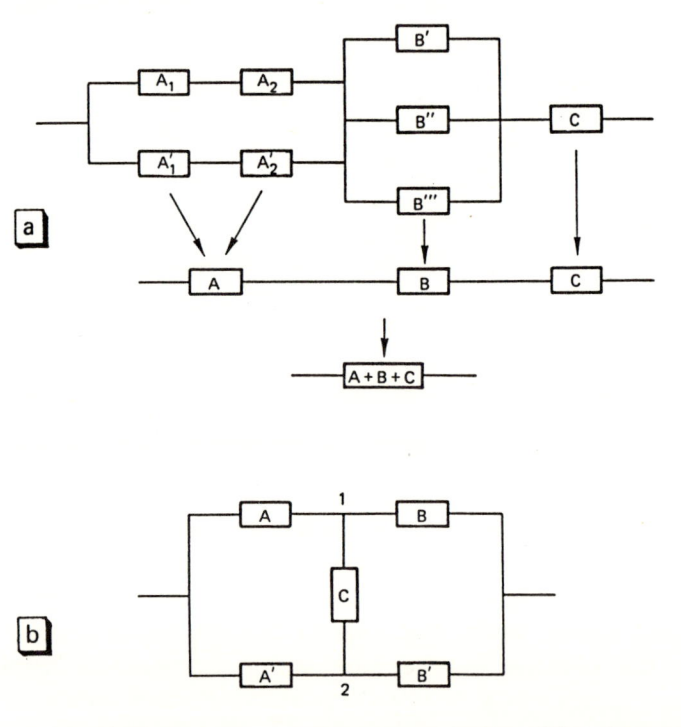

Figure 6.8 (a) Segregating complex systems in series and parallel subsystems. (b) Example of a system that cannot be simplified further (bridge structure).

If we assume that a unit can only function correctly (state 1) or has failed (state 0), so if we apply the catastrophic failure model, a simple calculation method is that which uses a *complete inventory list* of all states in which the system can possibly be. For the system of Figure 6.8b this inventory is shown in Table 6.3.

Unit states					System state
A	A'	B	B'	C	
1	1	1	1	1	1
1	1	1	1	0	1
1	1	1	0	1	1
1	1	1	0	0	1
1	1	0	1	1	1
1	1	0	1	0	1
1	0	0	1	1	1
1	0	0	1	0	0

– – – and so on (32 states in total)

Table 6.3 *Inventory list of all states of the system of Figure 6.8b.*

Subsequently, based on this list of all possible states one calculates the probability that the system is in state 1, expressed in the reliabilities of the units A, A', etc. Though simple, this method is laborious.

Another obvious method is the *inspection method*. It requires some insight into the system operation. For the example of Figure 6.8b this methods goes as follows: the system is reliable if one or more of the paths AB, $A'B'$, ACB' and $A'CB$ function. This is the complete set of all possible totally or partially different paths from the input to the output of the system. The reliability is now simply:

$$R = P(AB \cup A'B' \cup ACB' \cup A'CB).$$

After working this out and simplifying the result is:

$$R = P(AB) + P(A'B') + P(AB'C) + P(A'BC) +$$
$$- P(AA'BB') - P(ABB'C) - P(AA'BC) - P(AA'B'C) +$$
$$- P(A'BB'C) + 2P(AA'BB'C).$$

The notation $P(AB)$ denotes the probability that both A and B function correctly.

6.8 Optimisation

The pursuit of a higher reliability for a given application involves expenditure. The costs increase because the development costs are higher as a result of the necessary life tests, reliability simulations and design revisions. The production costs, too, will be higher due to extra or more stringent inspection procedures, training of production personnel and things like that. Furthermore, the initial capital outlay for the system will also be higher, due to the use of high-reliability components, redundant components procured from different manufacturers and so on.

However, the above is only one side of the coin. The future user/owner of the system is not only interested in the acquisition costs C_a of the system, but also in the maintenance costs C_m during the useful life of the system. The system owner will strive for the lowest possible *overall cost of ownership* C_t, with regard to the entire life cycle of the system: design, production and use. In Figure 6.9 is sketched how C_a and C_m for typical technical systems depend on the desired MTTF of the system.

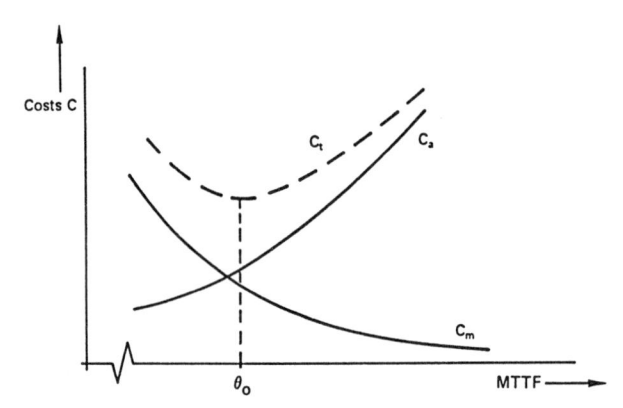

Figure 6.9 Acquisition costs C_a, maintenance costs C_m and total costs C_t of a fictitious technical system versus the MTTF of the system.

It will be clear that at a certain MTTF = θ_o the total costs with regard to the entire life cycle of the system are lowest. To the left of this minimum it may be useful to except higher initial costs because one will later save more than this extra expense on reduced maintenance.

Not all systems are at or in the vicinity of the minimum in the C_t curve of Figure 6.9. To the right of the optimum MTTF = θ_o the requirement was apparently that very little maintenance or none at all should be performed, for example, because maintenance is simply not possible or because system failure will have very costly consequences. What one can also find is often that the buyer has overspecified the system's MTTF and has not been cost-conscious enough.

The other case where the actual MTTF is lower than optimal is often seen in consumer goods. For products in the typical 'consumer field' most buyers practically only pay attention to the initial costs, the price C_a, to obtain the system. Usually, very large systems made to the order of institutions such as the government (for space travel, telecommunication or military purposes) are to the right of the optimum θ_o in Figure 6.9. The systems in the industrial field (available in the professional market) are mostly very close to θ_o because of the prevailing competition between the producers/manufacturers and the cost-consciousness of their customers.

Optimisation of the reliability R of a system is an important problem area in reliability engineering. We shall discuss this problem in the following. At the inception of a system, and also while keeping it in an operational state, a certain amount of costs C are incurred. Costs are here understood to mean *generalised costs*, i.e. not just money but also other necessary scarce matters such as weight or volume in the case where the total weight or the overall size of the system is critically constrained.

The system can be divided into units (components, modules, subsystems). This *system partitioning* is usually done such that each unit performs an independent function. The reliability of the i-th unit is assumed to be R_i and its price is C_i.

We now have to distinguish two optimisation problems :
- *reliability constrained optimisation,*
- *budget constrained optimisation.*

In the first problem the designer wants a system reliability higher than a certain minimally required value for the least budget. In the second approach he can afford no more than a certain maximum budget and he wants the highest possible system reliability. We shall briefly discuss both problems:

■ *Reliability constrained optimisation*

It is assumed that for all n functions necessary to create the overall system function, a number of alternatives is available. For the i-th unit m_i alternative units realising the same i-th unit's function are available. These may be units from different manufacturers, but they may also be parallel circuited units (so redundant modules). Each alternative for the i-th function has a corresponding reliability R_{ij} and corresponding generalised costs C_{ij} ($j = 1,2, \ldots ,m_i$). Now we arrange the m_i alternatives for each of the n functions in order of increasing reliability:

$$R_{i1} < R_{i2} < \ldots < R_{im_i}.$$

The associated costs should now be in an increasing order as well, so:

$$C_{i1} < C_{i2} < \ldots < C_{im_i}.$$

If one or more of the alternatives does not meet this requirement there are alternatives that are too costly given their reliability and they can be simply disregarded. After this elimination procedure further optimisation is done as follows:

For the system reliability of a certain realisation (choice of alternatives) we write:

$$R = \prod_{i=1}^{n} R_{ij}.$$

For the system costs holds:

$$C = \sum_{i=1}^{n} C_{ij}.$$

The values of j have now to be found such that $R \geq R_{min}$ and C is as low as possible. This is done as follows:
- For each function take the least expensive alternative.
- Calculate the corresponding system reliability R.
- For each function determine the relative contribution to the system's reliability per extra invested unit of cost of the next more expensive alternative. This contribution is:

$$\frac{\Delta R}{\Delta C_i} = \frac{R}{R_i} \cdot \frac{\Delta R_i}{\Delta C_i}.$$

- Select that unit to be part of the next system realisation for which this relative reliability increase is largest.
- Repeat the previous three points until: $R \geq R_{min}$.

110 Non-Maintained Systems

Now a disappointment: this method is relatively fast, but does not necessarily give the correct solution. If the differences in $\Delta R/\Delta C_i$ between the various functions are large, and if the number of functions (n) is large the only right method is to search all possible combinations by computer, calculate the associated reliability and cost and select the alternative with the lowest cost just exceeding the minimum reliability requirement.

- *Budget constrained optimisation*
  Again the system is partitioned into n functions that can be realised separately and for which there are a number of alternatives, each with its own reliability and costs. Again it holds that all functions are needed:

$$R \prod_{i=1}^{n} R_{ij},$$

and

$$C = \sum_{i=1}^{n} C_{ij}.$$

What are now the values of j for which: $C \leq C_{\max}$ and R is as large as possible?
The optimisation is, of course, identical to the preceding optimisation (the problem is symmetrical in R and C). This time we stop if the lastly chosen alternative realisation results in an budget overrun. Here, too, it holds that this is a practical method for a quick solution. The optimum solution is not necessarily found, to that end an exhaustive search of all possibilities has to be performed.

A calculation method sometimes found for improving the system reliability by means of adding redundancy is the so-called proportional assignment (proportional to the failure rate). This method introduced by ARINC* does not start with a consideration of the costs, but with the consideration that addition of a (passive) redundant unit increases the mean life with a certain factor (2). If the n units form a series system again, and the i-th unit has a failure rate λ_i, then:

$$\lambda = \sum_{i=1}^{n} \lambda_i \geq \lambda_{\text{spec}}.$$

The new allocation of failure rates λ'_i to the n units is done according to the assignment procedure:

$$\lambda'_i = \frac{\lambda_i}{\sum_{i=1}^{n} \lambda_i} \lambda_{\text{spec}} = \lambda_i \frac{\lambda_{\text{spec}}}{\lambda}.$$

So the new failure rates λ'_i are assigned in proportion to the old failure rates λ_i. All failure rates are reduced by the same factor $\lambda_{\text{spec}}/\lambda$. The addition of redundancy increases the mean life with a certain factor (although the failure distribution also changes into a non-

---

* ARINC: Aeronautical Radio Incorporated.

exponential distribution). So λ_{spec}/λ gives a measure for the required redundancy in the system. Once more: this method of proportional assignment is not based on a (cost) optimisation.

6.9 Analysis methods

In the next four sections we shall discuss a number of systematic calculation methods to find reliability of mixed systems.

6.9.1 Network reduction method

Applying the network reduction method requires the structure of a system to be taken together into substructures of which an analytical expression for the reliability is known. Examples of such substructures are: series structures, parallel structures, m-out-of-n structures, and so on. After this reduction only non-reducible structures are left, such as that in Figure 6.8b. One can subsequently apply *delta-star transformations* to these non-reducible structures. (Of course, also the more intuitive calculation methods already given in Section 6.7 can be used.)

In Figure 6.10 the delta-star transformation for reliability models is shown. Starting with a model with the delta structure indicated on the left with component reliabilities R_{AB}, R_{BC} and R_{AC} the following equations transform this structure into a star structure with element reliabilities R_A, R_B and R_C:

$$R_A = \sqrt{\frac{\{1 - (1 - R_{AC})(1 - R_{BC}R_{AB})\}\{1 - (1 - R_{AB})(1 - R_{AC}R_{BC})\}}{\{1 - (1 - R_{BC})(1 - R_{AC}R_{AB})\}}},$$

$$R_B = \sqrt{\frac{\{1 - (1 - R_{AB})(1 - R_{AB}R_{BC})\}\{1 - (1 - R_{BC})(1 - R_{AC}R_{AB})\}}{\{1 - (1 - R_{AC})(1 - R_{BC}R_{AB})\}}},$$

$$R_C = \sqrt{\frac{\{1 - (1 - R_{AC})(1 - R_{BC}R_{AB})\}\{1 - (1 - R_{BC})(1 - R_{AC}R_{AB})\}}{\{1 - (1 - R_{AB})(1 - R_{AC}R_{BC})\}}}.$$

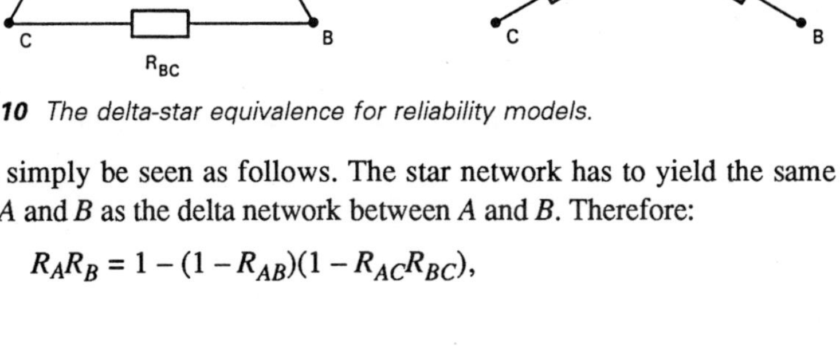

Figure 6.10 *The delta-star equivalence for reliability models.*

This can simply be seen as follows. The star network has to yield the same reliability between A and B as the delta network between A and B. Therefore:

$$R_A R_B = 1 - (1 - R_{AB})(1 - R_{AC}R_{BC}),$$

112 Non-Maintained Systems

and also:

$$R_B R_C = 1 - (1 - R_{BC})(1 - R_{AC}R_{AB}),$$
$$R_A R_C = 1 - (1 - R_{AC})(1 - R_{BC}R_{AB}).$$

These are three equations with three unknowns. Solution yields the previously given expressions. An interesting exercise is to try this method out on the network of Figure 6.8b.

6.9.2 Tie set and cut set method

It was explained in Section 5.1 that a catastrophic failure model may be regarded as a transfer model; as long as there is at least one connecting path from the input to the output the system functions correctly. The tie set-cut set method starts therefore with a complete inventory of all entirely or partially different tie sets that can be followed from the input to the output through the reliability model. A tie set is the set of components, elements (or branches) that has to be traversed going from input to output; as it were tying the input and output together.

In Section 6.7 this was not done in a systematic way. In order to trace all tie sets systematically we shall, as promised in Section 5.1, use the graph theory. In Figure 6.11 the graph representation of the structures in Figures 6.8a and 6.8b is given. The transfer gain of the element or branch is the reliability of the respective component.

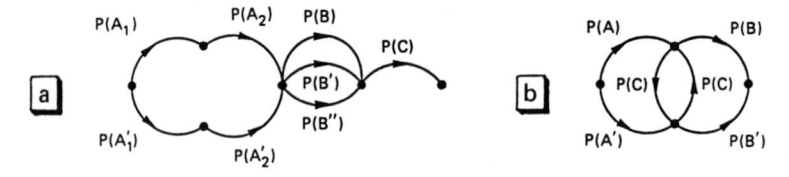

Figure 6.11 Reliability graphs of both systems of Figure 6.8.

There are at least as many branches as there are components. There may be more branches however (see Figure 6.11b). This occurs if a branch transfer occurs in several tie sets and has to be repeated for that reason. As we have seen, a tie set is a set of the branches that together form a link between the input and the output. Especially interesting are the *minimum tie sets*, i.e. the tie sets consisting of the smallest number of branches. If no node is passed more than once, the corresponding tie set is a minimum tie set. If there are n of these minimum tie sets T_i ($i = 1,2, \ldots ,n$) the system reliability is:

$$R = P(T_1 \cup T_2 \cup T_3 \ldots T_n).$$

A cut set is a group of branches which would, if they were removed from the graph, interrupt the transfer from the input to the output. A *minimum cut set* contains a minimum number of branches; a system failure already occurs if a minimum cut set is removed from the graph. If there are m of these minimum cut sets C_i ($i = 1,2,3 \ldots ,m$) the system failure probability is:

$$F = P(\overline{C}_1 \cup \overline{C}_2 \ldots \overline{C}_m),$$

where $\overline{C}_1$ indicates that C_1 has failed.

The tie set-cut set method provides an efficient way of calculation if there are no dependent failures in a system. We shall illustrate this method in the following example. We start with the reliability graph of Figure 6.11b. We can indicate several tie sets and cut sets (see Table 6.4).

	Tie sets		Cut sets
T_1:	AB	C_1:	AA′
T_2:	A′B′	C_2:	BB′
T_3:	ACB′	C_3:	ACA′
T_4:	A′CB	C_4:	ACB′
T_5:	ACCB	C_5:	A′CB
		C_6:	BCB′

Table 6.4 List of tie sets and cut sets of graph in Figure 6.11b.

It will be clear that tie set T_5 is not a minimum set, neither are the cut sets C_3 and C_6 minimum sets.

From the above follows:

$$R = P(T_1 + T_2 + T_3 + T_4) =$$
$$= P(AB + A'B' + ACB' + A'CB),$$

and for the system unreliability:

$$F = 1 - R = P(\overline{C_1} + \overline{C_2} + \overline{C_4} + \overline{C_5}) =$$
$$= P(\overline{AA'} + \overline{BB'} + \overline{ACB'} + \overline{A'CB}).$$

N.B.: If one or more non-minimum tie sets or cut sets are inadvertently added to this analysis, the final result remains the same: non-minimum tie sets and cut sets do not add to R or F respectively.

It is important for a correct expression for R or F that *all* minimum tie sets and *all* minimum cut sets have been accounted for. So we might state that R and F are obtained from the *complete set of minimum tie sets or cut sets* of the reliability graph representing the system.

The above expressions may generally be written as:

$$R = P(\sum_{i=1}^{n} T_i) = 1 - F = 1 - P(\sum_{i=1}^{m} \overline{C_i}),$$

$$P(\sum_{i=1}^{n} T_i) = P(T_1) + P(T_2) + \ldots + P(T_n) - [P(T_1 T_2) + \ldots$$
$$\ldots + P(T_{n-1} T_n)] + \ldots (-1)^{n-1} [P(T_1 T_2 \ldots T_n)].$$

By analogy one can easily derive the expression for the system unreliability F.

The above-mentioned expressions for R and F can be written out further. If it is assumed that the basic events A, A', B, B' and C are stochastically independent, the probability of

114 Non-Maintained Systems

the intersection may be written as the product of the probabilities. For the present example this results in the same expressions as we have already found in Section 6.7.

Writing out the above expressions goes as follows:

$$R = R_0 - R_1 + R_2 - R_3 + \ldots$$

Here R_0 is the sum of the reliabilities of all forward minimum tie sets T_i (without loop). It is equal to:

$$R_0 = P(AB) + P(A'B') + P(ACB') + P(A'CB).$$

R_1 is the sum of the reliabilities of all subgraphs with *one loop* (see Figure 6.12):

$$R_1 = P(ABB'C) + P(ABB'A') + P(ACA'B') + P(ABA'C) + P(A'B'CB).$$

Similarly R_2 is the sum of the reliabilities of all subgraphs with *two loops*:

$$R_2 = P(ABA'B'C) + P(ABA'B'C).$$

In our example three loops do not occur. So $R_3 = 0$. With:

$$R = \sum_{i=1}^{n} (-1)^i R_i,$$

in which n is the maximum number of loops and R_i is the sum of the reliabilities of all subgraphs with i loops, the system reliability R is simple to derive.

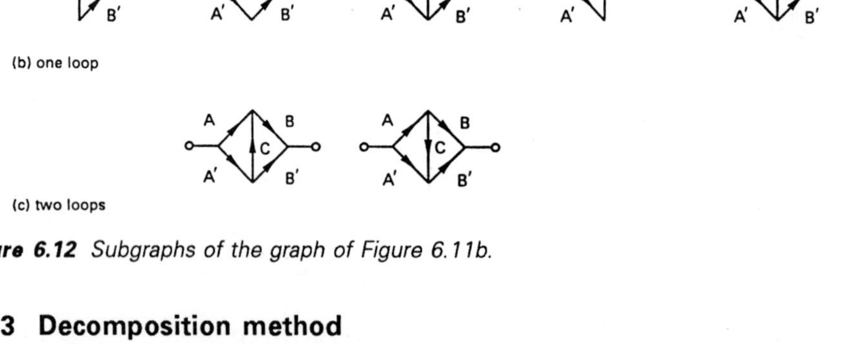

(a) no loop

(b) one loop

(c) two loops

Figure 6.12 Subgraphs of the graph of Figure 6.11b.

6.9.3 Decomposition method

Bayes' theorem for two events A and B is:

$$P(A) = \frac{P(AB)}{P(B|A)},$$

or alternatively:

$$P(AB) = P(A)\,P(B|A) = P(B)\,P(A|B).$$

This *conditional probability theorem* may be used when determining the reliability of complex systems. To that end, a component (or a group of components) with a key position in the structure of the reliability model of that system is singled out. The event that this component functions correctly is indicated by X; the complementary event by $\overline{X}$. Similarly S is the event that the total system functions well, while $\overline{S}$ is its complementary event. The reliability is then:

$$R = P(S) = P(X)\,P(S|X) + P(\overline{X})\,P(S|\overline{X}).$$

Likewise, the unreliability is given by:

$$F = P(\overline{S}) = P(X)\,P(\overline{S}|X) + P(\overline{X})\,P(\overline{S}|\overline{X}).$$

The proof is simple. Applying Bayes' theorem yields:

$$\left.\begin{array}{l} R = P(XS) + P(\overline{X}S) = P(S) \\ F = P(X\overline{S}) + P(\overline{X}\,\overline{S}) = P(\overline{S}). \end{array}\right\} \text{Q.E.D.}$$

In the bridge network of Figure 6.8b, for example, we advantageously can isolate the component C as the key component. We find for the reliability:

$$R = P(S) = P(C)\,P(S|C) + P(\overline{C})\,P(S|\overline{C}).$$

Here $P(\overline{C}) = 1 - P(C)$. It is easy to determine the two conditional probabilities $P(S|C)$ and $P(S|\overline{C})$ from Figure 6.13.

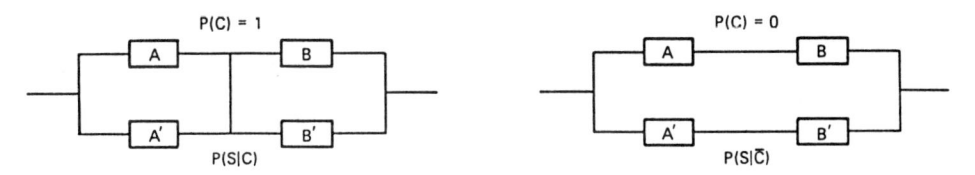

Figure 6.13 *Decomposition method applied to the bridge structure of Figure 6.8b.*

We find now:

$$P(S|C) = \{P(A) + P(A') - P(AA')\}\,\{P(B) + P(B') - P(BB')\},$$

$$P(S|\overline{C}) = P(AB) + P(A'B') - P(AA'BB').$$

The decomposition method can be used repeatedly and eventually leads to simple series and parallel networks. The difficult part is to make the right choice of the key components. In general, these are the components that have to be removed in order to yield simple, calculable structures.

6.9.4 State-space method

The state-space method (which assumes that the failure process can be regarded as a Markov process) is a very common method that can handle a wider variety of cases than any other method. The method can be used for systems with components that fail stochastically independent, but also for systems exhibiting dependent failures. Common-

116 Non-Maintained Systems

cause failures are therefore simple to represent. In addition, this method can also easily account for repairs. Furthermore, one does not meet with any difficulties of a theoretical nature if the system components fail in more than one distinguishable mode (multi-mode failures). It should be noted that, although all these different cases constitute no theoretical limitation of this method, the representation of such a general system in practice can become quite complex.

Without computer-aided synthesis and analysis of these Markov models for practical systems, the method would be *de facto* limited to almost trivially simple systems. As we have seen in Section 5.3 on the introduction of the Markov model, the system to be modelled is assigned a discrete and finite number of states that are mutually exclusive (exclusive, disjunct). All of the states which the system may 'visit' in practice have to be represented in the model. By introducing dummy states, time-dependent hazard rates can be simply represented. This can only be done for the hazard rates $z(t)$ that can be viewed upon as describing the hazard rate of a fictitious system composed of elements with a constant failure rate λ. One can then simply insert this entire dummy sub-system into the main Markov model.

We have already demonstrated this calculation method in Section 5.3. When we restrict ourselves to Markov models with constant failure rates λ_i the solution of the Laplace transform of the probability $P_{S_i}(t)$ that the system is in state S_i at time t, may be found as follows. For convenience, we start with the system drawn in Figure 6.14 which has only three states.

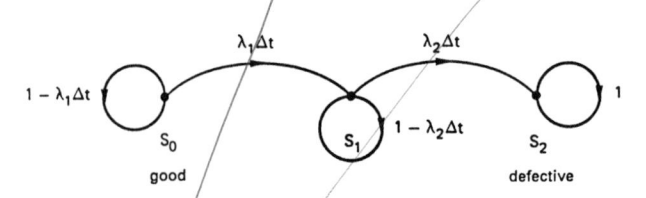

Figure 6.14 Markov diagram of a system with three states and constant failure rates.

The corresponding differential equations are:

$$\frac{dP_{S_0}(t)}{dt} + \lambda_1 P_{S_0}(t) = 0,$$

$$\frac{dP_{S_1}(t)}{dt} + \lambda_2 P_{S_1}(t) - \lambda_1 P_{S_0}(t) = 0,$$

$$\frac{dP_{S_2}(t)}{dt} - \lambda_2 P_{S_1}(t) = 0.$$

The transition matrix U is now:

$$\begin{bmatrix} \lambda_1 & 0 & 0 \\ -\lambda_1 & \lambda_2 & 0 \\ 0 & -\lambda_2 & 0 \end{bmatrix}.$$

State-space method

The Laplace transforms in which the initial conditions $P_{S_0}(0)$, $P_{S_1}(0)$ and $P_{S_2}(0)$ have been substituted, give:

$$(s + \lambda_1) P_{S_0}(s) + 0\, P_{S_1}(s) + 0\, P_{S_2}(s) = P_{S_0}(s),$$
$$-\lambda_1 P_{S_0}(s) + (s + \lambda_2) P_{S_1}(s) + 0\, P_{S_2}(s) = P_{S_1}(s),$$
$$0\, P_{S_0}(s) + (-\lambda_2) P_{S_1}(s) + s\, P_{S_2}(s) = P_{S_2}(s),$$

or in matrix notation:

$$\begin{bmatrix} s+\lambda_1 & 0 & 0 \\ -\lambda_1 & s+\lambda_2 & 0 \\ 0 & -\lambda_2 & s \end{bmatrix} \cdot \begin{bmatrix} P_{S_0}(s) \\ P_{S_1}(s) \\ P_{S_2}(s) \end{bmatrix} = \begin{bmatrix} P_{S_0}(s) \\ P_{S_1}(s) \\ P_{S_2}(s) \end{bmatrix}.$$

The first matrix in the above expression will be denoted A. When it is assumed that the initial condition is:

$$P_{S_0}(0) = 1 \text{ and } P_{S_1}(0) = P_{S_2}(0) = 0,$$

only the first column of the inverse A-matrix is important. This gives:

$$A^{-1} = \begin{bmatrix} \dfrac{1}{s+\lambda_1} & - & - \\ \dfrac{\lambda_1}{(s+\lambda_1)(s+\lambda_2)} & - & - \\ \dfrac{\lambda_1 \lambda_2}{s(s+\lambda_1)(s+\lambda_2)} & - & - \end{bmatrix}.$$

So the result is:

$$P_{S_0}(s) = \frac{1}{s+\lambda_1},$$

$$P_{S_1}(s) = \frac{\lambda_1}{(s+\lambda_1)(s+\lambda_2)},$$

$$P_{S_2}(s) = \frac{\lambda_1 \lambda_2}{s(s+\lambda_1)(s+\lambda_2)}.$$

Inverse transformation results in:

$$P_{S_0}(t) = e^{-\lambda_1 t},$$

$$P_{S_1}(t) = \frac{\lambda_1}{\lambda_1 - \lambda_2} (e^{-\lambda_2 t} - e^{-\lambda_1 t}),$$

$$P_{S_2}(t) = 1 - P_{S_0}(t) - P_{S_1}(t).$$

The reliability $R(t)$ of the system is the sum of the probabilities that the system is in a state where it is operating correctly (after all, the states are exclusive and disjunct). If we assume that S_0 and S_1 are such correct states, we find for the system reliability:

118 *Non-Maintained Systems*

$$R(t) = \sum_{i=0}^{1} P_{S_i}(t) = \frac{\lambda_1}{\lambda_1 - \lambda_2} e^{-\lambda_2 t} - \frac{\lambda_2}{\lambda_1 - \lambda_2} e^{-\lambda_1 t}.$$

The mean life or MTTF of a system can be calculated in various ways from the Markov model. When both the mean life and the reliability have to be found, it is efficient to calculate the $R(t)$-curve first and to integrate subsequently according to:

$$\text{MTTF} = \lim_{T \to \infty} \int_0^T R(t) \, dt.$$

However, if one only wants to know the mean life, the procedure may be shortened considerably. To begin with, it is possible to avoid the inverse Laplace transform. We can do that by invoking the initial-final value theorem:

$$\text{MTTF} = \lim_{T \to \infty} \int_0^T R(t) \, dt = \lim_{s \downarrow 0} R(s) = \lim_{s \downarrow 0} \sum_{i=k}^{m} P_{S_i}(s).$$

The summation is over all good states S_k through S_m out of the system.
It further turns out to be possible to avoid the Laplace transform completely. To prove this, we start with the set of Laplace equations governing the Markov model, represented in matrix notation as:

$$[s \cdot I + U] \, [P_{S_i}(s)] = [P_{S_i}(0)].$$

If we then integrate both the left-hand side and the right-hand side of this equation (multiply by $1/s$) and subsequently apply the initial-final value theorem the above equation changes into:

$$\lim_{s \to 0} [sP_{S_i}(s)] + U \cdot \lim_{s \to 0} [P_{S_i}(s)] = [P_{S_i}(0)].$$

Because it holds that:

$$\lim_{s \to 0} [sP_{S_i}(s)] = \lim_{t \to \infty} [P_{S_i}(t)] = [P_{S_i}(\infty)],$$

and:

$$\lim_{s \to 0} [P_{S_i}(s)] = \lim_{T \to \infty} \int_0^T P_{S_i}(t) \, dt] = [\theta_{S_i}],$$

in which θ_{S_i} represents the mean sojourn time the system remains in state S_i, the set of equations may be rewritten as:

$$U \cdot [\theta_{S_i}] = [P_{S_i}(0)] - [P_{S_i}(\infty)].$$

However, the final value vector $[P_{S_i}(\infty)]$ is known only if all absorbing (failure) states can be merged into one new down-state. For reliability and MTTF calculations this is to be recommended because the number of equations is then much smaller. It is possible to avoid the final value vector $[P_{S_i}(\infty)]$ as well, if we realise that the absorbing states can have no influence whatsoever on the good states of the system: after such a failure the system can never reach a good state again. For that reason we may omit those rows and columns from the transfer matrix and those rows from the vectors in which these absorbing states occur. This results in the following reduced equation:

$$U'\,[\theta_{S_i}]' = [P_{S_i}(0)]'.$$

This equation can also be derived directly from the Markov model. For the system of Figure 6.14, for example, holds:

$$\begin{bmatrix} \lambda_1 & 0 \\ -\lambda_1 & \lambda_2 \end{bmatrix} \cdot \begin{bmatrix} \theta_0 \\ \theta_1 \end{bmatrix} = \begin{bmatrix} 1 \\ 0 \end{bmatrix}.$$

From this follows:

$$\theta_0 = \frac{1}{\lambda_1} \text{ and } \theta_1 = \frac{1}{\lambda_2},$$

so that this system has a mean life of:

$$\text{MTTF} = \theta = \theta_0 + \theta_1 = \frac{1}{\lambda_1} + \frac{1}{\lambda_2}.$$

Of course, the assumption is again that S_0 and S_1 are correctly operating states. This information comes from the actual physical system that is being modelled.

Problems

6.1. What is understood by 'dependence' in reliability engineering?
Name two kinds of dependent failures and give an example of each.

6.2. Prove that the following equation holds for an active redundant m-out-of-n system, consisting of identical units with failure rate λ:

$$\text{MTTF} = \frac{1}{\lambda} \sum_{k=m}^{n} \frac{1}{k}.$$

6.3. A certain component has a failure rate of 4×10^{-8}/h in the on-state and a failure rate of 4×10^{-9}/h in the off-state. On average, over the life of this component, it is only 25 % of the time in the on-state.
What is the effective failure rate of this component?

6.4. a. Derive the reliability $R(t)$ of an n-fold, active redundant system with a redundancy degree $\eta = 1$, if the units are all identical and have a reliability $e^{-\lambda t}$. Failures occur stochastically independent.
b. Determine the mean life θ of this system from the expression for $R(t)$ found above.

6.5. A system consists of n identical units in series that can fail independently of each other. The reliability of one unit is R.
The system is made redundant as indicated in the configurations a and b below. The system reliability is R_a and R_b respectively in that case.

a. What is $\lim_{R \to 1} \frac{R_a}{R_b}$?

b. What is $\lim_{R \to 0} \frac{R_a}{R_b}$?

c. Give $\frac{R_a}{R_b}$ for $n = 3$ as a function of R.

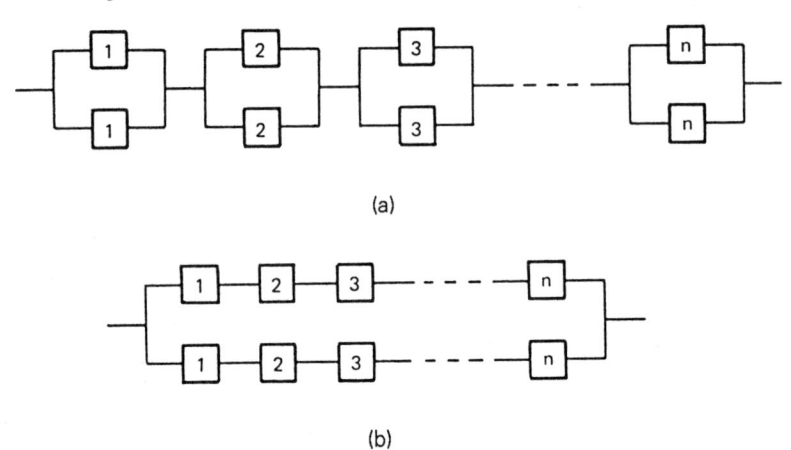

(a)

(b)

6.6. a. Calculate the failure probability of a system with the catastrophic failure model shown below. (In the rectangles the *failure* probability is indicated; failures occur stochastically independent.)

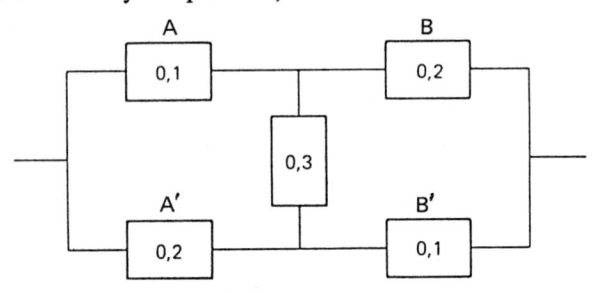

b. Does the presence of the central interconnection branch in the above failure model always result in an improvement of the system? (The failure probability associated with this branch is always smaller than one.)

6.7. One has four mutually identical water pumps, each with a reliability R_o ($0 < R_o < 1$) and a capability to pump up water 5 meters (16.5 ft). These pumps are used in a system in which the water has to be pumped up 7 meters (23 ft).

Six different configurations are depicted below. Arrange the configurations in order of increasing system reliability $R_a, R_b, ..., R_f$.

N.B.: Each rectangle represents one pump.

(a)   (b)

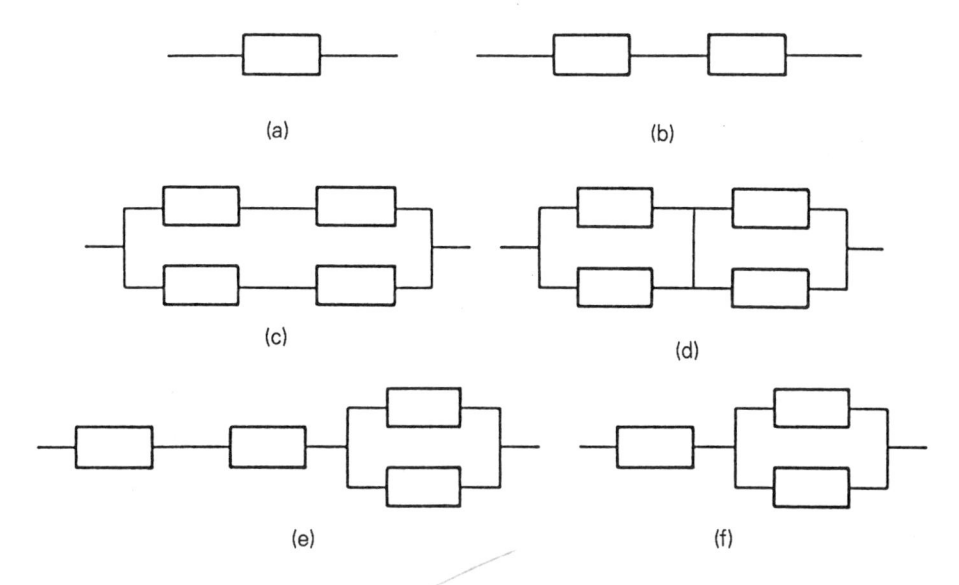

(c)   (d)

(e)   (f)

6.8. A redundant system without repair consists of two identical units with a constant hazard rate λ.
   a. Draw the Markov diagram for the *active* redundant and for the *passive* redundant system.
   b. Using both Markov diagrams reason how large the mean life of each of the systems is.

6.9. A computer system for process control consists of two active redundant computers, each of which is capable of controlling the process all by itself. Both computers are connected to the public power line. The following independent failures may occur in this system:

Failure mode	Failure rate
computer I fails	λ_1
computer II fails	λ_2
power line fails	λ_N

Draw the associated Markov diagram. Assume that repair is not possible. Clearly indicate the various states and the initial and 'down' state(s).

6.10. A system consists of n identical units in series which can fail independently. For each unit holds: $C = f(R)$, where f is a monotonically increasing function (C is generalised costs, R reliability). It is asked to make the reliability larger than or equal to the value R_s for minimum costs. How large is the reliability R_i of the i-th unit ($i = 1,2,...n$) after this optimisation and what are the total costs incurred for this system?

6.11. A stand-by redundant system is composed of two equal units. An activated unit has a failure rate λ_a. A unit in the stand-by mode has, as long as it is in this stand-by mode, a failure rate λ_o. The failure detection and switching unit of this system may fail to activate the second unit with a failure rate λ_s. The events are stochastically independent.

122 *Non-Maintained Systems*

- Draw the Markov diagram.
- Define the various states.
- Indicate the transition probabilities.
- Give the simplest Markov diagram of this system (also define its states and indicate its transition probabilities).

6.12. To guarantee a certain reliability it turns out to be necessary to implement that system in a dual redundant fashion. One can only choose from two possibilities, namely a two-unit active redundant system or a two-unit passive redundant system. In the passive redundant system it is necessary to install a switch which enables the system to switch to the other unit. This switch has one failure mode, namely that it fails to switch to the other unit (gets stuck). If the failure rate of the active unit is λ, a passive unit cannot fail and the failure rate of the switch is λ_2, what is then the minimum value of λ_s at which the active redundant system is better than the passive redundant system?

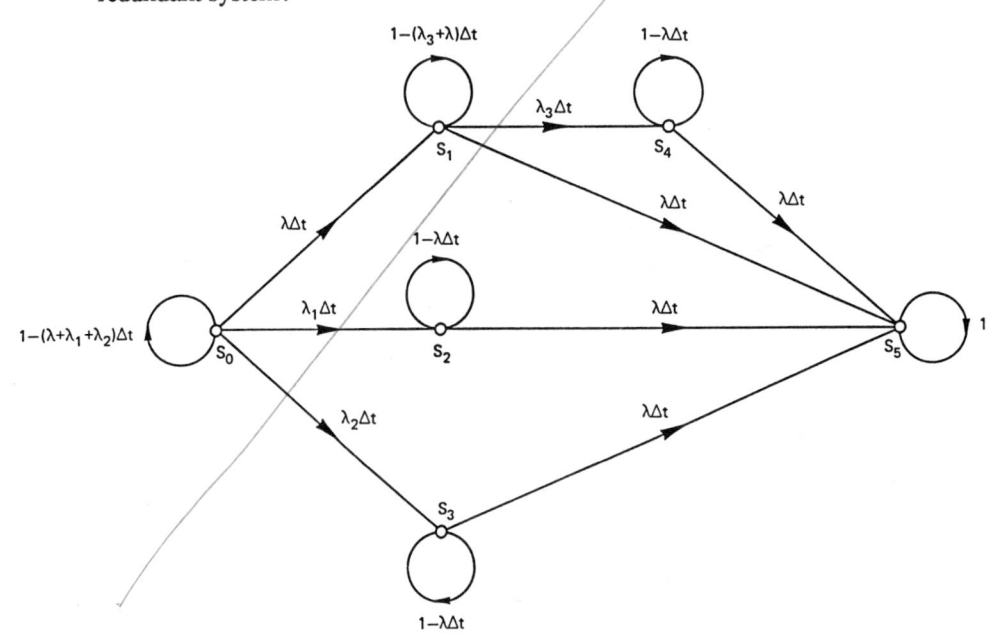

6.13. A system without repair is described by the Markov diagram above, where S_0 is the initial state and S_5 is the 'system down' state. What is the mean life (MTTF) of this system? *Hint: first reduce the Markov diagram as far as possible.*

6.14. A certain technical system is a 2-out-of-3 active redundant system consisting of three identical units with failure rate λ. The units fail independently.

It is given that this system is three times more expensive than a single unit and has a mean life that is only 5/6 of that of a single unit. Refute the paradox that this system is worse than a single unit system; after all, the costs are three times as high, the mean life is shorter and always at least two units have to be on if the system is to function properly. In your solution sketch the three $R(t)$ curves of 1 unit, of 2 units in series and of the 2-out-of-3 system.

6.15. An electrical power generator supplies a maximum power of 10 kW and has a failure rate of $\lambda_m = 4 \times 10^{-3}$/hour. If the generator is loaded with its nominal power (7 kW) or less, it has a failure rate of $\lambda_n = 10^{-3}$/hour. A certain application requires 10 kW. To that end two identical generators are used in an active redundant configuration (*The units fail stochastically independent*).
   a. Give the corresponding Markov diagram and determine the mean life of this system (MTTF).
   b. How large is the reliability, sketch it as a function of time. Also draw the reliability curve of the system if it would have been realised with only one generator (in the same figure).

6.16. A safety system consists of three monitors connected to one process under surveillance. For safety reasons the process is always switched off after a fixed operation interval to conduct maintenance. The process is also switched off if an unsafe situation is indicated.
   The mutually identical monitors can fail in three ways:
   1. The monitor incorrectly indicates an unsafe situation (the probability of this happening in the period mentioned is 0.05),
   2. The monitor does not detect an unsafe situation in the process (the probability of this happening in the period mentioned is 0.01).

   The costs of an undetected, unsafe situation are $ 25,000.– per event and the costs of incorrectly switching off the system (and restarting it again) $ 10,000.– per event.
   Assume that the failures are stochastically independent. Show, by means of a calculation, which of the following strategies is optimal:
   a. If at least one monitor indicates an unsafe state the process is switched of;
   b. If at least two monitors indicate an unsafe state the process is switched off;
   c. Only if all three monitors indicate an unsafe state is the process switched off.

6.17. In slurry transport it is necessary that the mixture in the transport pipes is kept in motion, otherwise clogging may result. Therefore, three pumps are used in a 2-out-of-3 configuration. In order to obtain the required capacity it is necessary that *two* pumps function. In the event that only one pump functions the capacity is not obtained, but the pipes do not clog, though. The pumps are identical and fail stochastically independent. The failure rate λ is constant for normal use; when still only one pump functions the failure rate of this pump becomes 2λ.
   a. What is the MTTF, in the event of the failure that the capacity is not realized?
   b. The second function of the system is to keep the mixture in motion. One pump suffices to do this. What is the MTTF of this function in the above configuration?

6.18. Six computer systems are interconnected by means of the network below. Each connecting line of the network can be used in both directions and has an independent failure probability $p_0 = 0.1$. Calculate the *probability* that a successful information exchange can occur between the computers 1 and 4. *Hint: first reduce the network as much as possible and subsequently use the decomposition method.*

124 *Non-Maintained Systems*

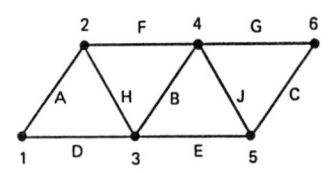

6.19. In an electronic circuit a diode function is necessary. To increase the reliability of this diode function one wants to apply active redundancy. One can afford, however, no more than three diodes. The diodes used may exhibit both open failures and short-circuit failures; the associated probabilities are:
 – open failure $p_o = 0.02$,
 – short-circuit failure $p_s = 0.01$.
The diodes fail stochastically independent. Indicate for which of the circuits below the *reliability* is *maximal* and motivate your answer.

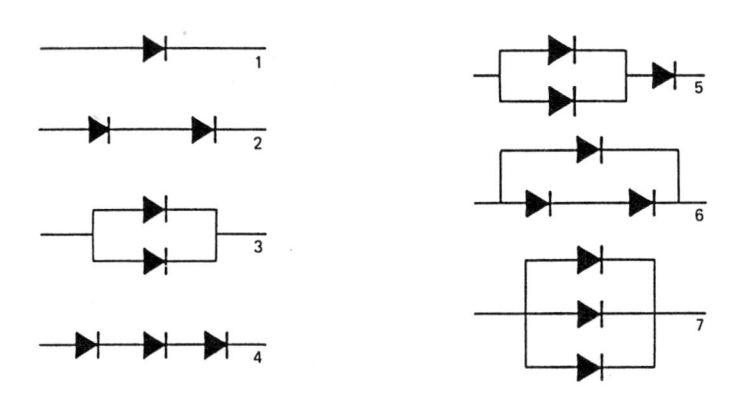

6.20. A passive redundant system consists of two units and a switch, which can all fail independently. As soon as the first unit fails the system switches to the second unit; both units have a failure rate λ. The switch can fail in the following ways:
 – the switch may get stuck, due to which switching over is not possible anymore. The failure rate for this event is λ_s.
 – the switching mechanism may be activated by interference pulses, which causes the switch, if unit 1 still functions, to spontaneously switch to the second unit. The (failure) rate with which these interference pulses occur is λ_i.
 a. Draw he Markov diagram. Clearly indicate the transitions and the states.
 b. Calculate the MTTF of the system.

6.21. The undercarriage of an aircraft is operated hydraulically. If the pressure drops, a warning system signals this condition in the cockpit. On the one hand the warning system may fail to signal the pressure dropping (with a probability 0.1) and on the other hand an incorrect signal may be given (also with a probability 0.1). One wants to apply non-adaptive, majority voring redundancy, whereby the alarm system may be implemented fourfold at the most; in case of doubt (a tie between defective and good indications) no alarm is given.
What is the best choice of system if one wishes to acquire a certainty of 94 % that an hydraulic pressure drop is reported?

6.22. In the figure below a series system composed of three components is depicted. The reliabilities of the components are indicated in the boxes. The system reliability is considered too low, and one wants to increase it to 0.8 or higher for a minimum investment. Active redundancy is therefore applied at components level.

If it is given that all components are equally expensive and that there are no extra costs for parallelling components, determine the optimum configuration (catastrophic failure model).

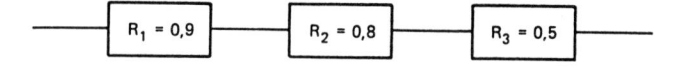

——[$R_1 = 0,9$]——[$R_2 = 0,8$]——[$R_3 = 0,5$]——

7
Maintained Systems

In Section 6.1 a maintained system was defined as a system that is firstly maintainable, and secondly, is actually maintained. Maintenance is conducted by constructive human intervention. Before dealing with maintenance in more detail, it is useful to look at a maintained system in a somewhat broader perspective.

As indicated in Figure 7.1, the total *life cycle* of a system can be divided into a number of stages or phases. The 'life' of a technical system begins with the design and manufacture of that system. The designer/manufacturer usually defines the system and its specifications based on market research and a competitive analysis. At this stage important matters are laid down: the system performance is projected, the price determined, as well as the quality and the reliability. Subsequently the stage of operational use begins which eventually, after a hopefully long and useful life, ends with the replacement and removal of the system.

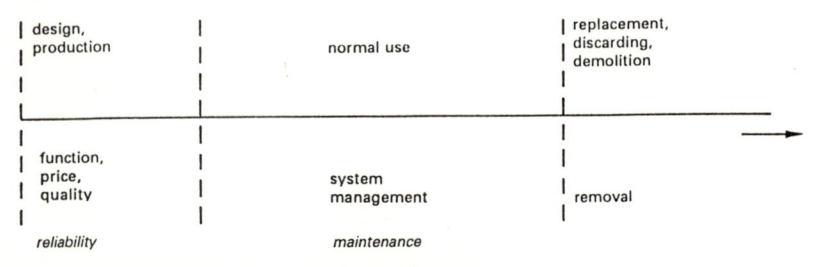

Figure 7.1 *The complete life cycle of a technical system.*

Now, many systems are so valuable, that is the invested initial or acquisition costs C_a from Figure 6.9 are so high, that it is economically attractive to perform *maintenance* during the functional life of the system. This allows the invested capital C_a to be written-off over a longer period. The reverse side of the medal is that this also involves expenses (maintenance costs C_m in Figure 6.9).

Especially for the more complex technical systems *maintenance* is labour-intensive and requires well-trained and, therefore, well-paid personnel (not just anyone is able to repair a video cassette recorder, let alone a radar system!). The maintenance expenses C_m accrued over the total functional life span of a system may therefore very well be a multiple of the replacement value of the system. To keep the maintenance expenses down, the system will have to be designed with maintenance in mind. This may be done by partitioning the system into easy to maintain and/or replace modules, by guaranteeing a good accessibility of the system components, incorporate automatic failure detection and location circuitry and also by supplying test programs, documentation and the like.

A maintenance friendly design is aimed at simplifying repairs and shortening the duration of the repair. In addition, other aspects come to the fore in the design phase. Usually some graceful degradation of the system is tolerated, as long as the associated failures do not generate a hazardous situation. *Fail-safe* conceptions and the corresponding constructions and circuitry must already be anticipated in the design stage of a system.

In order to use a system without too much trouble in the operational use stage, *well-planned* maintenance is required. There have to be sufficient spare parts, there have to be qualified repairmen that can come into action speedily, etc. Such a planning may, for example, result in multi-stage maintenance. This means that at the site of the system only modules are replaced by spare ones. These failed modules are subsequently sent away to be repaired at a central location. The first stage is *replacement*; it can be done fast, is simple and requires little training. The second stage, the actual repair, takes longer (if it is at all economical to repair the failed modules). Besides, valuable measuring equipment and well-trained personnel are required, which will be utilized more efficiently at this central location.

We see that there are clearly two aspects to the maintenance of a system; a *design aspect* embedded in the physical design of the system, and a *management aspect*, which is the responsibility of the system owner during the entire functional life of the system. The following sections will primarily be restricted to the latter aspect of maintenance. We shall look at the relative merits of a number of different maintenance strategies the system owner may pursue.

From the above it will be clear that maintenance may cost much, but no maintenance may cost a fortune. The optimal situation will be where the total (acquisition *and* maintenance) costs taken over the total functional life of the system are lowest. This optimum is called *minimum life-cycle cost*. In practice, one comes across extremes away from this optimum far too often. On one hand, one finds systems that are far too expensive and are designed under the motto 'invest now, save later' without proper regard to the later savings. One finds typical examples in areas where government institutions specify systems that become obsolete before they have degraded appreciably. On the other hand, one finds systems that are made cheap to lure customers but that require so much maintenance that it is hard to keep them up. Examples of this category one finds in consumer goods (cars, electronics, etc.) but also in post-war social housing and civil works in which too much wood and metal have been used instead of concrete.

7.1 Introduction

All models treated so far for the determination of system reliability have in common that they did not assume human intervention during the operational stage of the system's life cycle. Systems for which maintenance is not an economically sound concept are an exponent of this subset of non-maintained technical systems.

Example
Nowadays, when a repairman charges $ 40 per hour for labour, no one will even consider repairing an electronic watch that costs only $ 20.

128 *Maintained Systems*

A second class of non-maintained systems consists of systems for which maintenance is not possible. Exponents of this class are: space vehicles used only once, measuring equipment built into foundations of buildings, and cable amplifiers left on the ocean bed.

The remaining systems are the so-called maintained systems. As we have seen maintenance is defined as follows: *Maintenance is any form of human intervention in a system with the intent to keep the system in an operational state or (after failure) restore it to an operational state.*

In this definition of maintenance there are a number of striking things:

- *Human intervention* is essential, as we have already seen in Section 6.1. The mere fact that a system repairs itself (as it were) by engaging internal redundant sub-systems does not necessarily make this system a repairable or maintainable system.
- After the repair is completed, the system, nor the replaced components, necessarily have to be 'as new' again. The components used in the repair process may have deteriorated by, for example, prolonged storage. The repairman may also make mistakes (soldering too hot, fastening a nut too tight or too loose and so forth). Owing to these mistakes a time interval with an increased hazard rate may occur after a repair. The repairman may also have restored the system to a usable state, but not, for example, for all functions of that system or for the full capacity of the system. An illustration is replacing, along the roadside, a flat tire by a spare that is not 'full size' as most cars are equipped with today. This severely restricts the maximum speed and distance the car can be driven.
- In addition to restoring a system to a usable state or keeping it in that state, maintenance also serves to guarantee the *safe operation* of the system.

For a human intervention to qualify as maintenance, the system need not always fail. After all, if it concerns a redundant system, the redundancy can usually be repaired without loss of the system function. This is called *preventive maintenance*. Strictly speaking the maintenance is in this case preventive at the system level, but corrective at the unit level, since at least one of the redundant units must have failed. So, preventive maintenance is maintenance conducted before a system failure occurs. It is prophylactic insofar as it prevents future failures. This preventive maintenance will not suffice: not every system failure can be seen coming in advance. A practical system will therefore need corrective maintenance as well.

One distinguishes two categories of preventive maintenance, namely scheduled maintenance and condition-based maintenance.

With *scheduled maintenance* human intervention occurs in accordance with a predetermined plan or schedule. For example, every time after a certain interval of the life parameter has elapsed. An example is changing oil and the oil filter after every 7500 km (4500 mi) or after every 500 operational hours.

With *condition-based maintenance* the maintenance need of a system is measured. Based on the result, possible intervention is determined. For example, changing oil on the basis of the combustion and wear products in the oil; changing ball-bearings or rebalancing a machine when the vibration level becomes too severe.

In this way, future calamities in the form of a system failure (and corresponding damage

7.2 Systems with preventive maintenance

In this section it is assumed that only preventive maintenance will be performed on the system under observation. Eventually such a system will fail because preventive maintenance alone does not offer a waterproof guarantee against the eventual failure of a system. The time necessary for this preventive maintenance will be neglected here. After all, preventive maintenance can (usually) be conducted at a predetermined time when the system is down for other reasons anyway.

7.2.1 Scheduled maintenance

By using a simple scheduled preventive maintenance model, we shall illustrate the importance of the correct choice of the interval T, between two successive overhauls at $t = iT$.

The model involves the assumption that the maintenance at $t = iT$ is done so thoroughly that the system is *as new* afterwards. So the availability $A(t)$ of such a system is formed by a periodical repetition in time of the interval $(0, T)$ of the reliability $R(t)$ of that system without the scheduled maintenance (see Figure 7.2). We are now interested in the mean life θ, i.e. the time until the first failure occurs. This parameter is usually referred to as Mean Time To First Failure (MTTFF):

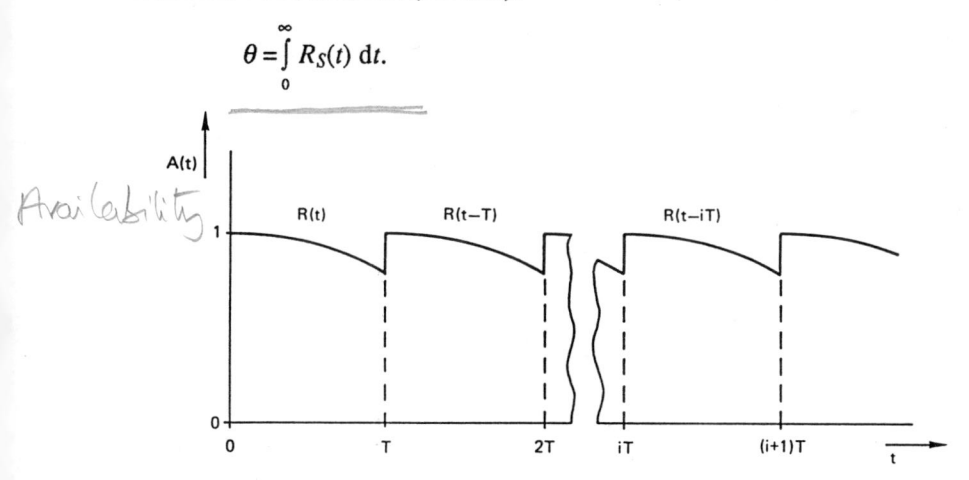

$$\theta = \int_0^\infty R_S(t)\, dt.$$

Figure 7.2 *The availability A(t) of a system with scheduled maintenance which restores the system to the 'as new' state.*

Here $R_S(t)$ is the reliability of the system *with* scheduled maintenance. For the first time interval $(0,T)$, $R_S(t)$ is given by:

$$R_S(t) = R(t), \qquad (0 \leq t < T).$$

For a point t in time in the $(i + 1)$st time interval the system has to have survived the first i intervals in addition to the subinterval (iT, t). We therefore find:

$$R_S(t) = R(T)^i R(t - iT), \qquad (iT \leq t < (i+1)T).$$

The integral that yields the mean life θ of this system with maintenance can be split into the sum of integrals over the various maintenance intervals:

$$\theta = \lim_{n \to \infty} \sum_{i=0}^{n} \int_{iT}^{(i+1)T} R_S(t)\, dt.$$

Since $R_S(t)$ is a simple periodic repetition of $R(t)$ in time we may rewrite the integral as:

$$\theta = \lim_{n \to \infty} \sum_{i=0}^{n} \int_{0}^{T} R(T)^i R(t)\, dt.$$

We may now remove the constant $R(T)^i$ from the integral. This yields with:

$$\sum_{i=0}^{\infty} x^i = \frac{1}{1-x},$$

for the mean life of the system with scheduled maintenance:

$$\theta = \frac{\int_0^T R(t)\, dt}{1 - R(T)}.$$

We already know that in a system with a constant failure rate λ, so with a reliability $R(t) = \exp(-\lambda t)$, scheduled maintenance is useless. After all, the (series) components of such a system that have not failed yet are as new. The MTTFF of such a system is therefore not changed by scheduled maintenance and remains equal to $1/\lambda$. This also follows from the above expression!

It should be noted that we have not made any assumption whatsoever with regard to $R(t)$ in the above derivation of the MTTFF: we have solved the general case. In Figure 7.3 we have plotted $R_S(t)$ for various maintenance periods T. We have done so for an arbitrarily chosen $R(t)$ function. For $T \to \infty$ we again get the underlying reliability function $R(t)$:

$$\lim_{T \to \infty} R_S(t) = R(t).$$

From this example we see that the maintenance interval T has a great influence on the reliability $R_S(t)$. In particular if the first maintenance interval ends past the nearly flat part of the $R(t)$ curve (read: if the redundancy of the system without maintenance is practically exhausted), scheduled maintenance is hardly useful. One has waited too long; the system is over the edge; the redundancy is used up.

Concludingly, a general remark about scheduled maintenance: after scheduled maintenance a practical system will not be 'as new'. Theoretically, the 'as new' assumption only holds for a (redundant) system consisting of units with a negative-exponential failure distribution.

A consequence of this so-called *non-perfect scheduled maintenance* is that the availability curve in Figure 7.2 of the system at the intervals iT (i = 1,2,...) is not equal to one, but remains lower than one. Without repair the curve will drop progressively lower and lower

Scheduled maintenance 131

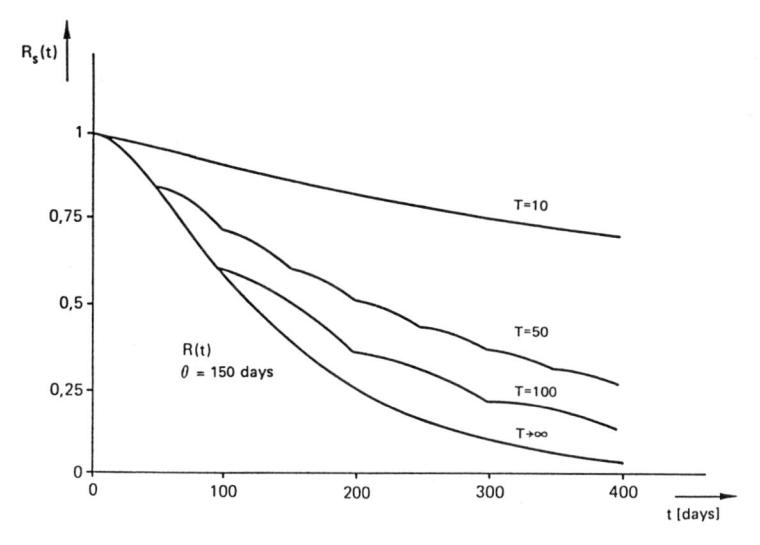

Figure 7.3 *The reliability $R_S(t)$ of a system with scheduled maintenance (every T days), after which the system is 'as new' again. R(t) is the reliability of the system without maintenance (when $T \to \infty$).*

with increasing i. Because of this the system reliability $R_S(t)$ in Figure 7.3 will drop sooner than depicted.

We will conclude this section about scheduled maintenance with a few words about the costs of scheduled maintenance with regard to repair. Suppose the costs of a scheduled maintenance are C_S on average. If no preventive maintenance would be conducted and one maintains the system by repair only, the repair costs are C_R on average per repair event. It will be clear that usually:

$$C_S < C_R,$$

because scheduled maintenance can be planned, owing to which a better utilisation of the available maintenance capacity is possible and there is no unplanned system-down cost involved. Moreover, usually more than just the primary failing components break when a system is operated till it fails.

Per unit of time the costs of scheduled maintenance with period T are C_S/T and for repair the costs are C_R/MTBF.

When the ratio ρ is introduced:

$$\rho = \frac{C_S}{C_R} \frac{\text{MTBF}}{T},$$

the requirement for a financially healthy scheduled maintenance scenario is:

$$\rho < 1.$$

To obtain a small cost ratio ρ between the two kinds of maintenance the maintenance period T cannot be chosen (too) small with regard to the MTBF of the system. Scheduled maintenance at times too far apart (T large with respect to the MTBF) results in low system reliability $R_S(t)$ and therefore leads to additional repair costs of the system in the field.

7.2.2 Condition-based maintenance

In the above it was discussed that for many systems it is not wise to allow the so-called *run-to-break* strategy (followed by corrective maintenance or repair). It could result in too much loss of production and unsafe situations, or be too expensive because the machinery would jam completely, for example.

Where a future failure of a system (or parts of it) can be predicted in advance, preventive maintenance is useful. In the preceding section we have discussed a special kind of preventive maintenance, namely scheduled maintenance, in more detail. With *scheduled maintenance* we predict future failure based on *statistical information* about the expected life of the system or its components.

Example
Under normal field conditions difficulties with a certain kind of jet engine are to be expected after 15,000 operating hours. For that reason, the hottest part of the jet engine is checked particularly and any suspect parts are replaced every 5,000 hours. Every 10,000 hours the entire engine is revised, except for the low-pressure compressor and every 30,000 hours a complete engine overhaul is performed.

It will be clear that preventive maintenance can be made to be more cost-effective if we can run the system longer, without endangering its correct operation. Rather than relying on *statistical* information one would like to gather *deterministic* information about the health of (the components of) a system. This is possible for some technical systems by the use of *condition monitoring* techniques. No longer do we need fixed intervals for maintenance, but we may adapt the maintenance intervals to the system's need for maintenance. This 'about to fail' information may (for many failure modes of a system) be obtained from a number of indicative parameters of such a system. These parameters are a measure for the internal condition of that system. Instead of scheduled or *time-based maintenance* we may turn to *condition-based maintenance*.

Unfortunately it can never be excluded that a failure mode escapes the system's condition measuring coverage. For that reason, time-based maintenance and condition-based maintenance are usually combined.

Example
The jet engine from the previous example has various forms of condition monitoring. During operation the engine temperatures, the fuel consumption and the vibration levels are constantly monitored. In addition, more indicative parameters are measured when testing the engine during a shop visit. If one or more of these parameters are off, the engine can be stopped in flight or can be run at a lower power to undergo extra maintenance immediately after the flight, in between the maintenance events planned above. If, while test running the engine, abnormal condition parameter values (e.g. oil contamination, etc.) are discovered, other additional maintenance can be conducted.

Condition measurement is conducted in practice not only to prevent further mishap after a failure (emergency shut down), but especially to see a failure approaching in advance. For the latter it is necessary that the *degree* of system deterioration can be measured (see Figure 7.4). Let us assume the system under consideration is a large hydropower turbine.

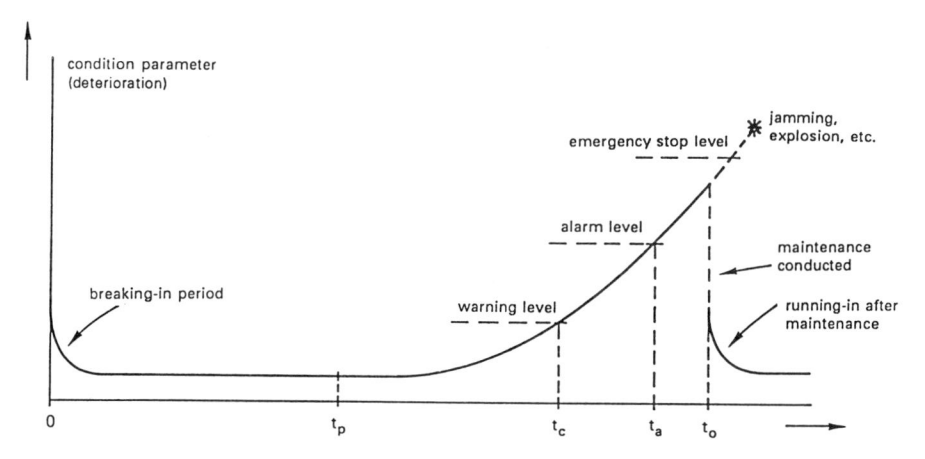

Figure 7.4 *The typical time dependence of the condition of rotating machinery. If there would be no intervention, the machine would completely jam after a certain period of time.*

As a condition parameter the system designer takes, for example, the vibration level of the turbine. After a run-in period, during which the mating parts of the machine wear in, the vibration level remains virtually constant. If one would (only) perform scheduled maintenance, one would already have to stop the machine at t_p for maintenance. Because this maintenance interval relies on statistical information, it needs a large built-in safety margin.

Now the machine is constantly monitored, for instance because of the large damage involved in failure. At t_c a warning signal will be given, because the vibrations level has increased, for example, due to turbine blade damage, ball-bearing wear or reduction gear wear. An analysis of the frequency spectrum associated with the vibration signal may show whether it concerns ball-bearing, gear box wear or turbine blade damage. In Figure 7.4, finally, the alarm level will be exceeded at t_a if no preventive action is taken. Beyond this time the turbine can, for example, no longer be operated at full power. If the operator would not intervene the turbine could destruct itself, its foundation and its direct environment.

From this example we see that a much larger maintenance interval t_a than with scheduled maintenance (interval t_p) will be allowable, if, based on the measured condition of the machine, we are able to extrapolate the condition into the future with a reasonable degree of certainty. This, of course, only holds for those failure modes of a system which announce themselves in the form of precursors for imminent failure.

The difficulty with condition-based maintenance is to find those system parameters that indicate a future failure. Besides this condition measurement often so-called *performance* measurements are conducted. This is useful if the desired function of the system is subject to graceful degradation. An example is given in Figure 7.5. With the help of an acceleration transducer the braking effectiveness of a vehicle is measured. The rise and fall of the deceleration gives a cursory impression of the functioning of the brake system. Based on this information one may renew the brake pads, turn and clean the brake drums, bleed the hydraulic brake lines etc. Other examples of performance measurement are: full

134 *Maintained Systems*

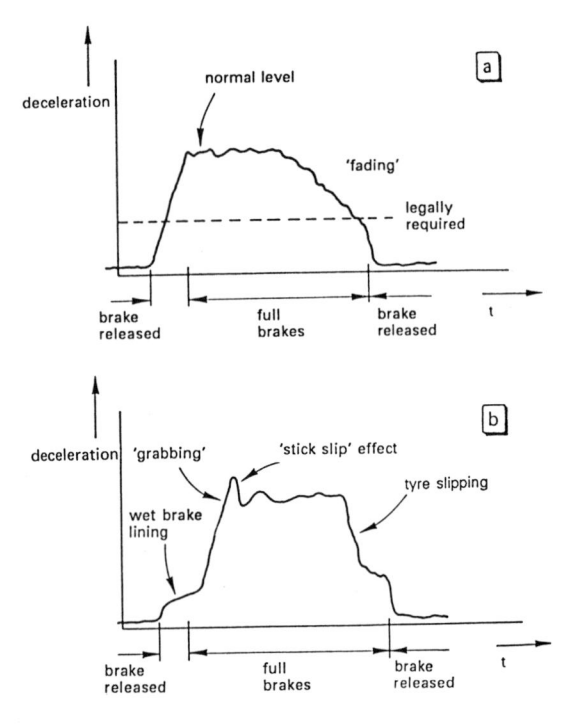

Figure 7.5 *A performance measurement of the brake system of a heavy transport vehicle.*

load tests for generators, fuel consumption measurements for vehicles, pump and transport capacity measurement of conveyor systems, etcetera.

Besides the incidental measurement of the condition or performance of a system, the continuous monitoring of a system in time to follow the trend of the condition or the performance is important. This so-called *trend monitoring* is used in complex systems to detect irregularities in the operation of the system at an early stage and to trace their cause. In Figure 7.6 an example is given. This case concerns a complex gear box. Frequency analysis of the vibration data shows that the greatest change in vibration level occurs in the spectrum below 1 kHz. Therefore a wide-band measurement (10 Hz – 1 kHz) of the root mean square value of the vibration velocity signal has been chosen for trend monitoring. The trend of this signal is illustrated in Figure 7.6b.

Condition-based maintenance has especially caught on in the maintenance of rotating and reciprocating machinery. To determine the condition, among other things the following parameters are measured, corrosion (contact resistance measurement, visual inspection), temperature (thermographic paint, infrared thermography), fatigue (detecting cracks by means of eddy current detectors, ultrasonic and X-ray techniques), unbalance, loose parts (vibration analysis) and wear products (oilfilters, cooling-water analysis, magnetic filters, spectrometric analysis).

In electronic systems future failures cannot be predicted (except, of course, in redundant systems). This is because most electronic components do not have wear-out behaviour. Nevertheless scheduled maintenance is desirable here, too. For example, measurement

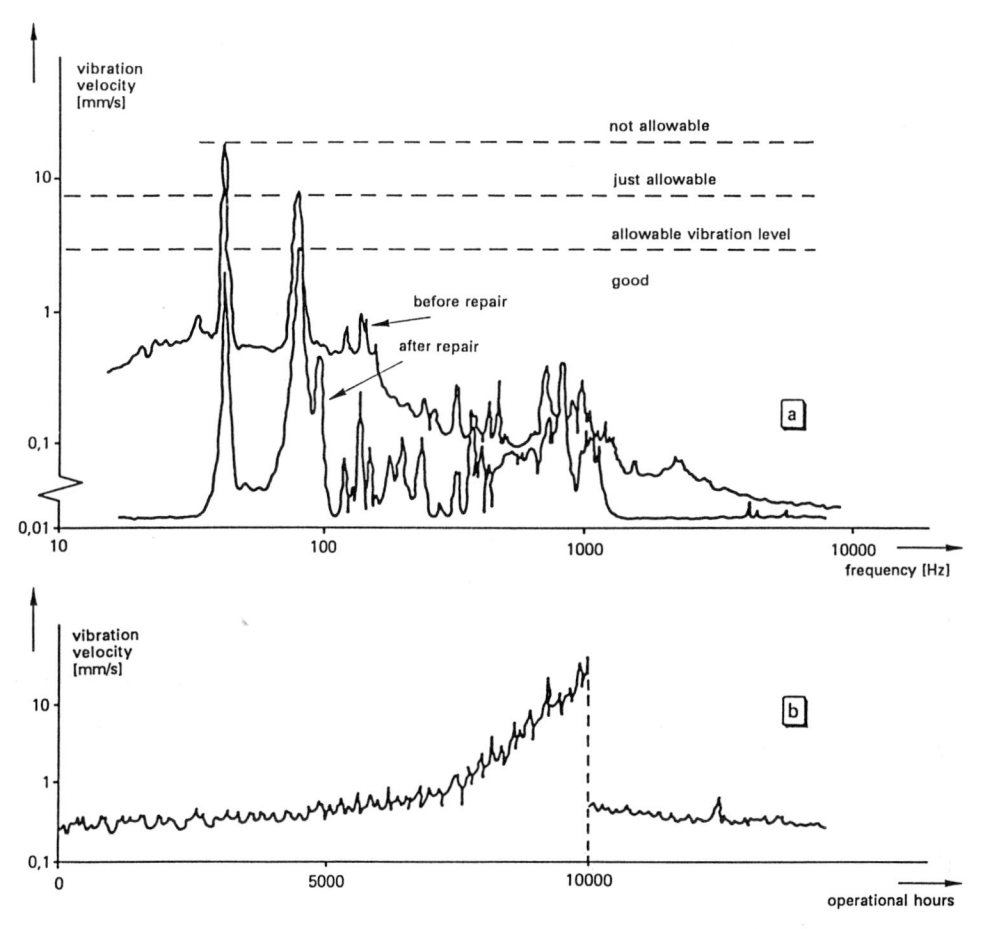

Figure 7.6 (a) The frequency spectrum of the vibration velocity (RMS value) of a high speed gear box before and after repair at t = 10,000 hours (3 % bandwidth spectrum); (b) the trend of the vibration velocity (from 10–1,000 Hz, RMS value) with the number of operational hours.

equipment needs to be checked periodically for accuracy. If necessary, the instrument has to be readjusted and components that have exceeded the tolerance much must be replaced.

7.3 Systems with corrective maintenance

The following sections will be devoted to the treatment of corrective maintenance. It is assumed here that a system or an essential part of that system fails before one intervenes. This human intervention (restoration, repair) is aimed at bringing the system back into a usable state.

Let us first assume that the time necessary for this restoration is neglectably small. This is (usually) the case with modular systems, if the failed module is simply replaced by a new or reworked one, without repairing it on the spot. Especially if the modules are also fitted with fault detection and indication, searching for and correcting the failure requires little time. With restoration by *replacement* we may therefore consider the repair time distribution to be a *Dirac function*. So, replacement is assumed to take effect immediately after the failure; the down-time of the system is neglible or not important.

In subsequent sections we shall discuss *repair* in which the repair process of failure reporting, failure location, ordering parts and actual repair takes so much time that it has to be accounted for in the form a *repair time distribution*.

7.3.1 Replacement

Under the assumptions made above about replacing broken modules, the availability $A(t)$ of the system is no longer an interesting quantity:

$$A(t) = 1.$$

For convenience it is also assumed that we are dealing with *ideal replacement*. This means that the replaced component is 'as new' and has not suffered while in storage or when it was fitted. Since replacement is a relatively simple operation this assumption is usually justified.

Examples are the replacement of thermal fuses in an electrical installation, the replacement of the head lights of a car or the replacement of entire component cards in a personal computer.

In practice one can never correct all the system failures that may occur by simply replacing one or more of the modules that make up the system. Even in a fully modular (electronic) system there will still be cable interrupts, connector failures and things like that. These failures will have to be repaired.

Suppose that $f(t)$ is the failure probability density of the system under consideration without replacement. After replacement the system will fail again. Suppose that the failure probability density of this second failure is $f_2(t)$, then we may find $f_2(t)$ with:

$$f_2(t) = \int_0^t f(t-\tau) f(\tau) \, d\tau.$$

It is assumed that both random variables (first survival time, second survival time) are independent and have the same distribution. The second failure time is then the sum of both survival times. It has the density $f_2(t)$ given in the expression above. The i-th replacement point in time is found by an i-fold convolution of $f(t)$ with itself. In recursive notation this gives:

$$f_i(t) = \int_0^t f(t-\tau) f_{i-1}(\tau) \, d\tau.$$

Next we introduce the *renewal density* $h(t)$, in which $h(t)\Delta t$ is the probability of a renewal in the time interval $(t, t+\Delta t]$. The value of $h(t)$ is then easily found with:

$$h(t) = \sum_{i=1}^{\infty} f_i(t).$$

Let us give a simple example. Consider a collection of a large number of identical components N which have a normally distributed life with mean μ and standard deviation

σ (for instance light bulbs). All failed components are replaced by new ones. The resulting renewal density is illustrated in Figure 7.7. It shows that eventually the renewal density $h(t)$ assumes a constant value given by:

$$h_\infty = \lim_{t \to \infty} h(t) = \frac{N}{\mu}.$$

Despite the fact that we started with normally distributed failure times we eventually get renewals that occur at random in time. By mixing old and new components the system will eventually behave as if failing with a constant failure rate that is by sheer accident. The initial peaks in $h(t)$ may very well exceed the renewal capacity of the maintenance service, or, if this capacity is made large enough to cope with these peaks, it results in a quite inefficient use of the maintenance staff. How does the *renewal strategy* have to be changed to get the final value N/μ for $h(t)$ right from the beginning $t = 0$? Is one justified in calling $h(t)$ the hazard rate $z(t)$ of the system with renewal?

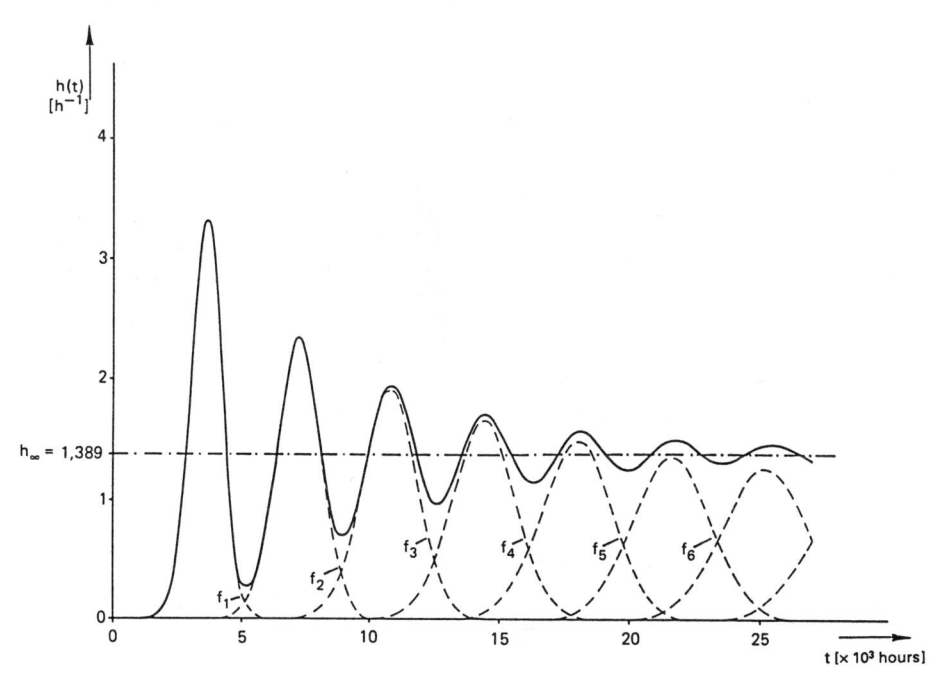

Figure 7.7 The renewal density of a system consisting of 5,000 (active) components having a Gaussian life distribution with a mean μ = 3,600 hours and a standard deviation T = 600 hours.

Example

A large sky-sign in Las Vegas is put into use. The illumination is entirely by light bulbs with an average life μ = 1,000 hours an a standard deviation σ = 100 hours. The display contains 1,000 bulbs. We may now expect with Figure 7.7 that the greatest replacement peak load for the maintenance crew is after 1,000 hours. For a normal distribution 63.8 % of the total distribution is within a ± 1 sigma-interval around the mean value μ. So 638 light bulbs fail in a period of 200 hours. That is about three bulbs per hour! The final

138 *Maintained Systems*

value, after infinitely many renewals, is $N/\mu = 1$ bulb per hour. The replacement peak can be smoothed out by already starting to replace an average of 1 bulb per hour from $t = 0$ on, even though the bulbs to be replaced are still working. However, this might be argued to result in a waste of good bulbs. A more thrifty strategy is to replace the broken bulbs at an average rate of 1 bulb per hour from $t = 0$ on and to return the replaced, still good bulbs to the stock room. The light bulbs from the stock room are chosen at random: no distinction is made between new bulbs and used bulbs. If the fresh part of the stock is sufficiently large one may in this way, without wastage, manage to maintain a virtually flawless sky-light with a minimal size maintenance crew that has a virtually constant work load. (Virtually constant: because there will be statistical fluctuations in the number of broken bulbs which have to be replaced anyway.) Admittedly it may take some persuasion to have the maintenance people replace perfectly good light bulbs in a sky-sign which, in the beginning, may have all lamps operating!

This massive replacement problem is important where there are large numbers of new systems taken into service at approximately the same time and exposed to the same operating time. In that case the mortality will also show a large correlation.

7.3.2 Repair

In the following sections on repair, it is assumed that a system has failed before human intervention takes place. It is further assumed that the repair or corrective intervention takes a certain time t to complete. Here t is a non-negative stochastic variable, referred to as the *repair or restoration time*. The variable t has the distribution $M(t)$. This distribution is referred to as the *maintainability*. It is also assumed that the maintainability distribution is not a degenerate distribution as was the case with *renewal* (the maintenance duration density was a Dirac function here).

The operational-use phase of the life cycle of such a repairable system (see Figure 7.1) is given in Figure 7.8.

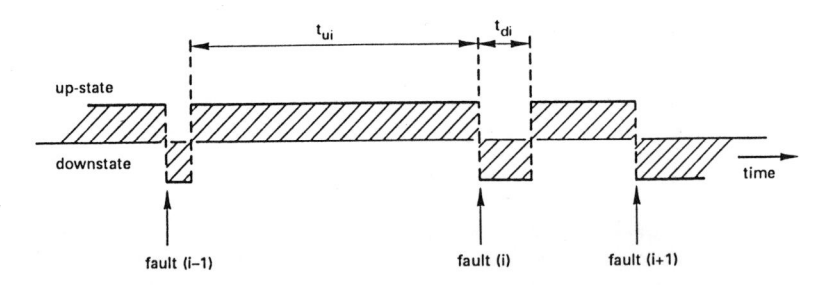

Figure 7.8 An example of the 'up' and 'down' intervals of a system or a unit versus calender time (see also Figure 3.2).

The time t_{ui} represents the i-th consecutive 'up' period or operational period, the time t_{di} the subsequent 'down' period or repair period.

The concept 'operational period' should not be interpreted too narrow here. The 'up' or operational period encompasses besides the time during which the system functions properly, also the time during which the system is ready for use, but is not used for whatever reason (other than a defective system).

The concept reliability as defined in Section 3.2 only is meaningful over the first continuous 'up' period t_{u1} of the system. The probability that this 'up' state lasts until the time t is $R(t)$. The mean time until the *first* system failure occurs (Mean Time To First Failure) is therefore:

$$\text{MTTFF} = \int_0^\infty R(t)\, dt.$$

It will also be clear that in a system like the one shown in Figure 7.8 the availability $A(t)$ is no longer equal to 1, as was the case with renewal.

In Section 3.2 the *availability $A(t)$* has already been defined: $A(t)$ is the probability that the system is in an operational state at time t, if it has been installed as intended and has been operated as specified during the interval $[0,t]$.

N.B.: It is striking that for non-repairable systems and also for the first time interval t_{u1} of a repairable system, the concepts *reliability $R(t)$* and *availability $A(t)$* are identical. It would therefore have been sufficient to only introduce the more general concept availability, also for non-repairable systems. Instead of 'reliability' the title of this book should have included 'availability'. Some authors do this consistently. However, the authors believe, because of the essential difference between maintained and non-maintained systems, that it is better to adhere to the generally used vocabulary.

Often one is interested in the average availability over relatively short periods of time. This measure of availability is important for airplanes and ships, which are especially not allowed to fail during their mission (flight, voyage). For that reason the measure '*mission availability*' was introduced, which is defined as:

$$A(t, T) = \frac{1}{T} \int_t^{t+T} A(t)\, dt.$$

This measure indicates the fraction of the time that the system was available (on average) during the time interval $(t, t+T)$ necessary for carrying out the required mission. In general the mission availability during the time T is a function of the absolute age t of the system; for an old system it will be different from that of a new system.

In analogy to mission availability *long-term availability* is defined as:

$$A_\infty = \lim_{t \to \infty} \frac{1}{T} \int_0^t A(t)\, dt.$$

The long-term availability A_∞ is often called *steady-state availability*. The latter expression is based on the observation that the availability $A(t)$ of a system may fluctuate very much directly after it has been put in use, especially if large numbers of components with (about) the same expected life have been employed. Eventually, however, for very large t, the age composition of the components has become rather uniform because of the many repairs. This makes the availability $A(t)$ of practical systems virtually constant for very large t; the system has then reached its steady-state availability A_∞. All this is illustrated in Figure 7.9. It should be noted that A_∞ is approached only after very many repairs. For that reason most technical systems will never reach A_∞ during the finite time they are used. Therefore it is better to regard A_∞ a limit of the real $A(t)$.

140 Maintained Systems

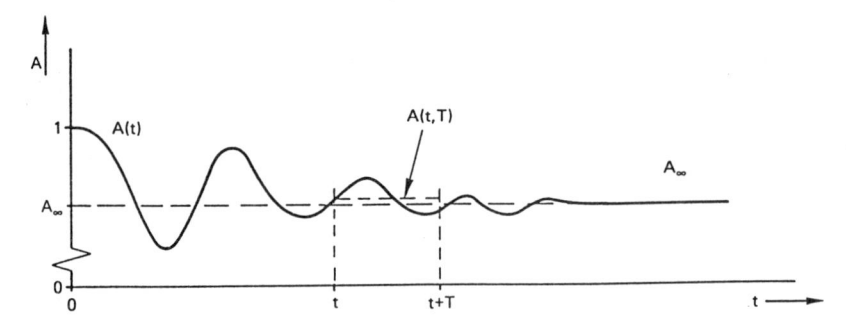

Figure 7.9 Example of the availability of a system as a function of the age t of that system.

For the long-term availability one may also write:

$$A_\infty = \frac{MTBF}{MTBF + MTTR}.$$

Here the Mean (up-) Time Between Failures (MTBF) is:

$$MTBF = \lim_{n \to \infty} \frac{1}{n} \sum_{i=1}^{n} t_{ui},$$

and the Mean (down-) Time To Repair (MTTR):

$$MTTR = \lim_{n \to \infty} \frac{1}{n} \sum_{i=1}^{n} t_{di}.$$

If, after sufficiently large but finite t, $A(t)$ has approached the value (A_∞) infinitesimally close, we may start executing the above summation from that time on. The first part in which $A(t)$ is not stationary does not matter anymore then. Accordingly, one may also relate the MTBF and the MTTR to the steady-state of a system only.

N.B.: The MTBF and MTTR are averages, they do not reveal information about the distribution of 'up' and 'down' times. The MTTR may, for example, be composed of a few very long repair times (which are dominant) and a large number of short repairs. Similarly a certain long-term availability (for example $A_\infty = 0.99$) contains no information about the number of failures. It may equally well be caused by a very large number of short failures as by a few long ones.

For a quantitative analysis of the availability of a system and the influence of the repair strategy, it is useful to make the following classification:

    A. *System*   1. Configuration
                         2. Homogeneity
                         3. Operational strategy
    B. *Units*   1. Failure modes
                         2. Failure distributions

C. *Repair*   1. Capacity
              2. Repair strategy
              3. Repair distributions

The *configuration* determines the structure of the system (parallel, series, etc.). The *homogeneity* determines whether the system may be modelled as a configuration consisting of statistically identical units. The *operational strategy* determines matters such as the use of redundancy in the active or in the passive mode, the continuation of the system operation after a failure, the complete or partial shutdown after a failure has occurred. In the later case no further failures will be developing during the down-time. Usually, however, restarting a complex system is accompanied by a temporary increase of the hazard rate. Therefore, systems with built-in redundancy or alternative functions may be allowed to continue operating.

The units which compose the system may display different *failure modes:* catastrophical, partial, complementary failure modes (open, short-circuited), dependent and intermittent failure modes. The failure modes may each have their own *failure distribution.*

The *repair capacity* is determined by the number of available *repair channels* and the number of repairmen available per channel. A repair channel is defined as a number of repairmen (with spare parts, measurement equipment and tools) who together work on restoring one failure. The number of repairmen determines the capacity of the relevant repair channel. A repair channel may also maintain more than one system. This is called *shared repair.* Also, two repair channels may work, consecutively, on the same system. The second channel repairs the next failure if it occurs while the first channel is still busy. The *repair strategy* determines which form of repair is chosen for a particular system. Finally, each form of repair has its own *repair time distribution.*

In the following we shall consider various configurations. Up front we assume a number of things:

- There always is at least one repair channel present. Such a channel comprises the repairman, his know-how, his measurement and testing equipment and the parts or modules he needs for the repair.
- The repair is ideal. This means that the used components are 'as new' and the system does not suffer from the repair; the repair does not introduce new early failures.
- There is an infinite supply of spare parts, unless otherwise stated.
- The time needed for repair is a non-negative stochastic variable that is distributed negative-exponentially and has a mean value $1/\mu$. The parameter μ is called the *repair rate*. This repair rate μ is the complementary concept of the *failure rate* λ.

7.3.3 Repairable systems without redundancy

We shall demonstrate the modelling of repairable systems without redundancy by taking a relatively simple system consisting of n units in series where the units are stochastically identical. This constitutes a simple homogeneous series system.

Let us first deal with the simplest repair strategy. It employs only one repair channel consisting of one repairman who immediately after a failure switches off the system and starts to repair it. Therefore, during the repair no more units can fail. Immediately after

142 Maintained Systems

completing the repair the system is switched on again. The failure rate of the units is λ, the repair rate for all units is μ.

We can model this simple maintainable system in a Markov model comprising only two states as indicated in Figure 7.10:
- State S_0 is the state in which all n units function properly, the repairman is on standby but idling.
- State S_1 is the state in which a unit has failed and the repairman is busy. The other $n-1$ units cannot fail now!

From this Markov model we can, with the rules given in Section 5.3, directly write the differential equations:

$$\frac{dP_{S_0}(t)}{dt} = -n\lambda P_{S_0}(t) + \mu P_{S_1}(t),$$

$$\frac{dP_{S_1}(t)}{dt} = n\lambda P_{S_0}(t) - \mu P_{S_1}(t).$$

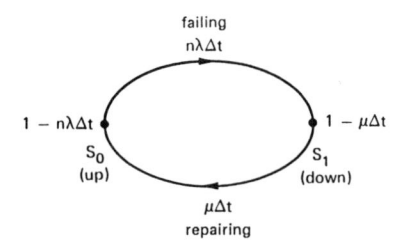

Figure 7.10 A state model of a series system with n units, in which no failures can occur during a repair (the units are switched off while the system is 'under repair'). To clarify the picture, the loops leaving from and returning to S_0 and S_1 have been deleted.

If we take as initial conditions $P_{S_0}(0) = 1$ and $P_{S_1}(0) = 0$ the following Laplace transforms result:

$$sP_{S_0}(s) - 1 = -n\lambda P_{S_0}(s) + \mu P_{S_1}(s),$$

$$sP_{S_1}(s) = n\lambda P_{S_0}(s) - \mu P_{S_1}(s).$$

Solving $P_{S_0}(s)$ yields:

$$P_{S_0}(s) = \frac{s+\mu}{s(s+n\lambda+\mu)}.$$

The availability $A(t)$ is found by inverse transformation of $P_{S_0}(s)$.
We then find:

$$A(t) = \frac{\mu}{n\lambda+\mu} + \frac{n\lambda}{n\lambda+\mu} e^{-(n\lambda+\mu)t}.$$

The mission availability $A(t,T)$ over the interval $(t, t+T)$ is then:

$$A(t,T) = \frac{\mu}{n\lambda+\mu} - \frac{n\lambda e^{-(n\lambda+\mu)t}}{T(n\lambda+\mu)^2} \{e^{-(n\lambda+\mu)T} - 1\}.$$

For the long-term availability A_∞ we find:

$$A_\infty = \frac{\mu}{n\lambda + \mu}.$$

With the expression given for the long-term availability in Section 7.3.2 we might have written down the latter expression directly:

$$A_\infty = \frac{\text{MBTF}}{\text{MBTF} + \text{MTTR}} = \frac{1/n\lambda}{1/n\lambda + 1/\mu}.$$

In Figure 7.11 $A(t)$ and A_∞ have been plotted. It is not assumed that $P_{S_0}(0) = 1$; various cases have been plotted. It will be clear that the transition phenomenon in the $A(t)$ curve only occurs if $P_{S_0}(0) \neq A_\infty$.

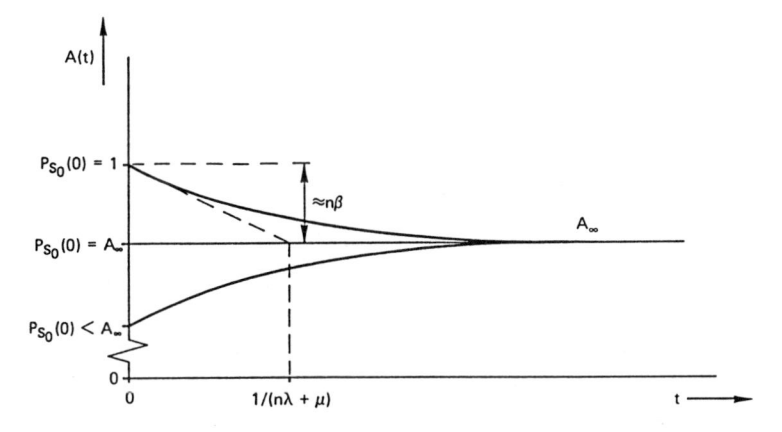

Figure 7.11 The availability $A(t)$ of a homogeneous series system consisting of n (identical) units for various values of $P_{S_0}(0)$. During repairs the system is disengaged and cannot fail.

Note: In repairable systems the failure-to-repair rate ratio is often introduced:

$$\beta = \frac{\lambda}{\mu}.$$

This ratio, which will always be small in the case of sound system management, indicates how much faster a unit fails than it is repaired. Inserting β gives the following approximation:

$$\overline{A}_\infty = \frac{1}{1 + n\beta} \approx 1 - n\beta, \qquad (\beta \ll 1).$$

The *unavailability* $\overline{A}_\infty$ is:

$$\overline{A}_\infty = 1 - A_\infty \approx n\beta.$$

The *renewal rate* is the average number of failures and therefore repairs (or renewals) per unit of time. The renewal rate is equal to:

$$h_\infty = n\lambda \frac{\mu}{\mu + n\lambda}.$$

144 Maintained Systems

For $n\beta \ll 1$ holds:

$$h_\infty = n\lambda \frac{1}{1 + n\beta} \approx n\lambda(1 - n\beta).$$

The quantity h_∞ determines the necessary capacity of the maintenance channel, so the number of required repairman hours.

We might also have chosen a more direct calculation of A_∞ than the one given above. In Section 7.3.2 it was defined that A_∞ is the steady-state value of the availability $A(t)$. In this static state the probabilities P_{S_i} of the states S_i ($i = 1,2,...,n$) become time-independent. So the derivatives in time of these probabilities are zero. Therefore these probabilities can be found from:

$$-n\lambda P_{S_0} + \mu P_{S_1} = 0,$$

$$n\lambda P_{S_0} - \mu P_{S_1} = 0.$$

Since the system is always in one of the states S_i, it holds that:

$$\sum_{i=0}^{n} P_{S_i} = 1.$$

For the n-unit homogeneous series system this results in:

$$A_\infty = P_{S_0} = \frac{\mu}{n\lambda + \mu}.$$

After this simple example of one repair channel applied to a disengaged series system, we shall analyse the effect of keeping the remaining $n - 1$ functioning system units in operation during a repair. This is a practical assumption because the switching off of a system during a repair is not always desirable. After the system is switched back on there is usually a new warming-up or running-in time, which should be added to the effective repair time. In addition, switching a system on and off is a form of failure accelerating stress which temporarily may increase the mortality immediately after the system is switched back on. Therefore, in the following it will be assumed that good units will continually remain switched on.

Now, after one failure has occurred, so during the repair of a unit, other failures may occur. It is also assumed that these failures occur with the same failure rate λ as under normal, uninterrupted operation. This assumption may not be quite fair since the system works now in a different operational mode, but its is a convenient assumption.

The following is the inventory of $n + 1$ possible mutually disjunct states:
- S_0: all n units function correctly.
- S_i: ($n - i$) units function, one unit is being repaired and ($i - 1$) units are queued up to be repaired ($i = 1,2,...,n$).

These states form the configuration of Figure 7.12.

Repairable systems without redundancy 145

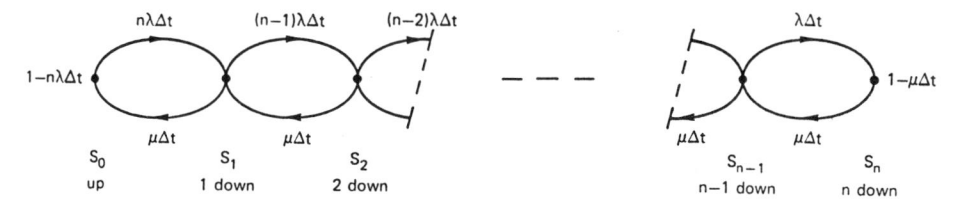

Figure 7.12 A state model of a homogeneous series system with *n* units that may fail in the course of a repair. (See remark in caption of Figure 7.10).

Based on the inspection of this Markov state model the following set of differential equations can be written if $\dot{P}(t)$ denotes $dP(t)/dt$:

$$\dot{P}_{S_0}(t) = -n\lambda P_{S_0} \qquad\qquad + \mu P_{S_1} \qquad\qquad\qquad = 0$$

$$\dot{P}_{S_1}(t) = +n\lambda P_{S_0} -[(n-1)\lambda + \mu]P_{S_1} \qquad + \mu P_{S_2} \qquad = 0$$

$$\dot{P}_{S_2}(t) = \qquad\qquad + (n-1)\lambda P_{S_1} -[(n-2)\lambda + \mu]P_{S_2} + \mu P_{S_3} = 0$$

. . . . . .
. . . . . .
. . . . . .

The probability that the system sojourns in state S_i in the steady-state is:

$$P_{S_i} = \binom{n}{i} i!\, \beta^i P_{S_0}, \qquad (i = 0, 1, \ldots n).$$

N.B.: We may also regard these probabilities as the fraction of the time the system (on average, over a sufficiently long period of time) dwells in the state S_i. Since the sum of all $(n+1)$ sojourn probabilities must be 1, we find for the steady-state availability:

$$A_\infty = P_{S_0} = \frac{1}{n!\sum_{i=0}^{n}\frac{\beta^{n-i}}{i!}} = \frac{1}{n!\sum_{i=0}^{n}\frac{\beta^i}{(n-i)!}}.$$

It can be simply seen that:

$$\lim_{\beta \to 0} A_\infty = 1.$$

For $\beta \ll 1$ the situation does not differ very much from that in Figure 7.11. If the repair rate μ is sufficiently larger than the failure rate λ, not switching the system off during the repair only introduces a small higher order correction term in the steady-state availability A_∞. In Figure 7.13 this long-term availability has been plotted versus failure-to-repair rate ratio $\beta = \lambda/\mu$ for a series system composed of three units and for one composed of ten units both with and without disengaging the system during repairs.

It can be seen that, for small values of β which are characteristic for a sound system management, it does not matter whether the system is switched off or not. Yet it is often useful to keep the system switched on for other than availability reasons, e.g. to prevent new early failures after the repair.

N.B.: This strategy may not be allowed to result in dangerous situations for the

146 Maintained Systems

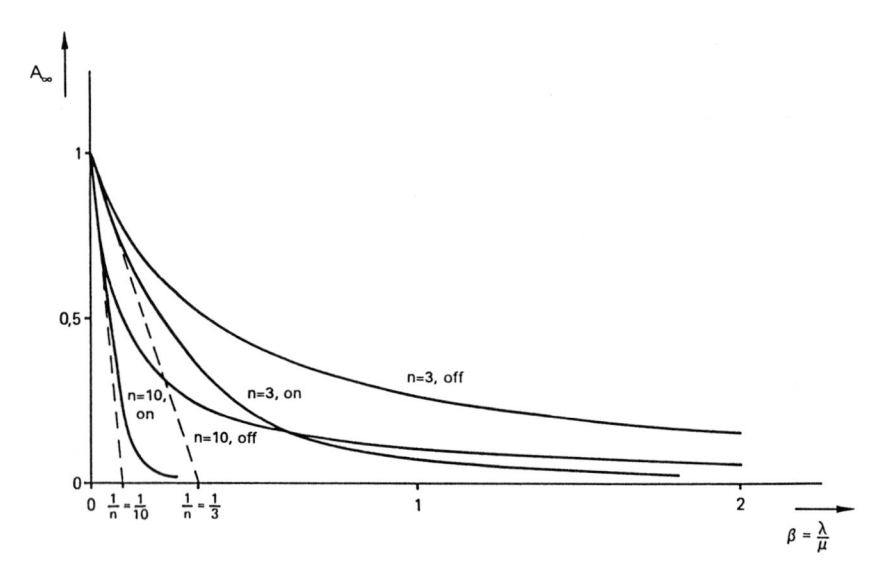

Figure 7.13 The availability A_∞ of an n-unit series system which remains in operation during a repair ('on') and the availability of the same system when it is shut down ('off') during a repair, as a function of the failure-to-repair rate ratio β.

repairman, nor may it give cause for longer repair times because of more elaborate repair to mitigate hazards associated with working on a still operative system.

So far we have considered only one repair channel consisting of one repairman. We shall now analyse the effect of more than one repair channel, each still consisting of one repairman. The good units of the homogeneous n-unit series system under observation will be always on. This can result in several failures. After the first failure occurs one of the m repair channels repairs the unit ($m \leq n$). If a second unit would break during this repair, a second repair channel becomes operative, etc. The repair channels are homogeneous too, all of them have a repair rate μ for a single unit repair.

In the associated Markov diagram the i-th state S_i ($i = 0,1,...,n$) will now mean that i units have failed, i of which are repaired if $i \leq m$ and m of which are repaired if $i \geq m$. The remaining $(n - i)$ units are still functioning properly but they may fail. The configuration of the Markov state diagram is as shown in Figure 7.14.

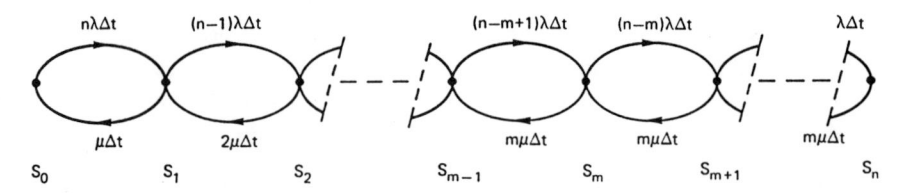

Figure 7.14 The Markov diagram of a homogeneous n-unit series system with m homogeneous repair channels and continuous operation.

If we look at this diagram in more detail we notice a number of things. If the system is in state S_0 at $t = 0$, the failure of a unit (n possibilities) will cause the system to transit to state S_1. From that state it is returned at rate μ back to state S_0 by the action of a repairman. The probability of yet another failure (state S_2) is smaller; there are only $(n-1)$ possibilities. Besides, the probability of correction of one of the two failures in the time interval Δt is twice as large, namely $2\mu\Delta t$. If we continue this line of reasoning we arrive at the conclusion that it is most unlikely that the m-th repairman will have anything to do at all. The probability of work for the m-th repairman decreases as m becomes larger (and for $m > n$ it is of course zero).

For $n = 2$, $m = 2$ and $\beta = \lambda/\mu$ the steady-state availability becomes:

$$A_\infty = 1/(1 + \beta)^2,$$

and for $\beta \ll 1$ holds:

$$A_\infty \approx 1 - 2\beta.$$

This is the same expression as we found for only one repairman. Apparently the repair time is now so short with regard to the life time of a unit that even the second repairman hardly ever has anything to do. In the plots of Figure 7.15 this is illustrated for a number of different cases.

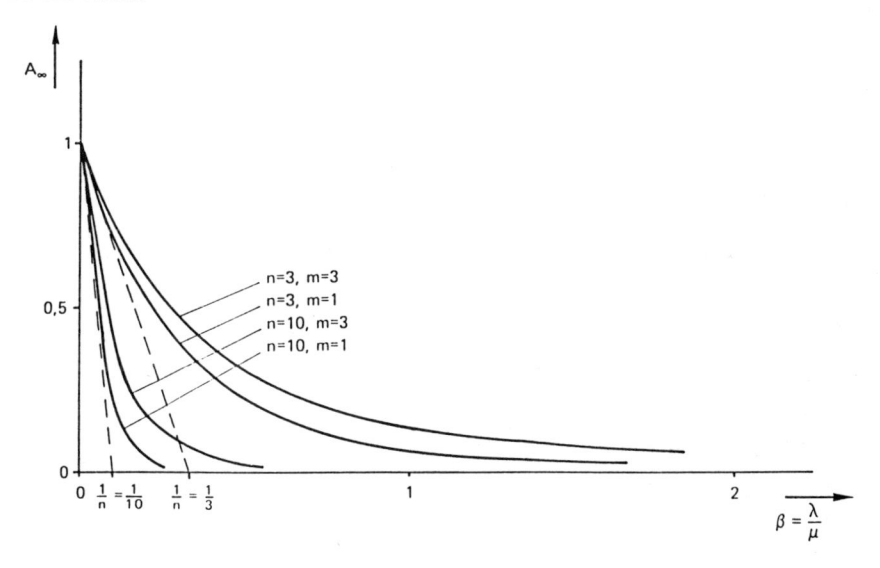

Figure 7.15 *The availability A_∞ of an n-unit series system with m repair channels when the system is continually 'on'.*

In the above, we have studied the effect of the number of repair channels m on the availability of an n-fold series system. We shall now analyse the case of only one repair channel, but one with more than one repairman. It is assumed that the k repairmen assigned to that single channel all are simultaneously busy repairing the occurring system failures. If only one unit has failed, all k repairmen are simultaneously working on that one unit. The repair will then, of course, be quicker, but, in general, not k times as quick. An efficiency factor α is introduced, such that the repair rate for k repairmen working on one unit becomes:

148 Maintained Systems

$$\mu_{\text{effective}} = \mu + \alpha(k-1)\mu, \quad 0 \le \alpha \le 1.$$

So, more precisely the factor α represents the effectiveness of the extra $(k-1)$ repairmen. α will depend on the job at hand; not all repairs are equally well suitable to be conducted by several people. Just think of using more than one person to fix a mechanical watch on the one hand ($\alpha \approx 0$), or using them to renew corroded pipes in an oil refinery on the other hand ($\alpha \approx 1$).

If more than one unit has failed, the group of repairmen is automatically split into a number of subchannels, so that there is always one unit being repaired per subchannel. With increasing numbers of simultaneous failures this continues until one repairman is working on one unit per subchannel. The subchannels are mutually independent, $\alpha = 1$ here; they are working on different units. The repairmen within one subchannel cooperate in the repair of that one unit ($0 \le \alpha \le 1$).

The state S_i in the associated Markov diagram is then characterised by:
- n units of which i have failed ($i \le n$),
- k repairmen prepared to cooperate ($k \le n$),
- i repair channels if $i \le k$, $\mu_{\text{eff}} = \mu + \alpha(k/i - 1)\mu$,
- k repair channels if $i \ge k$, $\mu_{\text{eff}} = \mu$.

The diagram is shown in Figure 7.16.

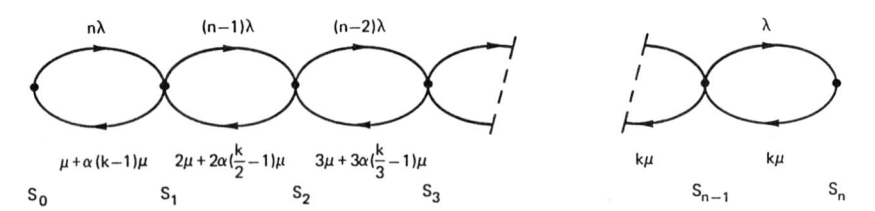

Figure 7.16 *The Markov diagram of an n-unit series system with k cooperating repairmen. (The factors Δt have been omitted here).*

On inspection of this Markov diagram we can directly see that (if for $t = 0$: $P_{S_0} = 1$) the first loop in the diagram, so the transitions between S_0 and S_1 are the most important, particularly for small values of $\beta = \lambda/\mu$. Therefore, it is clear that for $\beta \ll 1$ we may write:

$$A_\infty = P_{S_0} \approx 1 - \frac{n\lambda}{\mu + \alpha(k-1)\mu} = 1 - \frac{n\beta}{1 + \alpha(k-1)}.$$

This can be elucidated as follows. Let us consider only the first two states in the steady-state. If the efficiency α of the extra repairmen is zero, we will have the same situation as with k independent repair channels with one repairman each. If $\alpha = 1$ the repair is k times faster and consequently the availability larger.

Now that we have gained some insight into the maintenance behaviour of a series system, it appears on hindsight that instead of the availability A_∞ it is easier to determine the unavailability $\overline{A}_\infty$. The approximated expression above for $\beta \ll 1$ can be directly obtained from the Markov diagram. If $\beta \ll 1$ the likelihood that the system will ever reach the states S_2 and higher is negligible. We are only concerned with S_0 and S_1. For this two-state subdiagram the unavailability can be written down directly. It is the quotient of the failure transition probability ($n\lambda\Delta t$) and the repair transition probability $[\mu + \alpha(k-1)\mu]$ Δt. Thus:

$$\overline{A}_\infty = 1 - A_\infty \approx \frac{n\lambda}{\mu + \alpha(k-1)\mu}.$$

Example

Suppose $n = 2$ and $k = 2$, the previously explained method via the differential equations then gives:

$$A_\infty = \frac{1 + \alpha}{1 + \alpha + 2\beta + \beta^2}.$$

For $\beta \ll 1$ this may be approximated by:

$$A_\infty = 1 - \frac{2\beta}{1 + \alpha}.$$

So the unavailability $\overline{A}_\infty$ is about: $2\beta/(1 + \alpha)$. Owing to the efficiency α of the extra repairman this unavailability has apparently changed by 100 %; the sensitivity for the efficiency α is high in this case. (See also the example of Figure 7.17).

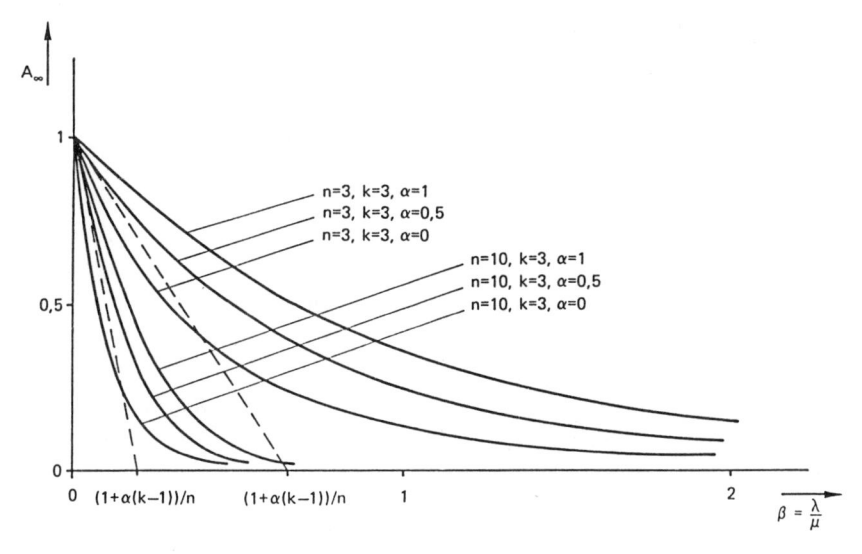

Figure 7.17 The availability of an n-unit homogenous series system with m repairmen with an efficiency α, as a function of the failure-repair rate ratio $\beta = \lambda/\mu$.

Concludingly, this section about maintained systems without redundancy has shown us:

- A system with sound maintenance management will have a failure-to-repair rate ratio $\beta \ll 1$ ($\beta = \lambda/\mu$).
- The effect of switching such a system off during repairs or leaving it on is small. This does not hold when β is not small; the system should then be switched off. The system should be left on when switching on or off is an important stressor. (See Chapter 2).

- Employing several, mutually non-cooperative repair channels has little effect on a system with otherwise sound maintenance.
- In the case of cooperating repairmen (effective within one repair channel) the improvement of the availability of the system depends strongly on the efficiency α of the extra repairmen.

7.3.4 Repairable systems with redundancy

It will be intuitively clear that a very powerful combination of maintenance strategy and system configuration is to preventively maintain a redundant configuration. The maintenance here already starts before all the built-in system redundancy is exhausted. It is possible to repair failed units and restore them to correct operation before the system as a whole has failed. Most of this is possible without having to even shut down the system (e.g. for the removal or insertion of the redundant unit). As the redundancy degree η (see Section 6.3) is greater and as the repair is done more expeditiously the probability of system failure will become less and less. After all, the system only goes down if the repair lasts till after the point in time at which yet another unit fails while all redundancy for that unit is already depleted. In general, the repair of failed units can be conducted during normal operation of the system. At the most a negligibly small break in the continuous system functioning is necessary in order to replace a failed redundant unit. In this section, therefore, a failure in a redundant unit and its repair shall not be considered a system failure, at least, not as long as there are sufficiently many well-functioning units left (think, for instance, of m-out-of-n redundancy).

We are interested in a number of reliability quantities of such repairable systems with redundancy, namely: the availability A_∞, the reliability $R_S(t)$ and the so-called Mean Time To First *system* Failure (MTTFF). For the calculation of the reliability and the determination of the MTTFF it is assumed that the repair stops if a system failure occurs. On the system level we treat the system as non-maintainable; on the unit level it is maintainable. As already said, this can only happen after all units added in a redundant fashion to a certain unit have failed. So we are then dealing with the repair of units only as long as the system function is still undisturbed; there are no repairs at system level. This assumption is only made in order to be able to utilise the concepts $R_S(t)$ and MTTFF; for the system availability A_∞ there are, of course, no objections to repairing the failed system. In fact, for determining A_∞ we must assume that a broken system will also be repaired.

Below we shall investigate the effect of the repair of systems with *active* as well as *passive redundancy*. For simplicity of calculation it is assumed that we are dealing with a homogeneous system, consisting of units that fail independently with a failure rate λ. The repairmen are also independent and individually realise a repair rate μ. The extra repairmen have an efficiency α.

Let us first look at the case of active redundancy. We want to determine the availability of an n-unit system, depending on the number of repairmen k (one repair channel). The Markov diagram is then very simple. It is exactly the same as that of Figure 7.16 which applies to a series system. The only thing left to settle is the redundancy degree η of the system; in other words, how many units within one system partition are allowed to have failed before the system fails, i.e. that partition fails. If it is assumed that we are dealing with an m-out-of-n system (system works correctly as long as at least m units are operating correctly), we have the most general case under observation. However, the problem then is that we can no longer arrive at manageable analytical expressions for A_∞. For a numerical analysis with a computer this is no problem, of course, as Figure 7.18 illustrates. Here not just A_∞ but the more general $A(t)$ is given for a number of values of the system parameters n, m, k, α, μ and λ.

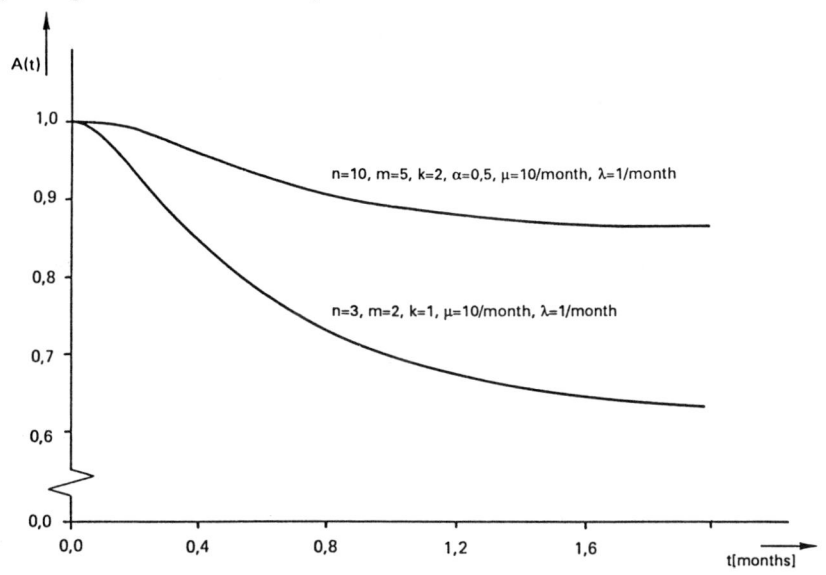

Figure 7.18 The availability of an m-out-of-n unit (active) redundant system composed of units with a failure rate λ and a repair rate μ, maintained by k cooperating repairmen with an efficiency α.

For the sake of analytical computability we shall now introduce a number of simplifications. Let us assume that we have just as many repairmen k as the number of system partitions n, so $k = n$. Also, the repairmen do not cooperate, but form $k = n$ independent repair channels. In fact, we in fact assume that $\alpha = 0$.
In an m-out-of-n system the availability is equal to the sum of the first $(n - m + 1)$ state probabilities P_{S_i}. Thus, in the steady-state:

$$A_\infty = \sum_{i=0}^{n-m} P_{S_i}.$$

Since every system partition has its own repairman and may therefore be considered to be independent of the rest, the problem may be formulated in another way. After all, each combination of a system partition and a repairman has a certain availability:

$$A_\infty = \frac{\mu}{\mu + \lambda},$$

and an unavailability:

152 Maintained Systems

$$\bar{A}_\infty = \frac{\lambda}{\mu + \lambda}.$$

For P_{S_0} all n partitions have to be available, so:

$$P_{S_0} = \binom{n}{n} \left(\frac{\mu}{\mu + \lambda}\right)^n.$$

For P_{S_1} there have to be $(n-1)$ partitions available and one unavailable:

$$P_{S_1} = \binom{n}{n-1} \left(\frac{\mu}{\mu + \lambda}\right)^{n-1} \left(\frac{\lambda}{\mu + \lambda}\right)^1,$$

so that:

$$\bar{A}_\infty = \sum_{i=0}^{n-m} \binom{n}{n-1} \left(\frac{\mu}{\mu + \lambda}\right)^{n-1} \left(\frac{\lambda}{\mu + \lambda}\right)^i.$$

Example

Consider an active redundant system consisting of two partitions ($n = 2$, $m = 1$) with a unit failure rate λ. Both partitions have their own repairman ($k = 2$). The repairmen do not cooperate ($\alpha = 0$). Based on the Markov model of Figure 7.16 it can easily be calculated that the steady-state availability now is:

$$A_\infty = \frac{\mu^2 + 2\lambda\mu}{(\mu + \lambda)^2}.$$

With $\beta = \lambda/\mu$ the unavailability may be written as:

$$\bar{A}_\infty = \frac{\beta^2}{(1 + \beta)^2}.$$

In the previous section we have seen that the unavailability of one single unit with repair was $\bar{A}_\infty = \beta/(1 + \beta)$. So, the improvement from applying redundancy *and* repair is a factor of $1/\beta$. Depending on the magnitude of λ with regard to μ the improvement can be very large.

Another simple, fairly general case is a truly redundant system with $\eta = 1$ (so $m = 1$), consisting of n units which are repaired by only one repairman ($k = 1$). We then find for the unavailability:

$$\bar{A}_\infty = \frac{1}{\sum_{i=0}^{n} \frac{1}{\beta^i i!}}.$$

Example

Consider an active redundant system consisting of two units ($n = 2$, $m = 1$) with one repairman ($k = 1$). For such a system it can be calculated that:

$$\bar{A}_\infty = \frac{2\lambda^2}{\mu^2 + 2\mu\lambda + 2\lambda^2} = \frac{2\beta^2}{1 + 2\beta + 2\beta^2}.$$

For a non-redundant system with one repairman we have seen that:

$$\overline{A}_\infty = \frac{\beta}{1+\beta},$$

so that the unavailability is reduced by redundancy by a factor:

$$1 + \frac{1}{\beta(2+2\beta)} \approx \frac{1}{2\beta} \qquad (\beta \ll 1).$$

So the improvement from introducing redundancy is about a factor $1/(2\beta)$. This is only a factor of 2 smaller than we would obtain by employing two (expensive!) repairmen (see the previous example).

We shall now determine the reliability $R_S(t)$ and the mean time to the first system failure (MTTFF) of active-redundant configurations. To that end, it is assumed that in the general Markov diagram of Figure 7.16 no repairs are performed anymore from state $S_{(n-m+1)}$ and higher states. For the rest the diagram remains the same. Between the MTTFF and $R_S(t)$ there is the previously given relationship:

$$\text{MTTFF} = \int_0^\infty R_S(t)\,dt,$$

or, if the Laplace transform of $R_S(t)$ is known:

$$\text{MTTFF} = \lim_{s\to 0} R_S(s).$$

To exemplify the calculation method we shall take the system, illustrated in Figure 7.19. The differential equations for this homogeneous 1-out-of-2, active-redundant system are:

$$\frac{dP_{S_0}(t)}{dt} = -2\lambda P_{S_0}(t) + \mu P_{S_1}(t),$$

$$\frac{dP_{S_1}(t)}{dt} = 2\lambda P_{S_0}(t) - (\lambda + \mu)P_{S_1}(t),$$

$$\frac{dP_{S_2}(t)}{dt} = \lambda P_{S_1}(t).$$

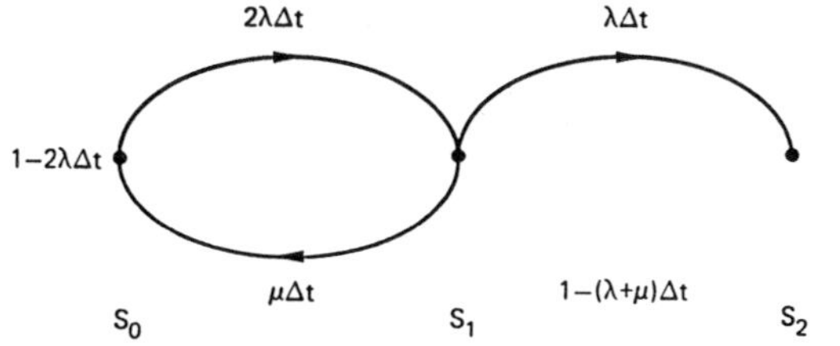

Figure 7.19 The Markov diagram of a homogeneous 1-out-of-2 active-redundant system with repair of the redundant units.

With the initial conditions $P_{S_0}(0) = 1$ and $P_{S_1}(0) = P_{S_2}(0) = 0$ the Laplace transform of the reliability becomes:

154 *Maintained Systems*

$$R_S(s) = P_{S_0}(s) + P_{S_1}(s) = \frac{s + 3\lambda + \mu}{s^2 + (3\lambda + \mu)s + 2\lambda^2}.$$

So the MTTFF becomes:

$$\text{MTTFF} = \lim_{s \to 0} R_S(s) = \frac{3\lambda + \mu}{2\lambda^2}.$$

Inverse transformation of $R_S(s)$ into the time domain results in:

$$R_S(t) = \frac{a\, e^{bt} - b\, e^{at}}{a - b},$$

in which:

$$2a = -3\lambda - \mu + \sqrt{\lambda^2 + 6\lambda\mu + \mu^2},$$

$$2b = -3\lambda - \mu - \sqrt{\lambda^2 + 6\lambda\mu + \mu^2}.$$

In general, taking the inverse transformation can be complicated. Therefore, it often has advantages to approximate $R_S(s)$ by an expression that can be inversely transformed in a simpler way. With $\beta < 1$, one of the poles of the expression for $R_S(s)$ turns out to be small with regard to the other. In what approximated expression for $R_S(t)$ does this result? (Answer: $R_S(t) \approx e^{-t/\text{MTTFF}}$ for $t \gg 0$).

So the improvement in mean system life by the application of repairs is in the above case equal to:

$$\frac{3\lambda + \mu}{3\lambda} \approx \frac{1}{3\beta}, \qquad \text{if } \beta = \lambda\mu \ll 1.$$

We shall now look into *passive-redundant configurations* and the effect of repair on them. Let us start with the Markov diagram for an *m*-out-of-*n* unit passive-redundant homogeneous system with *k* repairmen prepared to cooperate with an efficiency α. For this most general case the Markov diagram of Figure 7.20 is found.

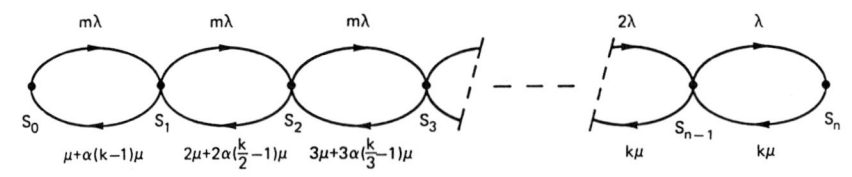

Figure 7.20 The Markov diagram of a homogeneous m-out-of-n passive-redundant system with k cooperating repairmen with efficiency α.

In the literature a number of simple analytical expressions can be found for the unavailability of such a system, provided that the system is made a little less general. If we imagine a redundancy degree $\eta = 1$ (so $m = 1$) and n independent repair channels with one repairman each (so $k = n$, $\alpha = 0$) we find:

$$\bar{A}_\infty = \frac{1}{n! \sum_{i=0}^{n} 1/(i!\, \beta^{n-i})}.$$

If, deviating from the above, it is assumed that the number of repairmen is not equal to n but equal to 1 (so $k = 1$) we find:

$$\overline{A}_\infty = \frac{1}{\sum_{i=0}^{n} 1/\beta^i}.$$

Example

For $n = 2$, $m = 1$ and $k = 1$ we find:

$$\overline{A}_\infty = \frac{\lambda^2}{\lambda^2 + \lambda\mu + \mu^2} = \frac{\beta^2}{\beta^2 + \beta + 1}.$$

The improvement over a similar system with active redundancy for the sound maintenance regime $\beta \ll 1$ is about a factor of two.

For our calculation of the reliability $R_S(t)$ and the MTTFF of the above passive redundant system we now assume k cooperating repairmen with efficiency α who lay down work at the moment the first *system* failure occurs. If we are concerned with an m-out-of-n system this will happen when $(n - m + 1)$ units have failed. This means that the repair transition probabilities associated with all states S_i in the Markov diagram of Figure 7.20 are zero for the states $i = n - m + 1$ and higher. So all these states can conveniently be taken together into one state: 'System Down'.

It is hard to give the expressions for the general case described above. Here, a simple example, namely $n = 2$, $m = 1$ and $k = 1$ will suffice. It then holds that:

$$R_S(s) = \frac{s + \mu + 2\lambda}{s^2 + (\mu + 2\lambda)s + \lambda^2}.$$

Consequently:

$$\text{MTTFF} = \lim_{s \to 0} R_S(s) = \frac{\mu + 2\lambda}{\lambda^2}.$$

If this is compared again with the corresponding active-redundant case, we see that keeping one unit at hand in a passive mode, results (as expected) in an MTTFF that is about a factor of two better. For the inverse transformation we find:

$$R_S(t) = \frac{a\,e^{bt} - b\,e^{at}}{a - b},$$

in which:

$$2a = -2\lambda - \mu + \sqrt{\mu^2 + 4\lambda/\mu},$$

$$2b = -2\lambda - \mu - \sqrt{\mu^2 + 4\lambda/\mu}.$$

Concludingly, we can make a simple comparison between a single unit system ($n = 1$) with one repairman ($k = 1$) and a double system ($n = 2$) with redundancy ($m = 1$) that is repaired by two non-cooperating repairmen ($k = 2$, $\alpha = 0$) or that is repaired by only one repairman ($k = 1$). We distinguish active and passive redundancy. The results are shown in table 7.1.

7.3.5 Shared-repair facilities

By shared-repair facilities we understand the repair strategy in which one repair channel has to maintain various systems. If the capacity of the repair channel consists of one repairman only, this single repairman has to trace and repair the failures in n systems; n systems will share one repairman.

For the availability calculation of this repair strategy we need not draw a new Markov diagram. For we may consider the system as a homogeneous series system consisting of n elements that can fail during repairs and that is maintained by one repairman. We will then get the Markov model of Figure 7.12. The availability of one out of these n systems is not equal to P_{S_0} in this case. It is:

$$A_\infty = \sum_{i=0}^{n} \frac{(n-i)}{n} P_{S_i}.$$

Now it also holds that:

$$P_{S_i} = \frac{n!}{(n-i)!} \beta^i P_{S_0},$$

$$P_{S_0} = \frac{1}{\sum_{i=0}^{n} \frac{n!}{(n-i)!} \beta^i}.$$

We then find for the availability of a particular system:

$$A_\infty = \frac{\sum_{i=0}^{n} \frac{n-i}{n(n-i)!} \beta^i}{\sum_{i=0}^{n} \frac{1}{(n-i)!} \beta^i} = \frac{1}{n\beta}\left[1 - \frac{1}{\sum_{i=0}^{n} \frac{n!}{(n-i)!} \beta^i}\right].$$

(a)

		one single unit	Improvement Factor			
			active redundancy $n = 2, m = 1$		passive redundancy $n = 2, m = 1$	
		$n = 1, k = 1$	$k = 2, \alpha = 0$	$k = 1$	$k = 2, \alpha = 0$	$k = 1$
	$\overline{A}_\infty$	$\frac{\beta}{1+\beta}$	$\frac{1}{\beta}$	$\frac{1}{2\beta}$	$\frac{2}{\beta}$	$\frac{1}{\beta}$
	MTTFF	$\frac{1}{\lambda}$	$\frac{1}{2\beta}$	$\frac{1}{2\beta}$	$\frac{1}{\beta}$	$\frac{1}{\beta}$

(b)

		Improvement Factor			
		without repair		with repair	
		active	passive	active	passive
	MTTFF	$\frac{3}{2\lambda}$	$\frac{2}{\lambda}$	$\frac{1}{3\beta}$	$\frac{1}{2\beta}$

Table 7.1 *Improvement of a system's MTTFF and steady-state unavailability $\overline{A}_\infty$ by the introduction of repair and redundancy.*

Shared-repair facilities

It is also simple to determine the probability P_b that the repairman is busy with one of the n systems. He is busy if at least one of the n systems has failed, so:

$$P_b = \sum_{i=1}^{n} P_{S_i} = 1 - P_{S_0}.$$

The probability that a system has failed and the repairman is busy fixing it (so, that system is *not* waiting for repair), can be written as:

$$P_r = \frac{1}{n} \sum_{i=1}^{n} P_{S_i}.$$

This probability is calculated per system; this explains the factor $1/n$. With this expression and that for A_∞ the probability that a system is waiting for repair becomes:

$$P_w = \sum_{i=2}^{n} \frac{(i-1)}{n} P_{S_i}.$$

The above may also be formulated differently. To that end let us consider the *fraction* of the time that a system functions (or is able to function) ρ_{up}, the fraction of the time that the system is waiting for repair ρ_{wait} and, finally, the fraction of the time that the system is actively repaired ρ_{rep}. We then get:

$$\rho_{up} + \rho_{wait} + \rho_{rep} = 1.$$

It will be clear that the above expression pertains to averages during the steady-state of the maintained system's life, so after very many repairs. It is now simple to see that:

$$\rho_{up} = A_\infty,$$
$$\rho_{wait} = P_w,$$
$$\rho_{rep} = P_r.$$

The probabilities calculated above may therefore also be regarded as average time fractions in the steady-state of the maintained system.

The mutual relationship between these time fractions all relating to one system is also simple, namely:

$$\frac{\rho_{rep}}{\rho_{up}} = \frac{1/\mu}{1/\lambda} = \frac{\lambda}{\mu} = \beta.$$

With the above expressions it easy to determine the average time fraction or the probability that the repairman is busy:

$$P_b = nP_r = n\rho_{rep} = n\beta\rho_{up} = n\beta A_\infty = 1 - \frac{1}{\sum_{i=0}^{n} \frac{n!}{(n-i)!} \beta^i}.$$

The availability A_∞ and the capacity utilisation of the repair channel P_b for different values of n and β are given in Table 7.2.

158 Maintained Systems

Failure-to-repair rate ratio $\beta = \lambda/\mu$

n	1/10 P_b	1/10 A_∞	1/30 P_b	1/30 A_∞	1/100 P_b	1/100 A_∞
1	0.090	0.910	0.032	0.968	0.010	0.9999
2	0.180	0.902	0.064	0.967	0.020	0.9992
5	0.440	0.872	0.161	0.963	0.050	0.9897
10	0.790	0.785	0.319	0.956	0.099	0.9892
20	0.998	0.499	0.620	0.930	0.198	0.9879
30	1.000	0.333	0.868	0.868	0.296	0.9863
40	1.000	0.250	0.986	0.739	0.394	0.9840
50	1.000	0.200	1.000	0.600	0.491	0.9814
60	1.000	0.167	1.000	0.500	0.587	0.9775

Table 7.2 *The availability A_∞ of a system and the probability P_b of being busy of the repair channel if this repair channel has to maintain n systems, as function of $\beta = \lambda/\mu$: the ratio of (average) repair to failure time.*

The availability A_∞ can also be expressed in ρ_{wait} and β:

$$A_\infty = \rho_{up} = \frac{1 - \rho_{wait}}{1 + \beta}.$$

N.B.: The availability of n systems with shared-repair facilities is smaller than that of one such system with repair. The deterioration is:

$$\frac{A_\infty(n)}{A_\infty(1)} = 1 - \rho_{wait}.$$

This expression can also be seen directly by ignoring the time intervals during which a system is waiting for repair in the total system time.

On average $\rho_{wait} \ll 1$, so that the availability hardly degrades by appointing one repairman to more than one system. A more optimal occupation of the repairman's time can now be realised. If we set the fraction of the time spent at work that the repairman can be occupied with actual repair work on average at 75 % of his daily task, the above expression gives how many systems, with a given β, may be appointed to such a repairman. It also allows us to determine the corresponding availability A_∞ associated with such a strategy. If this availability turns out to be too low, we will have to increase the capacity of the repair channel, so that β becomes smaller. This allows the repair channel to handle more systems. In this way, we can give a quantisation of the number of personnel necessary for maintaining a certain plant. We can also indicate how many people have to be immediately available at the location of the system(s) to be maintained.

The situation where a repairman is not immediately available is referred to as delayed repair. If the distribution of the delay time which is assumed to be a non-negative random variable is known, one can again determine quantities such as availability and MTTFF. Usually this has to be done numerically because of the complexity encountered.

If such a computation is done for a redundant system consisting of one active unit, one passive redundant unit and a repairman, and it is assumed that there is a constant queuing time τ, the following expression results:

$$\text{MTTFF} = \frac{2\lambda + 2\mu - \mu e^{-\tau\lambda}}{\lambda^2 + \lambda\mu - \lambda\mu e^{-\tau\lambda}}.$$

The constant queuing time may, for example, be caused by the logistic time of spare parts ordered from a central depot, or by the time it takes a repairman to travel from a central dealership to the plant site. In the previous section, we found that for $\tau = 0$ holds:

$$\text{MTTFF} = \frac{\mu + 2\lambda}{\lambda^2}.$$

The decrease in MTTFF caused by allowing a certain constant queuing time τ is approximately (the MTTFF for $\tau \neq 0$ divided by that for $\tau = 0$):

$$\frac{1}{1 + \mu\tau}, \qquad (\tau\lambda \ll 1).$$

The latter requirement means that the queuing time τ is supposed to be very small with regard to the mean life $1/\lambda$ of one system unit. If it is allowed to make this assumption it usually turns out to be possible to introduce an effective repair time $1/\mu_{\text{eff}}$ that is equal to:

$$1/\mu_{\text{eff}} = 1/\mu + \tau = (1 + \mu\tau)/\mu.$$

The degree to which the effective repair rate μ_{eff} is larger than the repair rate μ is exactly equal to the above-mentioned factor:

$$\mu_{\text{eff}} = \frac{1}{1 + \mu\tau}.$$

This practical approach drastically simplifies the calculations involved; one may simply use a corrected effective repair rate $\mu = \mu_{\text{eff}}$.

7.3.6 Inhomogeneous systems

In the previous sections it was always assumed that the systems under observation were homogeneous. This homogeneity means that the units (modules) of that system were mutually indistinguishable, statistically speaking. So they were assumed to have the same failure rate λ and repair rate μ.

It will be clear that this assumption is not always allowed. The units into which a series system can be partitioned perform different operational sub-tasks. It would therefore be highly coincidental if they would exhibit the same failure and repair rates: after all, they are physically different.

Even if we are concerned only with the redundant units within one partition of the system that have to perform the same operational (sub)task as the primary unit that was initially activated (passive 1-out-of-n redundancy), the units will often be purchased from different manufacturers and they will have been realised in different ways physically. This is done to avoid as much as possible the occurrence of dependent failures which, as we have seen in Section 6.4.1, may have desastrous effects on the redundancy.

160 Maintained Systems

The conclusion must therefore be that, in practice, we are concerned with *inhomogeneous systems*. However, because of the simplicity of the associated analysis, such systems are often approximated by a homogeneous system. What does happen in an inhomogeneous system that is essentially different from that is going on in a homogeneous system? In the previous sections we have seen that the calculated functions $R(t)$ and $A(t)$ were very monotonous functions. This is no longer the case with inhomogeneous systems. This can be best shown from the hazard rate $z(t)$ and the corresponding function $a(t)$ of a repairable system. In Section 3.2.1 we have already seen that:

$$z(t) = \frac{f(t)}{R(t)} = \frac{-1}{R(t)} \cdot \frac{dR(t)}{dt}.$$

In accordance with this we define for the repairable case:

$$a(t) = \frac{-1}{A(t)} \cdot \frac{dA(t)}{dt}.$$

What do these two quantities mean? To clarify this, $z(t)$ and $a(t)$ are rewritten as follows:

$$z(t) = -\frac{\Delta R(t)/R(t)}{\Delta t},$$

$$a(t) = -\frac{\Delta A(t)/A(t)}{\Delta t}.$$

We now see that these quantities indicate the relative decrease of $R(t)$ and $A(t)$ respectively per unit of time. If changes occur in these two functions, these changes will be seen even better in $z(t)$ and $a(t)$. Both of the latter functions can be used well for the analysis of the transition situation after $t = 0$ in a maintainable system (before the steady-state sets in). For, it is in this transition region where the differences with a homogeneous system occur, as we will see below.

Let us give an example of an inhomogeneous system and its analysis in view of $z(t)$ and $a(t)$. The system is illustrated in Figure 7.21. The input and output signals of this system

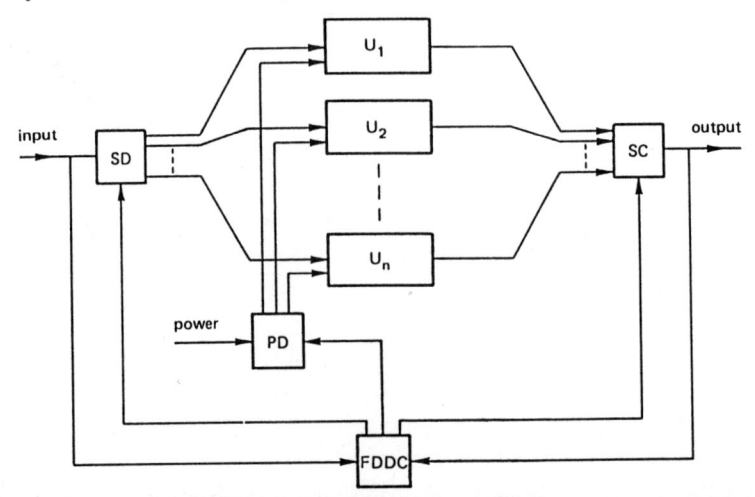

Figure 7.21 A general functional model of a 1-out-of-n redundant system. SD is a 'Signal Distributor' circuit, SC is a 'Signal Combination' circuit. PD is a 'Power Distributor' circuit. FDDC is a 'Fault Detection, Decision and Correction' circuit.

are switched over to the n different units U_i ($i = 1,2,...,n$) by SD and SC respectively. These two circuits also prevent that an input fault in one of the units (e.g. a short circuit to ground) or an output fault (e.g. an output shorted out to the supply voltage) would block the system input and/or output and thus constitute a common-cause failure. Therefore, besides *switching*, these circuits also perform *failure isolation*. The same holds for the routing of the supply energy by PD. The circuits PD, SD and SC are controlled by FDDC. This circuit *compares* the system's input and output signals and *detects* flaws in the system function, i.e. the performance which the system is supposed to produce. In response, the circuit *decides* which action has to be taken (switching off or on, averaging over more than one correctly functioning unit, and so on). Further, the circuit *corrects* the failure by generating the control signals for PD, SD and SC.

In Figure 7.22 the associated reliability model of this general redundant system is shown. If a unit is switched on, its failure rate is λ_i, if it is switched off the failure rate is δ_i. The switching function of the FDDC may fail by remaining 'stuck' in position, despite the failure of the unit (failure rate γ'). It may also fail by spontaneously switching to the next unit while the previous unit was still good (failure rate γ'). The first failure is only detected after the presently switched-on unit fails. The second failure can only interfere with the order in which the units are engaged. For, it is assumed that after the last unit has reportedly failed the first one is switched on again in an attempt to find a correctly operating unit that may still be available. All common-cause failures have been lumped in a series element as ε_0. The failures ε_i ($i \neq 0$) are caused by loose connections, decision errors, etc.

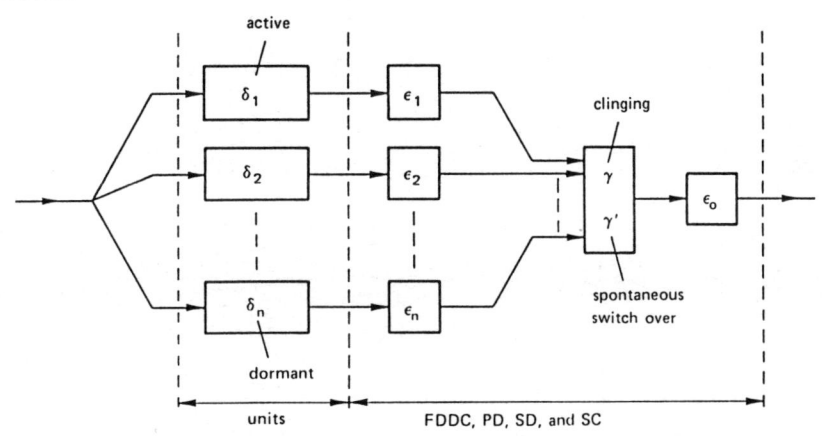

Figure 7.22 The general reliability model of the system of Figure 7.21. Here λ is the failure rate in the active state, and δ that in the non-active state. The other failure rates represent failures in the FDDC-, PD-, SD- and SC-circuits.

In Figure 7.23 the Markov model of the reliability model in Figure 7.22 is given. It has been assumed that we are dealing with a 1-out-of-3 redundant system. In Figure 7.23(b) we see how drastically this diagram would reduce were we to assume that the system is homogeneous, so if $\lambda_i + \varepsilon_i = \lambda$ and $\delta_i = \delta$. The complexity of the inhomogeneous model is so great that we had to take recourse to solving the associated equations numerically

162 *Maintained Systems*

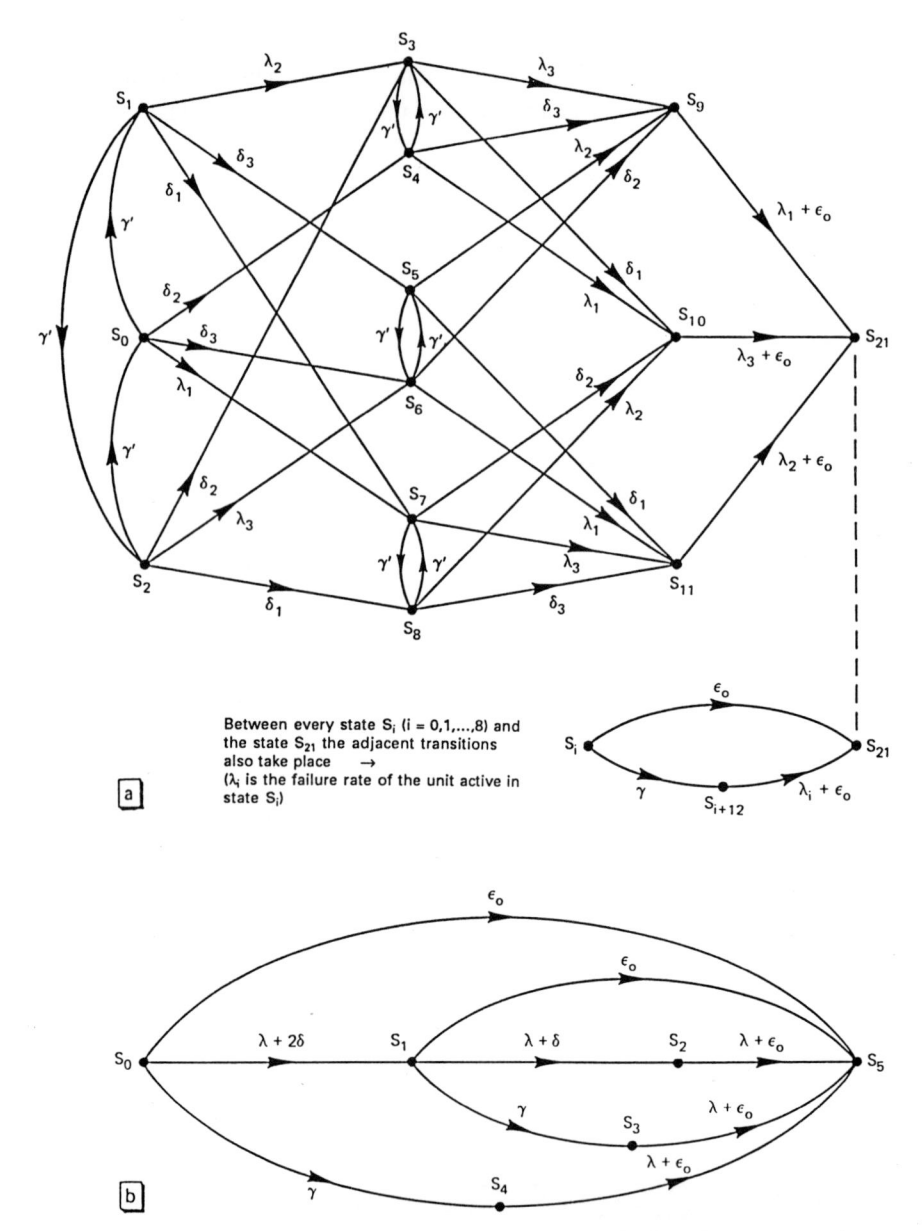

Figure 7.23 (a) The Markov diagram of the system of Figure 7.22, assuming that the system is equipped with one-out-of-three redundancy. (b) The system would become very simple if it were homogeneous (λ_i and ε_i have been taken together into λ).

with a computer. The results are presented in Figure 7.24. We have plotted a 1-out-of-4 passive-redundant system (so $n = 4$ and $\delta_i = 0$). Instead of rising monotonically to its final value, as is the case with homogeneous systems, $z(t)$ may become temporarily much larger than the final value and $z(t)$ may show one or more of such maxima (i.e. peaks in the fractional mortality per unit of time).

The only difference between the numbered eight curves in Figure 7.24 (drawn in solid

Inhomogeneous systems

lines) is the difference in the order in which the units are switched on. The four units in the system are functionally similar but have failure rates which spread a factor of 10. We can now see that the maxima develop in $z(t)$ because bad units are switched on first and, in addition, the 'clinging-to-a-unit' probability is different from zero. As a result the system may remain 'clinging' to a bad unit, which temporarily results in a larger breakdown. If no such clinging, sticking or adherence occurs during this bad unit's life, the system may still get stuck at the next unit. If this unit is also worse than the best one, this will cause one more peak in $z(t)$, and so on. If the likelihood of spontaneous or unjustified switching over to the next unit is large, each unit is switched on many times in its life which causes the $z(t)$ of the system to behave as the average of the other curves. This is the 'dashed curve' $\gamma' \to \infty$. Finally, the dotted curve illustrates the behaviour for zero 'sticking' probability ($\gamma = 0$).

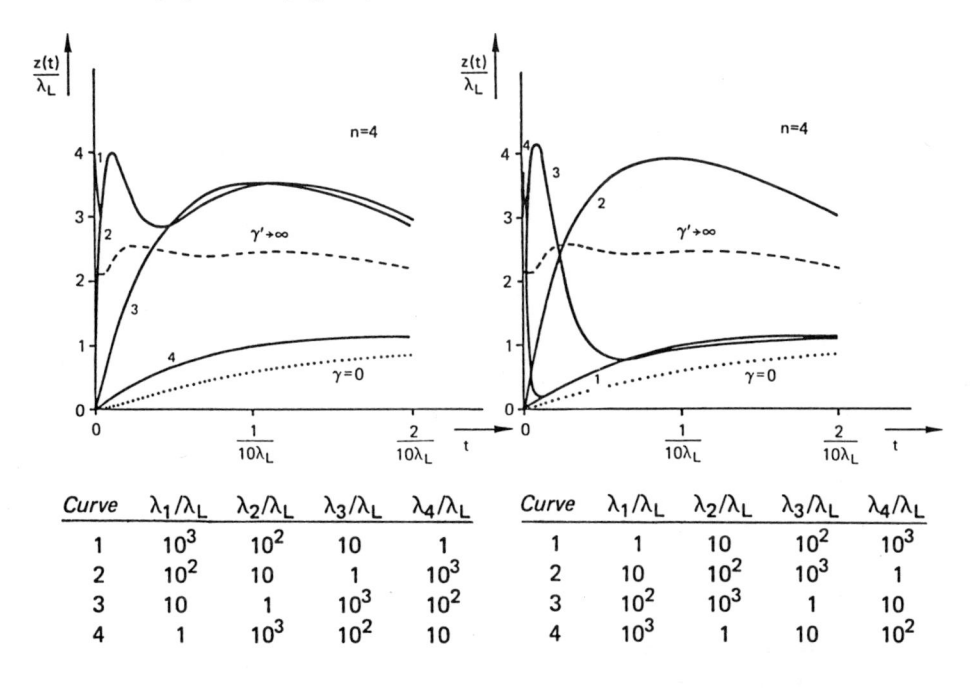

Curve	λ_1/λ_L	λ_2/λ_L	λ_3/λ_L	λ_4/λ_L
1	10^3	10^2	10	1
2	10^2	10	1	10^3
3	10	1	10^3	10^2
4	1	10^3	10^2	10

Curve	λ_1/λ_L	λ_2/λ_L	λ_3/λ_L	λ_4/λ_L
1	1	10	10^2	10^3
2	10	10^2	10^3	1
3	10^2	10^3	1	10
4	10^3	1	10	10^2

Figure 7.24 *The hazard rate $z(t)$ of an inhomogeneous 1-out-of-4 passive-redundant system with imperfect switching of redundant units. λ_L is the failure rate of the best unit. Spontaneous switching does not occur ($\gamma' = 0$). The switch gets stuck with a failure rate $\gamma' = 10\lambda_L$. There are no common-cause failures ($\varepsilon_0 = 0$).*

Let us now introduce repair to the general redundant system of Figure 7.21 and subsequently determine the function $a(t)$ numerically with a computer. We shall pursue various repair strategies. Thus we may, for example, start to repair a failed unit immediately with repair rate μ_i ($i = 1, 2, \ldots, n$). Repairing a 'sticking' switch is somewhat more difficult because this failure becomes manifest only after the corresponding unit fails. So for detection of such 'system stuck' failures, the repairman has to disturb the system on purpose, so that it is forced to engage a new unit. The question is if this is allowable and if a failed switch always can be repaired without disabling the system.

164 Maintained Systems

In general, failure such as 'system stuck' remain often undetected; the coverage of the system inspection is not 100%; masked failures will remain latently present in the system. We shall therefore assume that the system is allowed to break down and that all failed components are restored with the repair rate μ, after which the system is 'as new' again. What does this mean for the configuration of the Markov diagram? In Figure 7.25 $a(t)$ has been plotted for different situations. Explain the shape of the curves yourself. Why does $a(t)$ become negative?

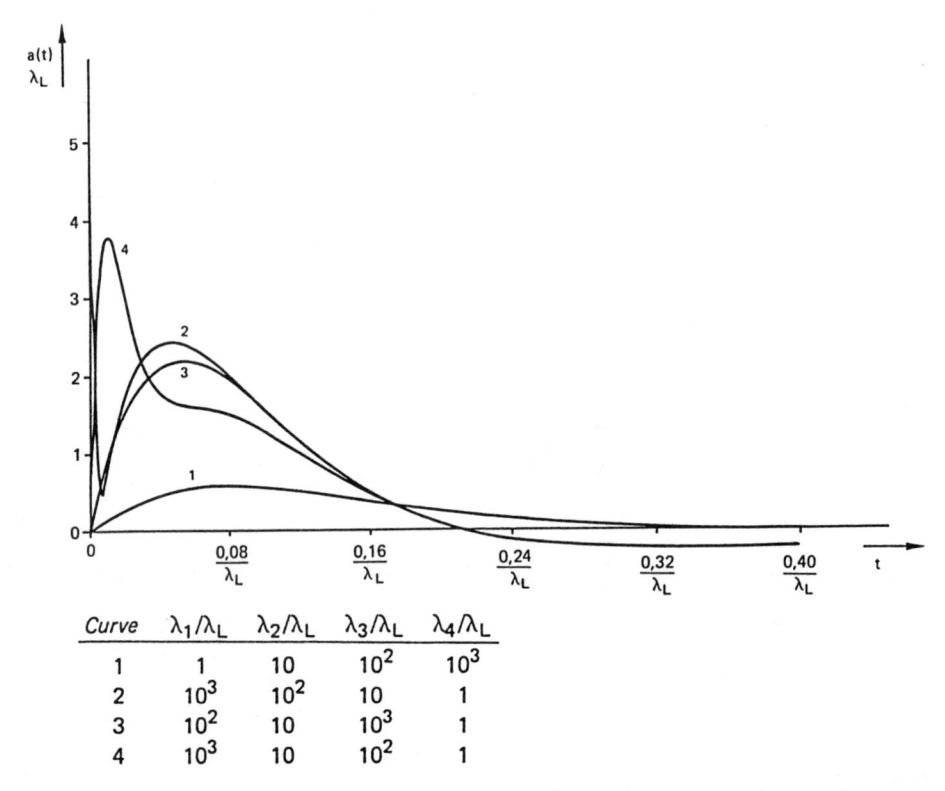

Curve	λ_1/λ_L	λ_2/λ_L	λ_3/λ_L	λ_4/λ_L
1	1	10	10^2	10^3
2	10^3	10^2	10	1
3	10^2	10	10^3	1
4	10^3	10	10^2	1

Figure 7.25 The $a(t)$-function of an inhomogeneous passive-redundant system with repairs (repair rate $\mu = 10$, $\mu = 4$, $\gamma = 10$ and $\gamma' = 0$).

7.4 Maintenance aspects

In this section we shall discuss a number of important aspects of technical maintenance. Here maintenance does, of course, not only mean the maintenance of the *function* of a system (*car*: transport function, *telephone*: communication function, etc.), but also the maintenance of the *safety* of that system (*car*: replace worn safety belts, *telephone*: repair of the two anti-parallel diodes across the microtelephone to prevent hearing damage caused by interference peaks on air lines).

First we shall inventory a number of different maintenance strategies.

7.4.1 Maintenance strategies

Already in the beginning of Chapter 7 it has been stressed that the only thing the owner of a system can do to increase the availability of a ready-for-use system is to optimise his maintenance strategy. Owing to its (maintenance-conscious) design the manufacturer should have given the system two inherent properties:

- *Low-maintenance operation.* Examples are: use of maintenance-free lead-cadmium batteries which do not have to be regularly refilled like ordinary lead batteries, self-lubricating closed bearings which have been filled with a lubricant for the duration of their life, self-adjusting clutches and brake shoes in cars, use of construction materials that do not need preservation by paint and the like, e.g. aluminium and some synthetic materials.
- *Maintenance friendliness.* Examples are: systems composed of modules which can be easily reached and replaced, systems composed of just a small number of different units (large numbers of mutually similar and interchangeable units), automatic failure detection and indication, check lights, test schemes with extensive failure diagnosis, and so on.

In view of these inherent system properties the owner (usually advised by the system designer or manufacturer) may set out a certain maintenance strategy based on cost and safety considerations.

We have also seen that corrective as well as preventive maintenance can be used. *Corrective maintenance* is often expensive because of the resulting damages (loss of production capacity, production of defective devices, jammed machinery, etc.). Just *preventive maintenance* alone is not sufficient, there will always be a certain probability of unexpected failures in a system. (Example: broken rear wind-screen.) Corrective maintenance will still be a necessity.

A very light form of preventive maintenance, which is usually performed by the lowest maintenance level (i.e. the system operator), is the so-called *normal care*. This comprises simple check-ups, inspections and adjustments. Examples of normal care by the system operator are: cleaning cassette recorder heads with a cotton bud dipped in alcohol, testing the ground current interrupt switch in a bathroom, lubricating simple constructions, and so on. In the case of *scheduled maintenance* all suspect components (burnt resistors) or failed components (redundancy) are replaced after a certain number of operational hours has elapsed, after a certain load has accumulated, after a certain distance (kilometres/miles) has been travelled, etc. In *condition-based maintenance* one interferes only if the measured condition of the system gives cause for it.

There is *direct* and *multi-stage maintenance*. Direct maintenance is performed locally on the site of the system, usually by the owner's own personnel. In multi-stage maintenance, parts of the system are replaced and sent to a central workshop for repair; for example to the dealer or the manufacturer. Examples of multi-stage maintenance are: the replacing and sending in for repair of punctured car tires, faulty measuring equipment, and so on.

Maintenance also has a *logistic aspect*. One should have sufficient *spare parts* for the correction of the problems encountered. For that purpose one may provide a supply at the location of the system (e.g. safety fuses). One may also use a central depot or stock room

166 Maintained Systems

(e.g. current electronic components) or order the spare parts from the manufacturer (e.g. electromotor of a disk memory). When doing so, one trusts that the manufacturer will keep in stock an adequately large amount of all the parts of previously produced systems for a sufficiently long period. Since this means an enormous capital investment some manufacturers may tend to skimp the subsequent delivery of spare parts. It can therefore be a paying proposition to have a second-source supplier for strategic parts.

A similar logistic problem occurs if the maintenance staff has to travel from one place to the other for performing the maintenance (e.g. computer maintenance as contracted by a maintenance agreement with the manufacturer).

System breakdown is a stochastic process in which large peaks may occur. If the *repair capacity* of the *maintenance channel* is saturated by a coincidental large number of simultaneous breakdowns, one has to cope with the resulting extra queuing time.

Many of these problems may be alleviated by applying systems with built-in *redundancy* and by having the system's *user* or *operator* taking simple *corrective action*. Possibly the system may not be restored to its nominally correct functioning by this user provided maintenance but it may still perform a sub-task to bridge the time to professional repair.

It will be clear that all aspects mentioned above, together determine the maintenance strategy for a particular system. This maintenance strategy is arrived at by recording of historic breakdown data, by maintenance experience gained with similar systems, by advise from the manufacturer and 'last but not least' by the good engineering judgement of the system user. It is very important for the maintenance staff that it is made clear which service is desired.

The following repair policies are distinguished:

1. First come, first served (first in, first out).
2. Last come, first served (last in, first out).
3. Random service sequence (sequential in, random out).
4. Priority service discipline.

The first policy will automatically occur due to of the pipe line effect of a repair channel with too low a capacity. With the second policy of maintenance it is agreed up front that a short period of time is always available per repair job; jobs that last longer are postponed until there is time. The third policy also is an automatic one (albeit a degenerate one). It is the result of 'favouritism' and/or 'hobbyism'. Some maintenance work is preferred over another; the 'nicer' jobs get priority or the highest priority is assigned to work for certain areas that have established good ties with the maintenance department. With the fourth maintenance policy one has several customers (or systems) each of which has its own pre-established priority; one works according to a predetermined priority list.

7.4.2 Spare parts supplies

Stocks of spare parts require a certain capital investment. Such an investment results in a financial burden: interest and the costs of space, etc. On the one hand it is therefore desirable to have as few spare parts in stock as possible (delivery on demand policy). On the other hand one needs to have sufficient strategic parts, in order not to hamper basic maintenance. The problem is how many *spare parts* to keep in stock in order not to block maintenance with a probability P_S during the time interval $(0,T)$. This is the so-called 'spare parts provisioning' problem.

Let us assume that we are concerned with a series system of n components. For convenience it is assumed that the components are mutually exchangeable. Further, we have a total of M spare parts. The number of accumulated failures at moment T in the series system is $N(T)$. This number will be equal to the number of replaced components, provided the stock is sufficiently large. We are now interested in the solution of:

$$P\{N(T) \leq M\} \geq P_S.$$

If it is assumed that the failure rate of the individual components is λ, we find:

$$P\{N(T) \leq M\} = \sum_{i=0}^{M} \frac{(n\lambda T)^i e^{-n\lambda T}}{i!}.$$

This expression can be derived easily. After all, effectively, we have a (series) system with a failure rate $n\lambda$, with M components in parallel in passive redundancy. The expression for $R(T)$ of this system has already been given in Section 6.4. It is equal to the above expression.

From this the number of required spare parts can be determined when n, λ, T and P_S are given, since:

$$P_S = e^{-n\lambda T} \sum_{i=0}^{M} \frac{(n\lambda T)^i}{i!}.$$

In Figure 7.26 this relationship is shown graphically for $P_S = 0.9$. In this way, for different operational hours T and effective failure rate $n\lambda$ the number of required spare parts M can be determined that is necessary to prevent the blocking of maintenance with a probability of 0.9.

Problems

7.1. In reliability engineering, what is meant by a maintainable system, a maintained system and maintenance?

7.2. What is meant by preventive maintenance and what does the hazard rate $z(t)$ of a system for which preventive maintenance is suggested have to comply with for this form of maintenance to be viable?

7.3. Which costs does the minimisation of a system's *life cycle cost* comprise?

7.4. A system consists of two active redundant units, each with a failure rate λ. The two units fail stochastically independent.
   a. How large is the MTTF?
   b. If scheduled maintenance is performed, with a time interval T, how large is then the MTFFF? (*Maintenance takes no time, the system is 'as new' afterwards.*)

7.5. A system with reliability $R(t) = 1 - t/2$ at the interval $0 \leq t \leq 2$ is maintained by perfect scheduled maintenance with a maintenance interval $T = 1$.

168 *Maintained Systems*

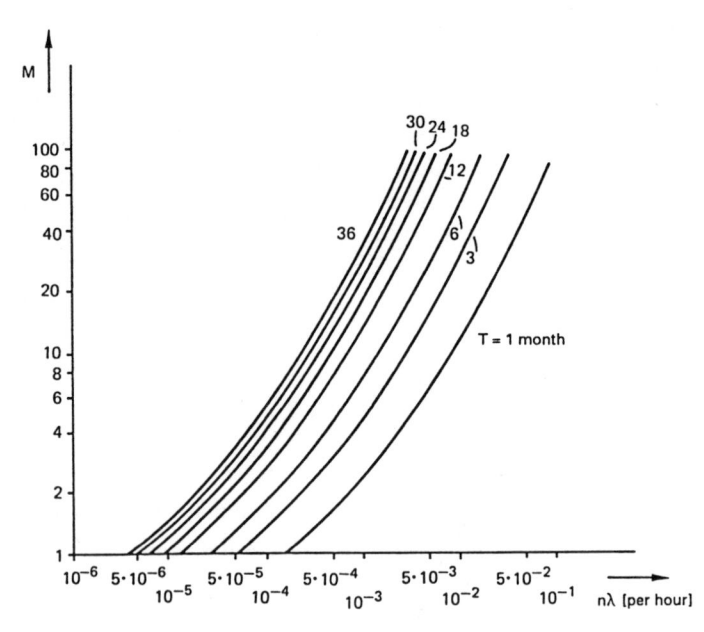

Figure 7.26 *The number of spare parts M necessary to bridge different operational intervals T as a function of the effective failure rate nλ. The probability P_S that the stock is not exhausted during the interval T is equal to 0.9.*

    a. Sketch the $R_S(t)$ function of this system versus the time t.
    b. How large is the mean life θ of this system (so the MTTFF)?

7.6. Machinery, consisting of n identical assemblies, requires that all assemblies function correctly for the correct operation of a process performed by the machinery. When one or more assemblies have failed, the other assemblies are kept in operation during repairs. For each assembly a repairman is available. The repairmen do not work cooperatively on one assembly. The repair rate is μ per repairman, the failure rate is λ per assembly and the assemblies fail stochastically independent.
    a. Give the Markov diagram of the above-mentioned maintainable system.
    b. Determine an expression for the steady-state availability A_∞ based on this Markov diagram.

7.7. Draw the Markov diagram of a homogeneous 2-out-of-3 passive redundant system with two cooperating repairmen with efficiency α. (The repair rate is μ for an isolated repairman, the failure rate is λ per energised unit).

7.8. A submarine has two identical electric propulsion motors, both of which are capable of driving the submarine at cruising speed. The motors fail independently of each other and always have a failure rate λ_m. In normal operation, both motors run. The required electrical energy is generated by a diesel gen-set with failure rate λ_g. Should the diesel gen-set fail, power is supplied by a battery which cannnot fail while it does not deliver energy. However, the battery can only supply current for a limited period. The time-dependent hazard rate of the battery is given by $z_u(t - t_0)$, where t_0 is the time the battery is switched in circuit. The total system fails if the propulsion fails.

Draw a Markov-diagram for this system, clearly giving the meaning of the states, the transitions and the begin and end states.

7.9. Draw the Markov diagram of a homogeneous 2-out-of-4 passive redundant system with two non-cooperating repairmen (efficiency $\alpha = 0$), if it is given that the repairmen only start repairing if at least two units have failed and continue until all units have been fixed again. (The repair rate is μ per repairman, the failure rate is λ per unit and the system is not switched off during the repair.)

7.10. A telephone cable comprises three lines; along each line one conversation can be held. The calls are placed independently with an average rate λ.
If there is no free line no more calls can be placed.
It is further given that the duration of the conversation obeys a negative-exponential distribution with mean $1/\mu$, compute the steady-state availability of this cable connection (i.e. the probability of one or more free lines).

7.11. The reliability of a system is given by:

$$R(t) = \frac{\tau}{t + \tau}, \text{ with } 0 \leq t \leq \infty \text{ and } \tau > 0.$$

It is decided to perform scheduled maintenance with a period T. The time necessary for this scheduled maintenance is assumed to be neglectably small and the system is as new again after maintenance.
a. What is the mean life (MTTFF) of this system for $T = \tau$, and for $T = 3\tau$?
b. Which of these two values of T gives the largest mean life, and how can this be explained?

7.12. A system consists of two units in active redundancy. The units have a constant failure rate λ of 10^{-3} per hour and fail stochastically independent.
a. How large is the MTTFF if no corrective maintenance at unit level is performed?
b. How large will the MTTFF be if there are repairs allowed at the unit level? The repair rate μ is 10^{-1} per hour.
c. What is the addressing frequency of the repair channel in case (b) if it may be assumed that $\lambda \ll \mu$?
d. The repair costs of a unit are $ 500 per event. However, if the system goes down the costs, mainly because of the loss of production, are $ 5,000 per event. Determine, based on the outcome of (a), (b), and (c) whether it is economically sound to perform corrective maintenance at the unit level.

7.13. A valve may fail in two ways; it may get stuck in the open or in the closed position. The 'open' failure mode has a constant failure rate λ_1 and a repair rate μ_1. Similarly the 'closed' failure mode has a failure rate λ_2 and a repair rate μ_2. What is the steady-state availability of this valve?

170 Maintained Systems

7.14. A tugboat has to tow a drilling platform from Norway to the Gulf of Mexico. For the correct functioning of its wall-to-ship communication system it is necessary that both the transmitting and receiving equipment are operating. Both have a failure rate of 10^{-3} per hour and cannot be repaired during the trip: only replacement is possible. The trip takes six weeks (1000 hours) without the need to refuel.

How many spare transmitters and receivers have to be taken along to reduce to smaller than 2 % the probability that the communication system fails during the trip, if it is further given that the transmitter and the receiver fail independently of each other and that a spare transmitter or receiver cannot fail while 'on the shelf'?

N.B.: The inverse transform of: $\dfrac{a^i}{(s+a)^{i+1}} = \dfrac{(at)^i}{i!} e^{-at}$.

7.15. a. A photoset machine consists of a mechanical and an electronic part. The electronic part has been duplicated to give active redundancy. The non-redundant mechanical part has a failure rate λ_m, while each of the electronic units has a failure rate λ_m. Failures occur stochastically independent.

Draw the catastrophic failure model of this system and compute its reliability.

b. It is decided to perform maintenance on the photosetter under (a). This requires two repairmen: one for the mechanical part and one for the electronic part.

The repair rate for the mechanical part is μ_m and the repair rate for each of the electronic units is μ_e.

The machine is not switched off during repairs.

Since the two repairmen cannot not work simultaneously on the photosetter, the order of repair is laid down as follows:

– The functioning of one of the electronic parts has the highest priority, because the proper functioning of the mechanical part can be tested with a functioning electronic unit.

– Subsequently, the functioning of the mechanical part is most important, because the photosetter is able to function correctly with only one functioning electronic unit.

– The redundant electronic unit has the lowest priority for repair.

Draw the Markov diagram of the system with such a maintenance strategy, specify the states used and give the initial and 'system-down' state(s).

c. Next it is assumed that the system is switched off during repairs when maintenance strategy (b) is used, as soon as the system is no longer able to function (so when the mechanical part has failed or when the two electronic parts have both failed). It is further assumed that those system parts which are still working at that moment cannot fail during the time the machine is switched off.

If it is assumed that the system is switched on again as soon as it is able to function again (so when at least the mechanical part and one electronic part function), what is then the steady-state availability A_∞ of this system, given $\lambda_e/\mu_e = 0.1$ and $\lambda_m/\mu_m = 0.01$?

7.16. Explain why a number of the functions $a(t)$ drawn in Figure 7.25 can become negative.

7.17. A printing shop uses a printing press with a failure distribution which for time intervals smaller than 10^5 hours, can be approximated by:

$$F(t) = 50 \times 10^{-12} t^2.$$

To correct drift in the adjustment of the printing press and to correct roller wear, scheduled maintenance is conducted, after which the press is 'as new' again. The time duration of the maintenance can be disregarded.
   a. Calculate and sketch the availability $A(t)$ for the event that maintenance is conducted once per 50×10^3 hour.
   b. Calculate the mean availability $A(t)$ of the printing press for the interval $[0,\infty)$.

7.18. A passive redundant system consists of three identical units and a switch (see figure). Every time the activated unit fails the next unit is switched in. The switch finally remains in position 3.

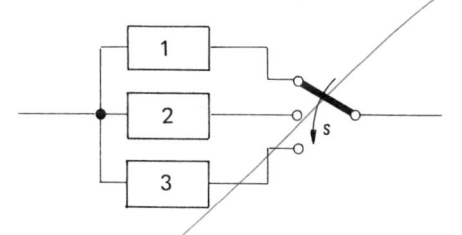

The switch can fail in two ways:
   - A 'stuck switch' which cannot engage another unit. The failure rate is $\lambda_k = 1$.
   - An intermittent failure caused by interference pulses from the mains which (incorrectly) causes the switch to engage the next unit. The time between two consecutive interference pulses is distributed negative exponentially with $\lambda_s = 1$.

As soon as the system has failed repairs are performed until all functions well again. During the repairs nothing can fail. If it is given that the units have a failure rate $\lambda = 2$ and that the repair rate of the system is $\mu = 100$, calculate the MTTFF and the steady-state availability.

7.19. An ore train, consisting of 20 wagons and 3 diesel locomotives is able to travel if at least 2 locomotives *and* all used couplings function correctly.
In the event that all locomotives function correctly they exhibit a failure rate λ_1 (load sharing). With one locomotive failed the remaining ones have a failure rate λ_2. If the system is down (the train comes to a halt) good components cannot fail. It holds that: $\lambda_1 < \lambda_2$. All couplings between the wagons and the locomotives may fail when the train rides and have a failure rate λ_c.
Both the locomotives and the couplings can be repaired, but one only starts to repair if the system has failed, after which one continues until everything has been repaired. One repairman is available for the locomotives, with a repair rate μ_l. A second repairman repairs the couplings with a repair rate μ_c.
Draw the Markov diagram and clearly indicate the transition probabilities and the (down) states.

8
Evaluation Methods

In the early 1960's a number of evaluation methods were developed for the analysis of the availability and safety of complex technical systems. These methods enable the system designer to trace possible design failures at an early stage in the design. This avoids the confrontation with such errors at a much later stage when the design is final and only costly design changes can remedy the problem. Or, worse still, it avoids changes in the system when it is already in production. An example of the latter is a manufacturer recalling cars for modifications to the steering or braking systems.

These evaluation methods are also used in retrospect to trace the most probable cause(s) of failure, if the system itself is not accessible or has been lost. It has, for example, been possible to prove in retrospect that the cause for the Surveyor-2 satellite's undesirable tumbling motion was caused by the failure of position control engine no. 3 tot start.

8.1 Introduction

The reliability engineering evaluation methods usually comprise a graphical representation of the causal connection between the events that may give rise to the occurrence of a certain undesirable event and that event. This representation takes the form of oriented graphs with a tree structure. The nodes in the structure represent events, the transition between the nodes represent the occurrence of such an event.

The evaluation methods discussed in this chapter are *event-oriented* in contrast to the *reliability models* used in Chapter 5 for the reliability evaluation of a system which were *structure-oriented*. These reliability models allowed us to tackle structural failures (hardware failures) that were the cause of an undesirable deviation from the system performance which, if large enough, would in turn result in the respective system function exceeding the tolerances stated in the system specification for that function.

With the above-mentioned *event-oriented evaluation methods* we cannot only model hardware failures but also undesirable situations that may develop due to errors in software, operator errors, operation or maintenance errors and even undesirable situations caused by a deviation from the specified environment for the system. An example of the latter is a large mechanical system containing many electromotors which are powered by the (external) line voltage. This electrical power supply is part of the environment. If the supply voltage drops significantly (undesirable environment; "brown-out") the electromotors may become too hot and burn out. The evaluation methods dealt with in this chapter are much more powerful than the evaluation by means of reliability models. All evaluations by means of reliability models can also be performed with the evaluation methods described in this chapter.

There are two essentially different groups of event-oriented evaluation methods. In the one group the relations between cause and event are set up in the *causal direction*. First the cause is plotted and subsequently the resulting event. These evaluation methods are therefore called *forward methods*. One starts with all possible failure events in the components of a system (of which there usually are many) and one finishes with the single event that the system fails (of which there is only one). The graphic presentation of such an evaluation gives a graph that starts wide and ends narrow (see Figure 8.1a). Such an evaluation is therefore often called *bottom-up evaluation*. Another name for this evaluation method stems from the logic reasoning of *induction* in which one derives from the particular (or detailed) information the general (or more global) information. Since this group of evaluation methods is based on such derivation by inductive reasoning they are also called *inductive methods*.

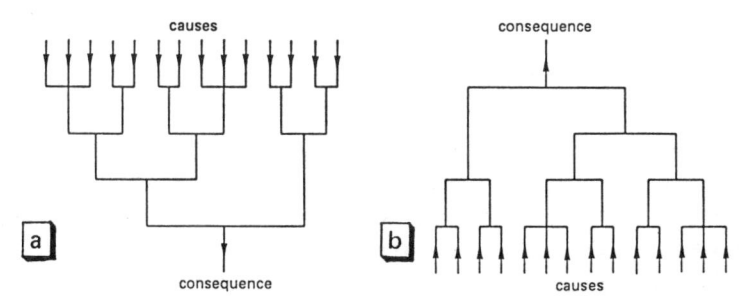

Figure 8.1 The structure and the orientation of graphs used for the evaluation of system reliability. (a) Inductive, forward or bottom-up graph. (b) Deductive, backward or top-down graph. The arrows indicate the direction of building-up these tree-like graphs.

The second group of evaluation methods is characterised by the fact that its basic reasoning is *anti-causal*: one starts with the resulting event and traces this back to all possible causes. After what has been discussed above, it will be clear that these methods (see Figure 8.1b) are also called *backward*, *top-down* or *deductive methods*.

In the following sections we shall discuss the most important members of each of these evaluation methodologies.

8.2 Causal evaluation

Causal evaluation is a systematic evaluation of an entire system or a subsystem beginning with the elementary or basic events occurring at the lowest useful system level.
Examples are: a particular weld in a pipe leaks ..., an obstruction in the oil supply of a certain slide bearing which causes ..., the failing of the high temperature safety device which causes ..., etc.
A disadvantage of this method is that one has to know the system in even the smallest detail before one can begin a causal evaluation. During the design of a system one usually starts with the overall specifications which, in turn, lead to specifications of subsystems. Not until the design has been completed at the higher complexity levels, are the details being filled in. The design at the level of units and modules only happens in the last place. Nevertheless, these details are necessary for a causal evaluation. In the design stage anti-

174 Evaluation Methods

causal evaluation methods are therefore mostly used, which due to their nature trace the design from the top down, in close harmony with the actual design process.

Causal evaluation is often used for hazard and risk analysis and for the identification of dangerous bottlenecks in a system, such as *single-point failures* (these are failures in one single component or module that will disable the entire system). The reason for the use of causal evaluation for this purpose is that, by starting at the lowest level, one can very easily write down all possible failure modes of each independent unit, component or module of a system and its consequences. In this way one easily acquires a *complete* causal tree in which no events are missed. This is far more difficult when using anti-causal methods.

8.2.1 FMEC Analysis

The most important member of the group of causal evaluation methods is the so-called 'Failure Mode, Effect and Criticality Analysis' (FMECA). We shall explain this forward evaluation method in more detail on the basis of a qualitative example.

Example

A petrochemical industry requires the investment of many hundreds of millions of dollars. Relatively short production breaks in an industry like this may amount to millions of dollars per year. For that reason an availability very close to 100 % is required. In this branch of industry which produces plastics and rubbers from crude oil, good process engineering is of key importance. For properly controlling the often complex process, the control engineers make intensive use of computers, transmitters, terminals and control equipment. Figure 8.2 gives an example.

The level labelled A is required to function most reliably since the instruments at this level are closest to the process and have to pass on data to higher hierarchy levels. These instruments are usually composed of modules and are standardised to a very high degree so that they are mutually interchangeable. In addition, at critical points these instrument have been implemented redundantly. Level B, that is the level of the process control computer, can do with a lower reliability. This is because the instruments at level A are also able to operate autonomously (for not too long a period of time). This will, of course, make the process more inefficient. This autonomous mode of operation will therefore only be used to bridge the time it takes to get level B back on its feet. The process computer at level B combines data from various parts of the process and converts this data into control data for the instruments at level A. Further, so-called data logging and reporting are important functions of this B level. For that reason, a modular composition and the use of some redundancy is usually prescribed at level B. However, the much greater complexity at this level lowers the availability realised. At the highest level, C level, the break down of a system is more a nuisance than a cause of catastrophic consequences. The availability may therefore be lower here than at level B.

We shall now focus in on the study of one of the control loops at the most critical level A. It has already been detailed in Figure 8.2. For the FMECA of this control loop we go about as follows:
– Draw a functional schematic diagram in which *all* relevant parts appear (see Figure 8.3).

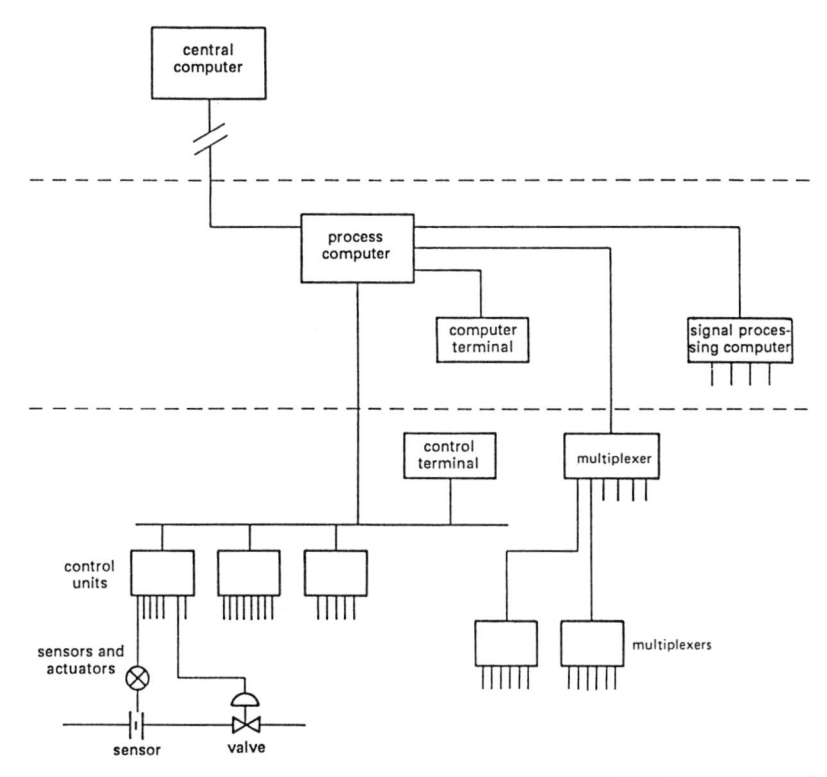

Figure 8.2 An example of the hierarchy in a control system used in the petroleum industry. Level A is the level with the highest required availability, level C is that with the lowest availability.

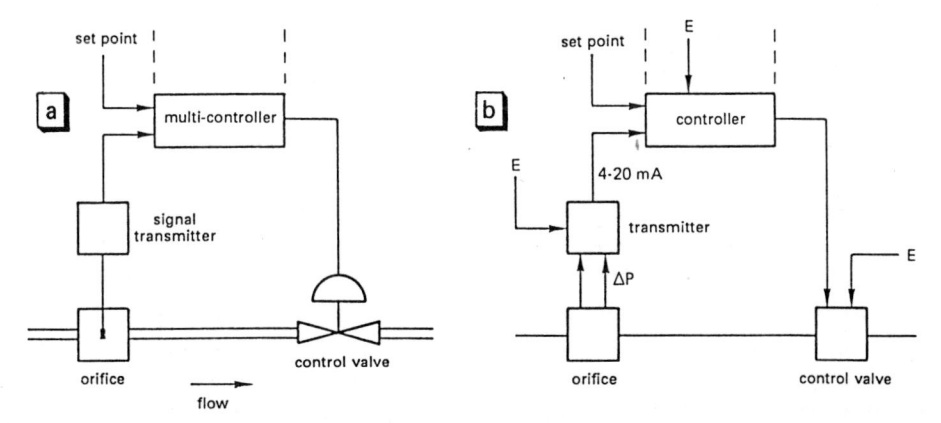

Figure 8.3 The control loop of Figure 8.2 (level A) shown in more detail. (a) Functional representation. (b) Electrical representation (E is the power supply).

- Catalogue *all* parts and indicate *all* failure modes of these parts (see Table 8.1).
- Determine the resultant *effect* on the system of a certain failure mode from the relation of that part to the rest of the system (see Table 8.1).
- Determine how *critical* the resultant *effect* is.
- Determine how *often* the failure mode *occurs*.

176 *Evaluation Methods*

Component	Failure mode	Effect	Criticality	Frequency
Measuring orifice	Contamination	Flow controlled to wrong level	Critical	High
	Nearly blocked	Instability	Marginal	Very low
Pressure transmitter	Out of callibration	Wrong flow	Critical	Low
	Zero offset failure	Wrong flow	Critical	Low
	Short circuit	Control stops	Negligible	Very low
Controller	No set point	Control stops	Negligible	Very low
	Amplification:			
	– Too high	Tends to become unstable	Marginal	Very low
	– Too low	Slow	Negligible	Very low
	Shorted	Process stops	Marginal	Very low
Control valve	Got stuck	Control stops	Negligible	Low
	Faulty:			
	– Fully open	Explosion	Catastrophe	Very low
	– Completely closed	Process stops	Marginal	Low
Power supply E	Defective	Process stops	Marginal	Very low

Table 8.1 Matrix of the failure mode, the consequence of failure, the degree of risk associated with that failure and the relative failure frequency of the components of the detailed control loop of Figure 8.3.

We may consider the FMECA method as being composed of a 'Failure Mode Analysis' (FMA) followed by a 'Failure Effect Analysis' (FEA) and a 'Failure Criticality Analysis' (FCA).

- FMA: We have considered *all* important system components and *all* failure modes of these components. In the example just given not all failure modes have been detailed, of course. In general, one is concerned with much larger matrices than given in Table 8.1. The computer is a very useful tool for these real life analyses.
- FEA: With this sub-analysis the consequences for the system operation are traced for the system components failing in a certain failure mode. It should be noted that in the example above the system apparently was already designed well, i.e. *fail-safe*. If the controller erringly would allow full flow an explosion would be likely to occur here. This must be avoided, of course. That is the case in this design: if the supply voltage E becomes too low the control valve (held open against a spring) closes completely, if an input of the controller fails it freezes the valve in the position it was in before the error, if the controller fails the valve closes also, etc.
- FCA: The risk associated with a failure is an important quantity. After all, a system failure may not result in expensive consequences or consequences that are hazardous for human beings or the environment. For that reason several *criticality levels*, negligible, marginal, critical and catastrophic, are distinguished. In the example an incorrect flow may result in an overdose of one ingredient in the chemical process which is dangerous. It has therefore been marked critical. In Section 8.4 we shall return to the safety of technical systems and the risk associated with operating them.

FMECA is a technique that can also be used well to determine the maintenance need of a system. In the example of Table 8.1 the measuring orifice, for example, has to be be regularly cleaned (with steam) to prevent intolerable clogging. Since a clogged-up condition results in a critical failure, scheduled maintenance has to be performed at relative brief intervals T. Furthermore, extra safety circuitry would be advisable in view of the high frequency of this contamination. (For example by measuring the turbulence in the flow resulting from oil and tar residues clogging the orifice plate.) In addition, the pressure transmitter should be regularly re-adjusted and the control valve should be checked for operability.

In this way one can already come up with the maintenance program during the design phase of the system.

In the above example we have only pictured one of the control circuits in Figure 8.2, at the lowest complexity level of the process control using the FMECA method. It will be clear that if the consequences of failure of this control circuit are to be regarded again as input events for a higher level FMECA this will result in a gigantic forward diagram that can only be constructed and evaluated with the help of a computer.

8.3 Anti-causal evaluation

As we have already seen an anti-causal evaluation of the availability of a system begins with the most complex event, that is the final event 'system down' at the highest level of complexity of the system under observation. Examples of this 'top' event are: The astronaut's life is in danger, the satellite has no solar power (because the solar panels of the spacecraft do not unfold after launching), etc.

By deductive reasoning this top event is split up into *all* other events that may give rise to the occurrence of this top event. This immediately introduces a weak point of anti-causal evaluations: one easily overlooks a possible cause of the consequential or top event.

A strong point of these evaluation methods is that they enable us to closely follow the progressing system design (from system level down to component level). Anti-causal evaluations, much like causal ones, may become very time-consuming and complex when it concerns real life systems. For these practical systems the evaluations are usually performed by a computer or the evaluation is restricted to only the most critical subsystem. Various computer programs have been written for this purpose. Some of these are in wide spread use.

The evaluation can be presented graphically with the help of standardised symbols. Representing a system in this manner can be a big help for designers, users and system managers to discuss design changes or maintenance strategies. If the overall system is too complex one usually splits the system up in several subsystems so as not to lose the overview.

Whereas we have given a qualitative example of a causal evaluation method, we shall now give a quantitative example of an anti-causal method. This will show how to determine the availability from such an evaluation. In doing so, we shall only discuss the most important anti-causal evaluation method.

178 *Evaluation Methods*

8.3.1 Fault tree analysis

'Fault Tree Analysis' (FTA) has become the best known anti-causal evaluation technique, predominantly due to the standardisation of this technique. The standardisation of the graphical symbols used enables everyone to learn FTA easily. Owing to the ease with which the designer can choose the degree of detail in complex systems (by entering into sub trees) and owing to the fact that many companies and government institutions require manufacturers to produce FTA's with the system to be procured, this analysis has become widely accepted.

The evaluation by means of a fault tree was developed by H.A. Watson of the Bell Telephone Laboratories in 1962 for the analysis of the availability of rocket launch control systems. Since its inception this technique has invaded all branches of industry and technology: aeronautics, telecommunications, (nuclear) power generation, transport systems, and so on.

The analysis by means of a fault tree consists of an oriented graph representing the causes and consequences (events) in relation to each other. These events (or states) can be global, such as: radioactivity leaking from a nuclear power station, but also very detailed, such as: warning lamp 3 has failed in subcircuit *V* of the core temperature measurement system. The relation between events is presented by means of standardised logic symbols such as shown in Table 8.2. In addition, Figures 8.4. (a), (b) and (c) give an example of the structure of the (small) fault tree associated with a water heater.

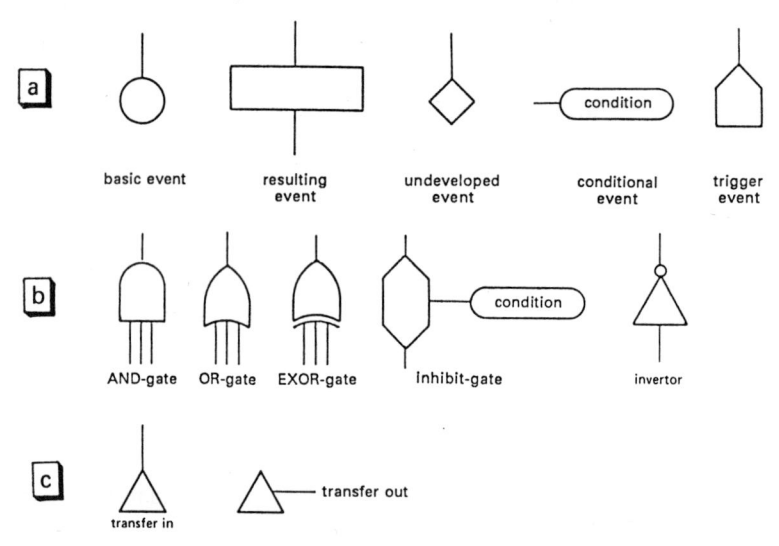

Table 8.2 *Standardised fault tree symbols. (a) Event symbols, (b) Gate symbols, (c) Transfer symbols.*

The meaning of the symbols used in Table 8.2 is as follows:
- *Basic event*. This symbol indicates one of the end points of a fault tree. The circle indicates a basic fault event: the failure of a most elementary component, an environmental failure, or a human failure (operation, repair). For a quantitative analysis the probability of this event or its probability distribution must also be given.

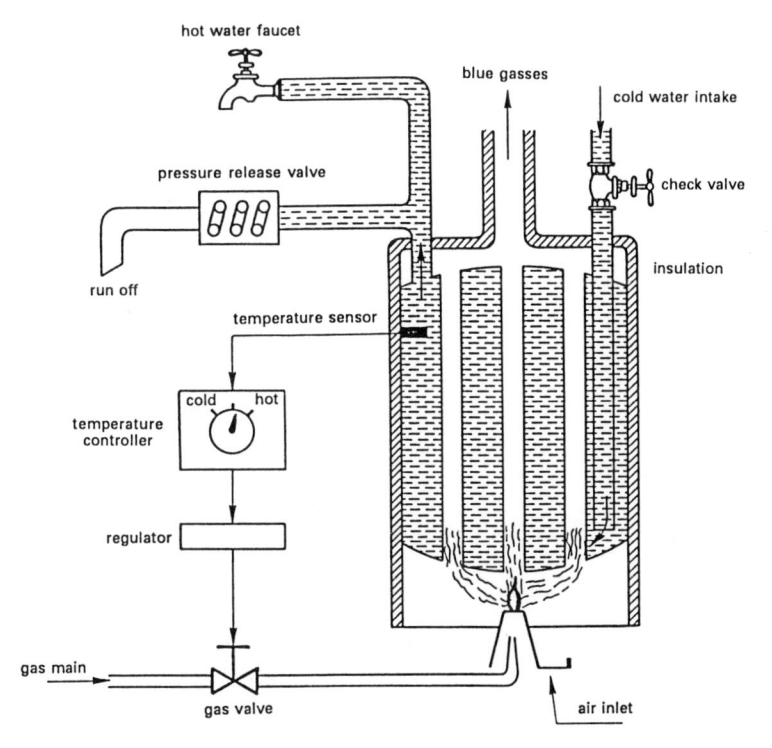

Figure 8.4 (a). Schematic cross section of a water heater. (The pilot light and the flame-out safety circuit have been omitted. This circuit usually consists of an extra valve in series with the main valve which is kept open against a spring by a current powering an electromagnet. This current is generated by the electromotive force of a thermocouple heated by the pilot light. This same light also ignites the burner if it is turned on by the controller.)

- *Resulting event*. The rectangle indicates an event resulting from a combination of events at the input of the associated logic gate.
- *Undeveloped event*. The diamond symbol indicates an event the causes of which have not been further developed. This event might be developed further if the required information and the necessity to do so are available.
- *Conditional event*. The ellipse indicates a condition or a restriction that holds true for the associated gate.
- *Trigger event*. The house symbol indicates an event that is implied during normal use, or an event that is expected not to occur during such use. It can function as the trigger for a number of other events.
- *AND-gate*. The output event occurs only and only if all the gate's input events occur.
- *OR-gate*. The output event occurs if one or more of the input events occur.
- *Inhibit-gate*. The output does not occur if the condition stated is met.
- *Transfer symbol*. This is used to avoid repetition of parts of the fault tree. A triangular symbol, for example, connects parts of fault trees that extend over several consecutive pages.

180 *Evaluation Methods*

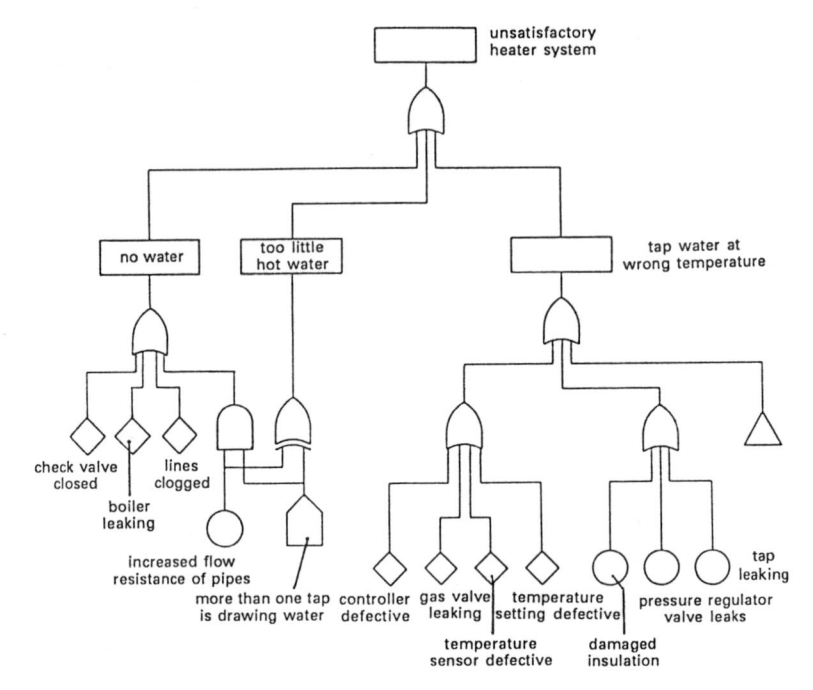

Figure 8.4 *(b). The structure of the fault tree representing the water heater of Figure 8.4 (a).*

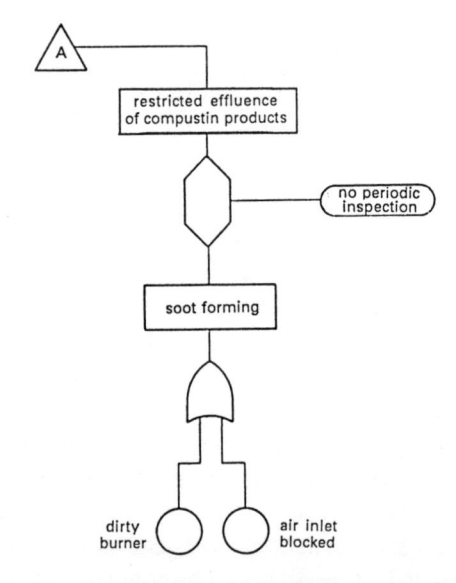

Figure 8.4 *(c). The sub fault tree associated with the transfer symbol A of the fault tree of Figure 8.4(b).*

Fault tree analysis

In this way, starting with the (top) event one wants to analyse, a fault tree can be detailed further and further down until one gets stuck at the basic events that can be developed no further, because of lack of information or necessity.

Let us now give a simple example of a fault tree. We will take the system of Figure 6.8a from Section 6.7.

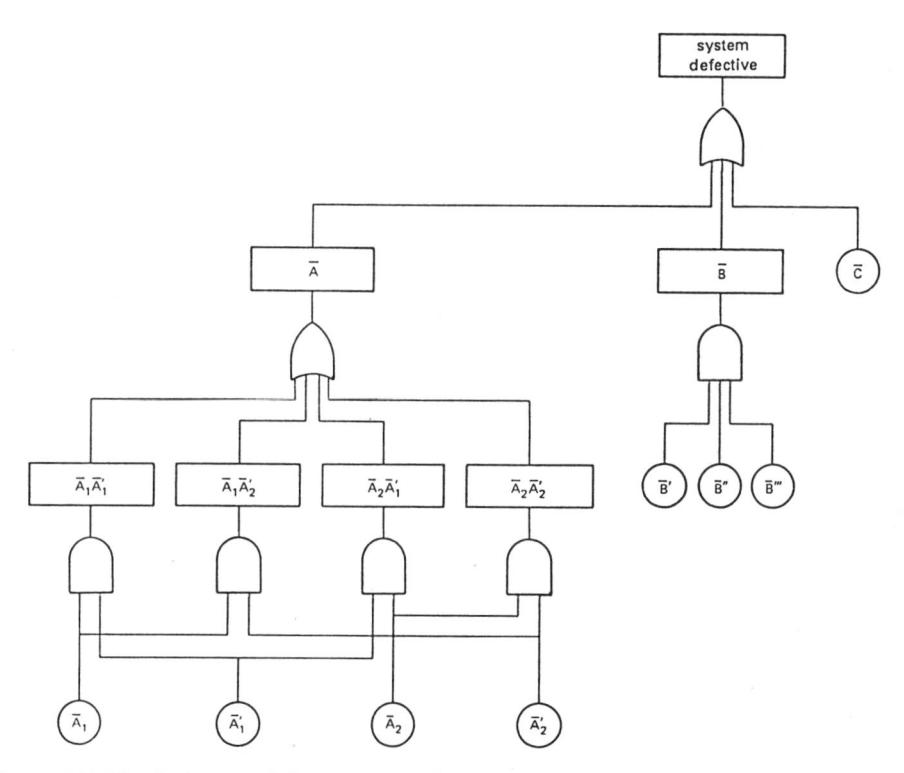

Figure 8.5 *The fault tree of the system in Figure 6.8a.*

It can be seen that occurrence of the event $\overline{C}$ constitutes a so-called *single-point failure*, because this basic event directly leads to the top event via a logic or-gate. It will further be clear that series systems apparently give cause for OR-gates and parallel systems result in AND-gates in the fault tree. In the fault tree of Figure 8.5 we have apparently stopped the development of the tree at the unit level; we regarded the failure of a unit as a basic event. If the probability P with which a unit fails, the failure distribution $F(t)$ or the unavailability $(1 - A(t))$ of the basic event are also known, one can determine the probability, failure distribution or unavailability of the top event. So, the construction of a fault tree is also adequate to determine the reliability or the availability of a system, just as the reliability model of Figure 6.8a.

The calculation on the basis of the fault tree of Figure 8.6 goes as follows: For the event $\overline{A} \cup \overline{B}$ holds:

$$P(\overline{A} \cup \overline{B}) = P(\overline{A}) + P(\overline{B}) - P(\overline{A} \cap \overline{B}).$$

If the events 'A fails' and 'B fails' would be disjunct (which is usually not the case), we may write:

182 Evaluation Methods

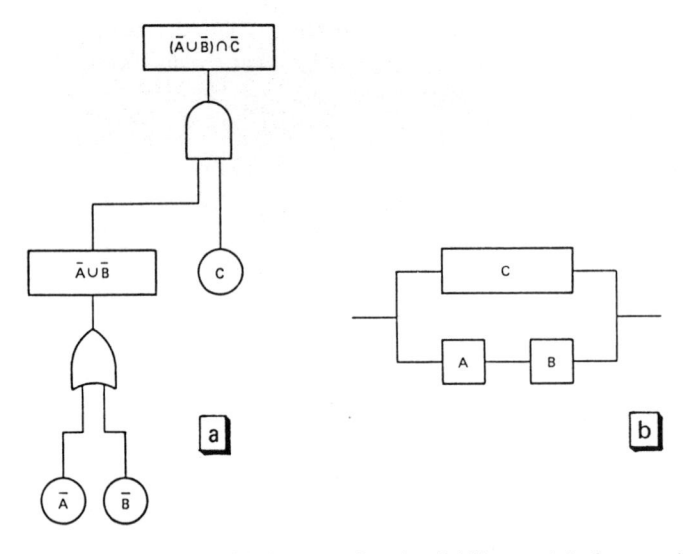

Figure 8.6 (a) The fault tree and (b) the associated reliability model of a certain system.

$$P(\overline{A} + \overline{B}) = P(\overline{A}) + P(\overline{B}).$$

With Bayes' theorem the probability of the intersection of $\overline{A}$ and $\overline{B}$ may be written as:

$$P(\overline{A}\overline{B}) = P(\overline{A}) P(\overline{B}|\overline{A}) = P(\overline{B}) P(\overline{A}|\overline{B}).$$

So if the events $\overline{A}$ and $\overline{B}$ are stochastically independent it holds that:

$$P(\overline{A}\overline{B}) = P(\overline{A}) P(\overline{B}).$$

Substitution results in:

$$P(\overline{A} + \overline{B}) = P(\overline{A}) + P(\overline{B}) - P(\overline{A}) P(\overline{B}).$$

Usually the failure probabilities are very small. We can then neglect the products of such small probabilities as being higher order small. This would give:

$$P(\overline{A} + \overline{B}) \approx P(\overline{A}) + P(\overline{B}).$$

The second step in the fault tree of Figure 8.6 necessary calculate the top event probability does not supply any new insight. Under which assumption does it hold that the 'output probability' of an AND-gate is approximately equal to the *product* of the 'input probabilities'?

For the total fault tree we find:

$$P_{top} = P((\overline{A} \cup \overline{B}) \cap \overline{C}) \approx [P(\overline{A}) + P(\overline{B})] P(\overline{C}),$$

provided that $P(A), P(B) \ll 1$ and A, B and C are stochastically independent.

Conclusion: under the assumptions made above we may simply determine the probability of the top event by *adding* the input probabilities of an *OR-gate* and *multiplying* those of an *AND-gate*.

Next, we assume that the basic events in a fault tree are distributed negative-exponentially and are independent. The output events of the OR-gates in the fault tree are then distributed negative-exponentially again with a failure rate equal to the sum of the failure rates of the n input events (series system). So:

$$z_{OR}(t) = \lambda_{or} = \sum_{i=1}^{n} \lambda_i.$$

For an AND-gate holds:

$$z_{AND}(t) = \frac{\sum_{i=1}^{n} \lambda_i(\alpha_i - 1)}{[\prod_{i=1}^{n} \alpha_i] - 1},$$

in which:

$$\alpha_i = 1/(1 - e^{-\lambda_i t}).$$

The fault tree is a rather 'negative' approach to the reliability problem; it is failure oriented. Yet, in reliability engineering we are concerned with increasing the likelihood of survival or *non*-failure. A failure-event-oriented evaluation, however, can be easily converted into a success-event-oriented evaluation. The *fault tree* is then replaced by its *dualogue* or *dual*, the *success tree*. With a number of simple conversions a tree can be converted into its dual form:
– Replace all AND-gates by OR-gates and also all OR-gates by AND-gates in the original tree.
– Replace all events in the original tree by their complementary event.
If we do this for the fault tree of Figure 8.6 we obtain the result shown in Figure 8.7.

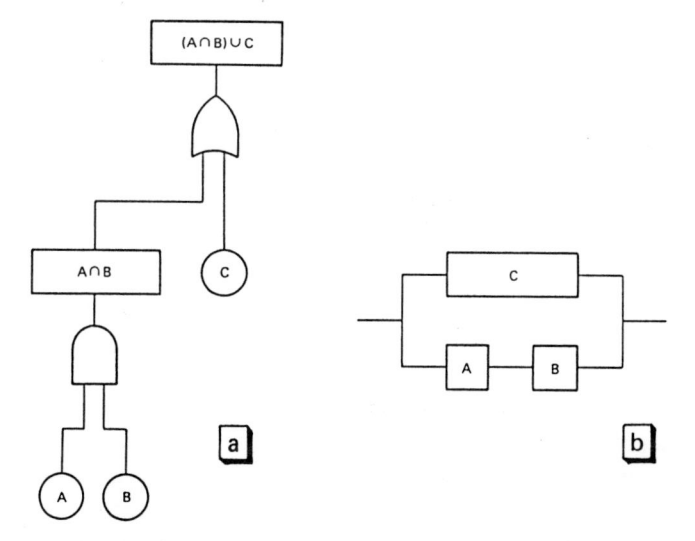

Figure 8.7 (a) The success tree and (b) the associated reliability model of a particular system (same as in Figure 8.6).

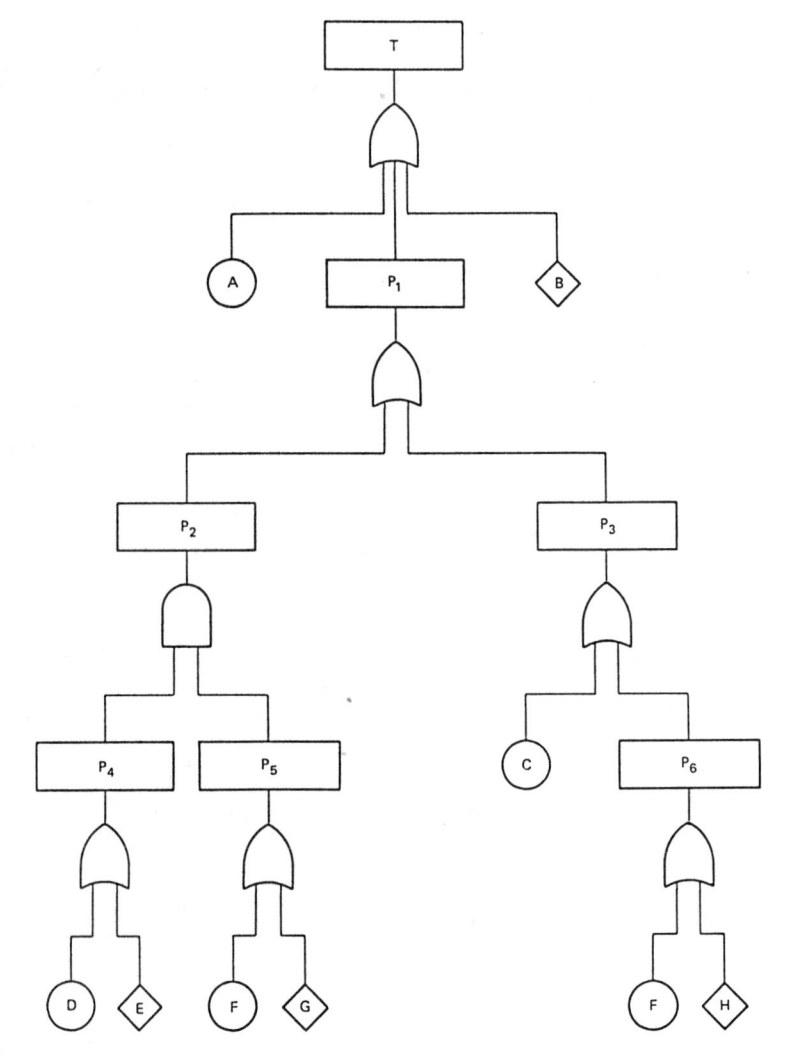

Figure 8.8 A fault tree serving as an exapmle to illucidate the minimum cut set algorithm.

The fault tree constructed for a practical system can usually be drastically reduced. This can be done by applying analysis methods discussed in Section 6.9.

A method for the analysis of a fault tree is for instance that which we have already come across in Section 6.9.2: The tie and cut set method. In the terms of a failure tree a *cut set* is any set of basic events the joint occurrence of which causes the top event. A *minimum cut set* is a set of basic events that can be reduced no further and which still (only just) cause the top event. A list of the minimum cut set of a system is very important for design purposes to identify the weakest links in the system. In larger fault trees these minimum cut sets can no longer be found by simply visually checking the fault tree. We shall therefore give an algorithm (the Fussel-Vesely algorithm) to determine these minimum cut sets. The algorithm is based on the observation that an AND-gate in a fault tree always increases the *size* of a cut set, whereas an OR-gate always increases the *number* of cut sets.

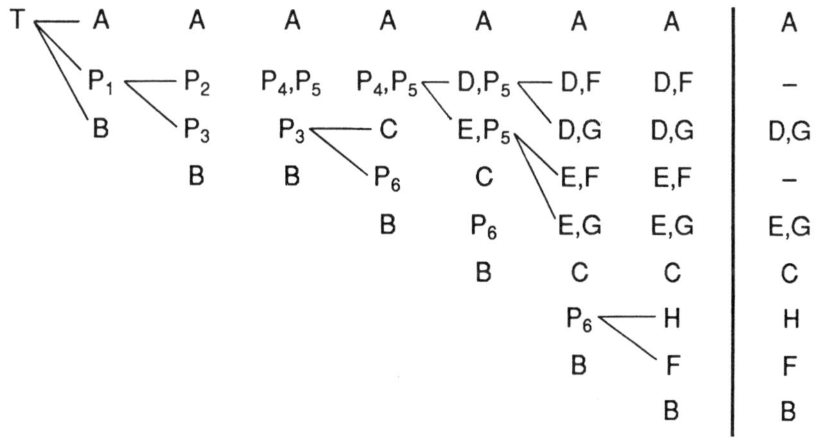

Fault tree analysis 185

```
T ← A      A       A        A         A        A      A  │  A
  ↖P₁ ←P₂  P₄,P₅  P₄,P₅ ← D,P₅ ← D,F   D,F  │  –
    ↘B    ↘P₃    P₃ ←  C    E,P₅ ↘  D,G   D,G  │  D,G
           B     B    ↘P₆   C    ↘E,F   E,F  │  –
                        B    P₆   E,G   E,G  │  E,G
                             B    C     C    │  C
                                  P₆ ← H    │  H
                                    ↘B  F    │  F
                                       B    │  B
```

Figure 8.9 *List matrices of the events resulting in the top event T of the fault tree of Figure 8.8.*

We shall explain this algorithm with the help of the fault tree in Figure 8.8. We trace the fault tree from the top to the bottom and in doing so we make a list matrix that is re-adjusted every time we come across a gate. The idea is to replace the output event of every gate we meet by the input events of that gate. This is continued until we have a list matrix that contains only basic events. If the gate we come across is an *OR-gate* the output event of that gate in the list is replaced by a *column* consisting of the input events of that gate. If we encounter an *AND-gate* the output event is replaced by the *row* of input events of that gate. This means that vertically in a column we have the *union* of events resulting in a top event. Horizontally in every row we have the events the *intersection* of which results in the top event. If this list matrix of events (eventually) contains only basic events, we have a *minimum* cut set on every row if none of the basic events occurs more than once. If the basic events occur more than once in the list matrix we still have to eliminate the non-minimum cut sets from the matrix to obtain one with minimum cut sets only.

Starting with the top event, we find the list matrices of Figure 8.9 for the fault tree of Figure 8.8. The basic event F occurs twice. So not all rows of the list matrix found are minimum cut sets. We can ascertain that F all by itself forms a minimum cut set, therefore the intersections DF and EF are not minimum cut sets. If they are eliminated as non-minimum cut sets we find the rows placed behind the vertical line in Figure 8.9. Based on this information we may now draw the *reduced fault tree* of Figure 8.10. From this fault tree we see that A, B, C, F and H each individually may cause a 'system down', whereas D, G and E, G are only capable of putting the system out of commission if they occur jointly. If the events in this fault tree would correspond to failures in system components we might think in terms of 'series' and 'parallel'. In a fault tree, however, human failures such as operator errors and incorrect maintenance will also show up as events, just as failures caused by an incorrect environment for the system (for example: condensation in an electronic system).

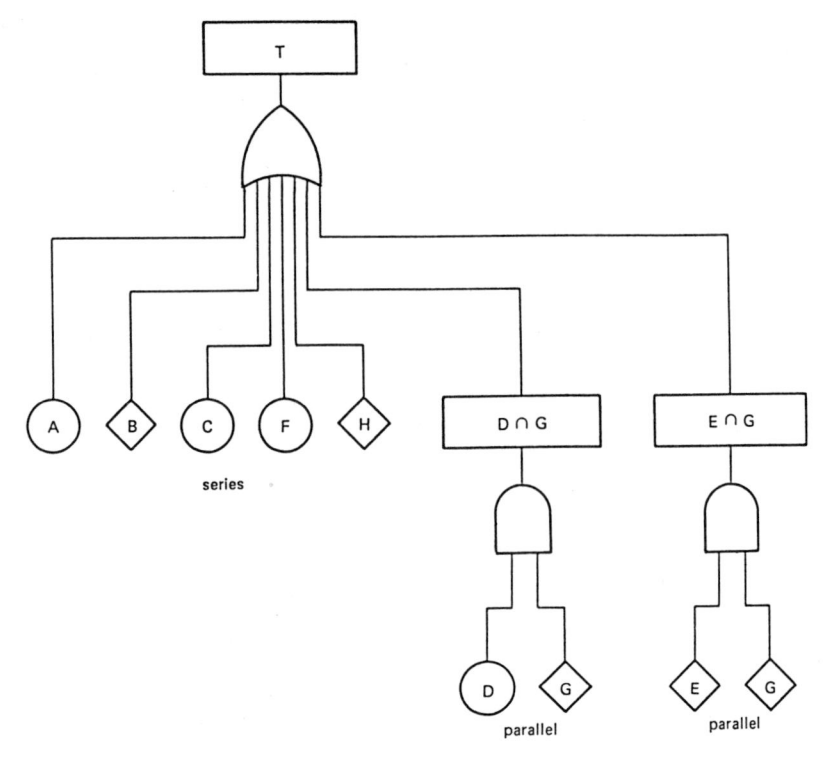

Figure 8.10 *The fault tree of Figure 8.8 reduced by means of the minimum tie set, cut set method as illustrated in figure 8.9.*

In the fault tree of Figure 8.10 *A, B, C, F* and *H* are so-called *single-point failures*. These are failures or events that may cause the top event all by themselves. For example, the cooling water of a system is missing, the power supply of an electronic fly-by-wire system used in a jet fighter is failing, etc. It stands to reason that in reliable systems these failures have to be avoided as much as possible. This can be accomplished by reducing their probability of occurrence. This may be achieved by forced combination with other events. An example of the latter is: the system breaks down only if both the main cooling and the auxiliary cooling systems break down, and, in addition, the system operator has disregarded the warning lights about the auxiliary cooling not being ready for use, and additionally if the operator fails to shut down the system within 5 minutes after the cooling system has failed. (The coolant available in the auxiliary system has a sufficient heat capacity to absorb the heat generated without causing excessive temperatures.) This results in a similar situation as with *D, G* and *E, G* in the fault tree in Figure 8.10. Only the intersection (AND-gate) of these events can cause failure. In addition one can keep the probability of these basic events as small as possible. This may be done by using quality materials, by performing good maintenance and by instructing and training the operating personnel well. When all this is done a greater certainty of non-failure is achieved than with a single-point failure of which one only tries to keep the probability of occurrence small.

8.4 Risk and safety

In Section 3.2 the concepts risk and safety were defined. We saw that risk is the probability of a system failing in an unsafe way, i.e. damaging to the system or its environment. Safety is a measure for the acceptability of the risk associated with operating that system.

Unfortunately all human enterprise involves a certain non-zero risk; there is no such thing as 100 % safety! In addition, any technical intervention, no matter how simple, involves risk. Just compare the use of a ladder for painting a house. In Figure 8.11 the risk of a fatal accident on a normal working day is shown as the day is passing. *Risk analysis* intends to recognise the nature of the various risks and to access the magnitude of the risks. In addition, risk analysis can form the input for measures to diminish the risk to such an extent that the remaining risk is acceptable, in view of the associated technical, economic, social, psychological and legal aspects.

N.B.: The above list of aspects is given in a random order and does not attempt to indicate the relative priority of these aspects!

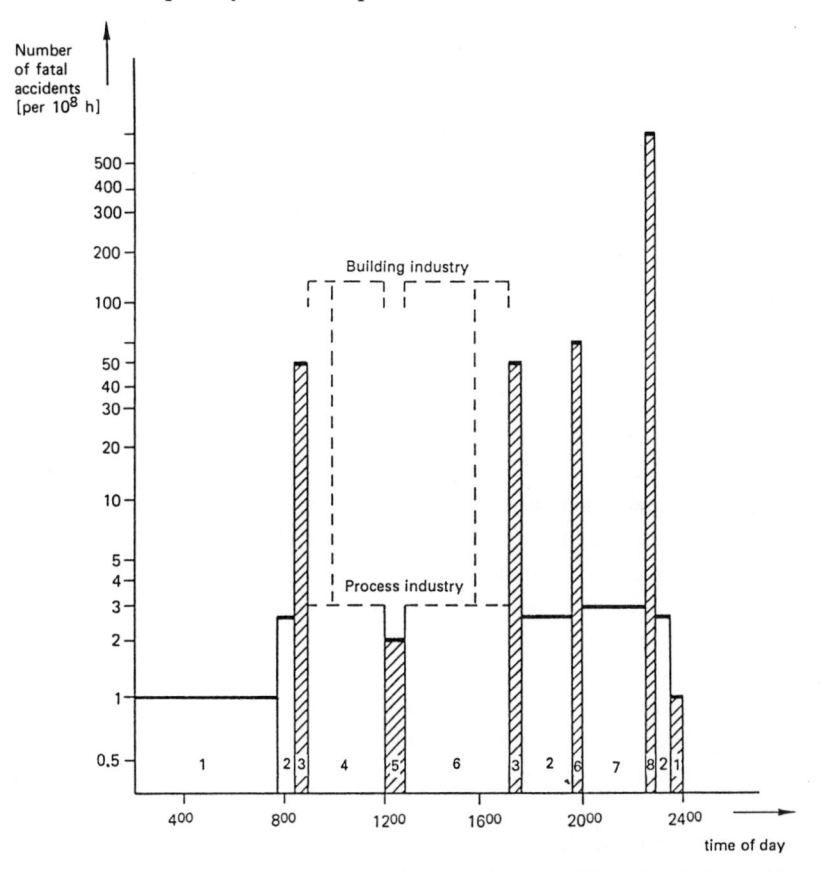

Figure 8.11 An example of the dependency of the probability of a fatal accident on the time of day: (1) Sleeping, (2) washing, dressing and eating breakfast at home, (3) commute to work, (4) risk at work, (5) lunch break, (6) driving to evening engagement in he dark. (7) visiting friends, a bar or other recreation, (8) driving back home at night (intoxicated?).

188 Evaluation Methods

Example
Living in a country below the sea level (which constitutes a considerable technological achievement) involves risks. These risks can be made acceptable by civil engineering works such as: well designed dikes that withstand severe winter storms, the use of redundancy: inner and outer dikes, and by timely warning the population to evacuate when a dike-burst is imminent.

As we, human beings, are using a host of technical systems on an ever larger scale, the risk also increases, at least, if no adequate counter measures would be taken. This is because:
- man is becoming more dependent on his technical systems (pacemakers, avionics);
- more dangerous processes are being used (nuclear reactor, chemical industry producing herbicides, for example);
- processing is becoming more critical (the use of higher pressures, temperatures and the like);
- processes are being scaled up (larger process units causing larger effects, e.g. larger passenger aircraft or larger cities built over an earthquake fault line).

Due to this, the potential dangers associated with further industrialisation and more intensive use of technical systems are becoming increasingly tangible (Plane crashes after a midair collision in a busy traffic area around an international airport over a densely populated area, radioactive discharge of nuclear reactors, but also the dangers brought about by failures in ground-fault circuit interrupters, pacemakers, and so on).

Besides the analysis of the extent of the risk involved in the use of a certain technical system, a risk analysis will (or should at least) also give an answer about the nature of the risk; it should especially make clear the *consequences* of an unsafe failure. One needs to know how critical a certain failure is with regard to the *consequential damage*. This damage may be incurred by *the system itself* (loss of function, damage or even destruction) but also by *the environment* (environmental nuisance, hindrance, destruction) and by *human beings* (endangering life, health risk, risk of injury, demotivation, etc).

Such an analysis is referred to as *Criticality Analysis* (CA). This analysis can be combined harmonically with the FMEA into an FMECA as we discussed in Section 8.2.1. Globally, this analysis entails:
- Describing the system at hand and drawing up an inventory of its component parts.
- Tracing all risks associated with the system.
- Quantifying the probability of the failure that brings about each risk.
- Determining the degree of criticality of the consequence of such a failure.

If counter measures are also to be included in this analysis the following steps are added:
- Drawing up alternative counter measures to reduce the risk.
- Effectiveness of the proposed alternative measures.
- Forecasting (by applying the above four steps to the alternatives) the result of each alternative.
- Making a choice from the alternatives.
- Checking the system with the selected alternative in place.

A few remarks are in order here. When determining the effects of failures in technical systems one should take into account not only material but also immaterial damage. The difficulty here is that the social, ecological and psychological aspects of system failure are hard to approach in a quantitative way.

In the choice of an alleviating solution the system designer is confronted with the difficult question whether the resulting risk level is now acceptable or not. (For example, if one were to design today's traffic system would one deem it acceptable that there are thousands of people getting killed each year in this traffic system?)

If legislation makes regulations with regard to the risk associated with the use of certain technical systems, the designer generally assumes simply that a risk below the limits set by the law is acceptable. An example is the unprotected use of voltages below 50 V in humid environments. Further, risks can be classified in occupational risks, voluntary and involuntary risks (see Table 8.3). Generally one may speak of a negligible risk (caused by a technical system) if this risk is negligibly small with regard to the risks already present if the system were not used. Keep in mind, however, that the exposure may vary greatly! People living closer by, or the system operator will be exposed much more.

		Risk per hour ($\times 10^{-8}$)	Exposure per life (hours)	Risk per life ($\times 10^{-3}$)
Occupational:	process industry (1970)	3	2500 × 40	3
	industry (1970)	2	2500 × 40	2
	industry (1930)	60	2500 × 40	60
	building industry (1960)	120	2500 × 40	120
	mining (1970)	150	1600 × 30	72
Voluntary:	rail travel	6	600 × 50	1.8
	air travel	9	100 × 50	0.45
	car travel	50	600 × 50	15
	cigarette smoking	10	?	–
	private flying	100	?	–
	motor racing	350	?	–
Involuntary:	natural disaster	0.2	10,000 × 70	1.4
	fire	0.5	10,000 × 70	3.5
	illness	1	10,000 × 70	7

Table 8.3 Example of the probability of a fatal accident (risk) per hour of exposure and per human life.

In an *ad hoc* analysis or *case history* of large calamities there invariably appeared to be a number of common factors involved:

- In serious accidents it turns out that usually numerous rules and regulations have not been observed correctly; it hardly ever occurs that one single violation results in a calamity (the designer, the user and the maintenance personnel all make mistakes which accumulate to generate an unsafe condition).
- After a serious accident the safety regulations are strictly observed again, even if the accident occurs somewhere else. One is aware once more that accidents do happen. However, discipline gradually slackens until the next accident or a near miss occurs.

Therefore, one has to see to it that during the entire life-cycle of a system, so during the:

190 *Evaluation Methods*

– design stage,
– normal use and maintenance stage,
– discarding stage,

the rules (for designing, using, maintaining and discarding) are not only made *reasonable* (the near impossible is not done anyway), but also that it is periodically checked how reasonable it is to circumvent or violate them under real life circumstances. In this regard it has to be remarked that: *'seeing is doing'*. Especially when no sanctions are imposed, or if the violator gains by short cutting (finished quicker, able to go home sooner and so on) it is very tempting not to abide by these rules. We have to realise well that we are dealing with people who, although of good will, cannot always muster the self-discipline it takes to suppress the urge to follow the tempting path of least resistance in daily practice. This holds for the designer (a not very well-thought through design that had to be finished quickly), the user (not reading the manual), the repairman (using other than the prescribed materials or parts) as well as the refuse collector (simply disposing of hazardous waste or lithium batteries in household refuse can be very polluting to the environment).

Habituation is frequently the cause of sloppy procedures, *'familiarity breeds contempt'* applies here. Gradually, safety procedures will not be taken seriously anymore. Especially employees who have been working with the same technical systems for long periods of time fail to recognise possible dangers; after all, so far all has gone well. In addition, use of drugs (and some medications) as well as alcohol abuse, even though taking place outside working hours, have a great influence on the alertness of the employees and their ability to make the right decisions.

Human errors occur regularly and are made in all areas:
– wrong labels on bottles (chemicals) and handbooks (unretrievable);
– overfilling, filling at too high a pressure and filling bottles, tanks, cylinders and the like with the wrong contents.
– leakage of pipes, hoses, but also of electric cables (electrocution);
– fumes and dust released during certain operations that cause explosions or fire;
– wrong position of switches, valves etc.;
– no longer paying attention to alarms which 'usually go off by accident anyway'.

All in all, factors generally recognised when working risk-consciously, but which, in practice, may very well cause catastrophic accidents due to unawareness, neglect or even ill will.

An important analysis that is usually performed is the *consequence analysis* of a technical accident. The consequence analysis is based on:

$$D = P \cdot E,$$

in which D is the damage potential arising from the incident, P the probability of the incident and E is the effect caused by the incident. The weak link in consequence analysis (weak in the sense that it is difficult to accept emotionally by most people) is that it does not matter in this consequence analysis whether 2,000 people die annually in dispersed traffic accidents or if 2,000 people would get killed in one big calamity. This is closely related to the *acceptability* of the risk, which is the *subjective* experience of that risk and its admissibility. Thus it might be said that one big disaster (in a certain length of time) is less acceptable than a large number of small disasters over the same length of time. With voluntary exposure one accepts greater risks than with obligatory or professional exposure (just think of sports as ski races, hang gliding, motor races). On the other hand one is prepared to take greater risks as an employee of a factory than when one just happens to live close to that factory. It might be put like this, the employee derives benefit from the exposure, whereas the exposure of people living close by is imposed (no benefit, exposure is experienced as a nuisance or worse, as imminent danger).

Problems

8.1. In addition to fault trees, are there success trees as well? If your answer to this question is affirmative, then convert the left-hand fault tree shown below with the failure events A, B, C and T into a success tree with the survival or success events A', B', C' and T'.
What is the corresponding catastrophic failure model?

8.2. Using the Fussel-Vesely algorithm determine the reduced fault tree of the right-hand fault tree below.

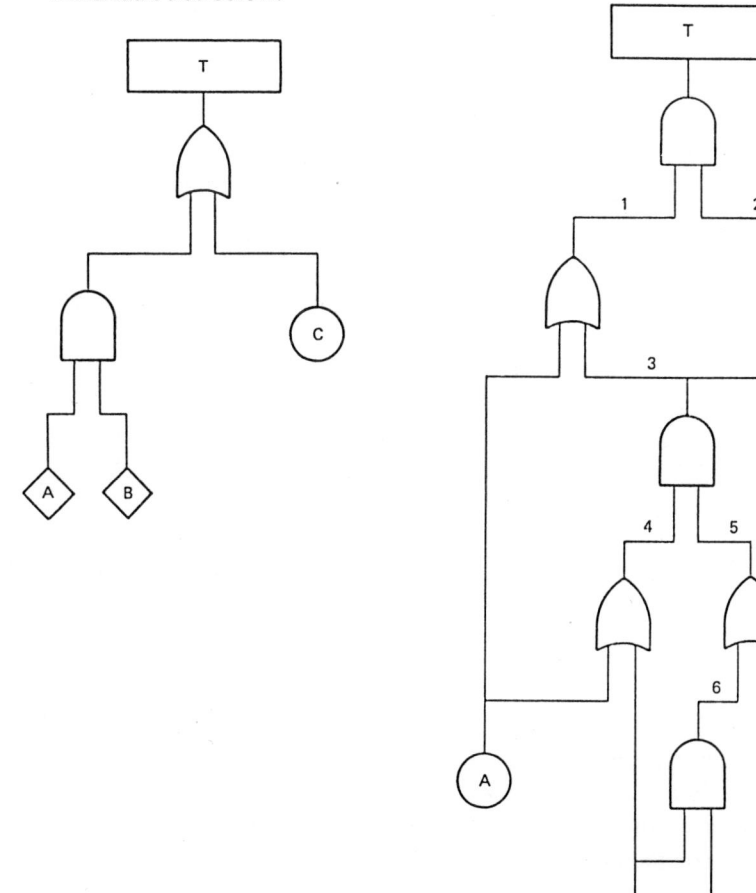

192 Evaluation Methods

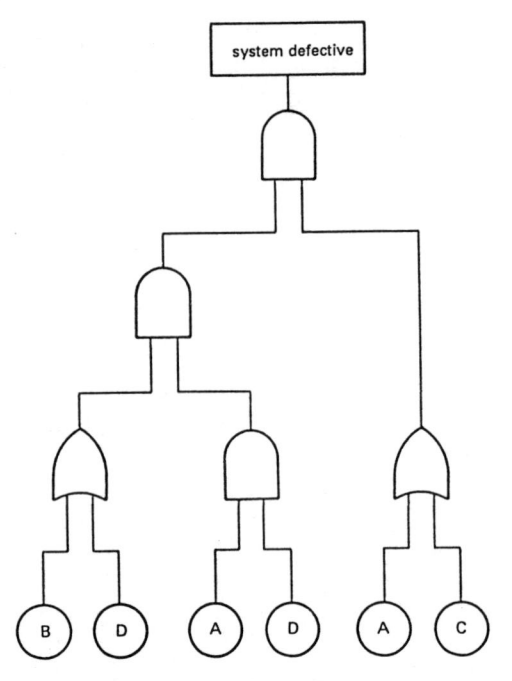

8.3. a. Of the fault tree shown above determine the minimum cut sets with the Fussel-Vesely algorithm and give the reduced fault tree.
  b. If it is given that the events A, B, C and D are stochastically independent and that the failure rates associated with these events are respectively $\lambda_A, \lambda_B, \lambda_C$ and λ_D, what is then the failure probability of this system at the moment t?

8.4. A four-engine airplane is able to continue its flight with two engines operating if necessary, regardless of the positions of these engines. The engines fail stochastically independent.
  a. If the reliability of each engine is 0.9 during the flight, how large is the probability that airplane arrives safely?
  b. Now, if it is also essential, in the event that still two engines function, that these engines have to be on either side of the fuselage (so one on each wing), then give the fault tree of this system.
  c. In case (b) how large is the probability that the airplane arrives safely if the reliability per engine is 0.9?
  d. If the failure rate of the engines is λ, then how large is then the mission reliability in case (b) if the flight time is T?

8.5. A solo yachtsman who wants to cross the Atlantic brings along two radio transceivers. One of those is the normal transceiver that can be plugged into the 24 V board power supply. The second system is for emergency use and can be connected to both the 24 V net and a 6 V battery pack.
Both sets consist of a transmitter part and a receiver part that can fail *separately*. For proper communication one has to be able to transmit as well as receive messages.

a. Draw the fault tree for the event that no proper communication is possible. Use the following notations:

N = board power supply B = battery pack
T_1 = transmitter 1 R_1 = receiver 1
T_2 = transmitter 2 R_2 = receiver 2

b. If all the components of the installation have a failure probability of 0.1, then calculate the probability of the top event.

8.6. The low-pressure injection system of a nuclear reactor consists of a water tank from which water can be pumped to a cooler by an active redundant pump. To enable maintenance of the pumps when they are in operation, a pump in the circuit can be isolated by shutting the valves K_{ai}, K_{a2} and K or K_{b1}, K_{b2} and K respectively. The valves and pumps are connected to the same electrical power network. A valve has one failure mode only, viz. being unjustly closed. The pumps can fail and, in addition, the electrical network can fail which causes all valves to shut automatically and the pumps to stop.

a. Using the tie set method, determine the reliability of this system if it is given that the reliability of a valve is R_k, that of a pump R_P and that of the electrical network R_E.

b. Give a fault tree of this system with the top event: The cooler not getting any water supplied.

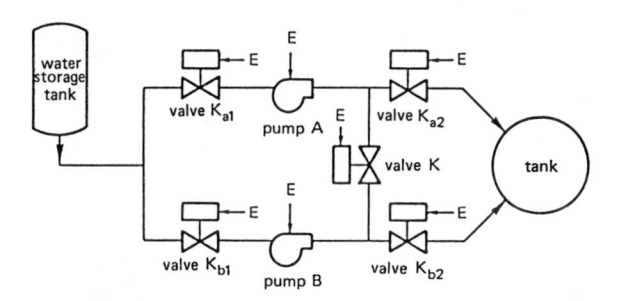

8.7. A process control computer system consists of two redundant computers, each of which is able to control the process individually. To bridge short power supply failures an emergency supply has been fitted between the public utility power network and the network connection of the computers, which, among other elements, incorporates batteries which are able to supply the energy required to both computers for a period of one hour.
This means that if the power failure has not been fixed after one hour, the process control will stop. Furthermore, the following four independent failures can occur in this system:
– computer I fails;
– computer II fails;
– the emergency supply has failed;
– the public utility power supply fails.

194 *Evaluation Methods*

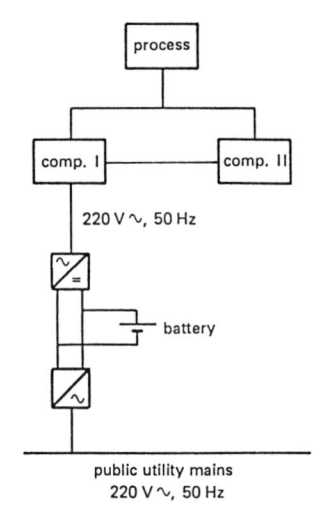

[Diagram: process connected to comp. I and comp. II, with 220 V ~, 50 Hz line to an inverter/rectifier with battery, connected to public utility mains 220 V ~, 50 Hz]

Draw a fault tree for this system which incorporates all the above-mentioned data. (The top event is that the process is no longer controlled.)

8.8. A process control computer system has to function for 1000 consecutive hours. The system consists of two redundant computers, each of which is able to control the process individually. To bridge short power supply failures an electric circuit has been fitted between the public utility power network and the power connection of one computer (configuration A) or both computers (configuration B). This electric circuit incorporates batteries which are able to supply the energy required to the computer(s) connected during a period of two hours for configuration A and during one hour for configuration B. This means that if a power failure has not been fixed after two hours and one hour respectively, the process control stops.

Furthermore, in this system the following four independent failures can occur (the probability that a failure occurs during the above-mentioned 1000 hours is also given):

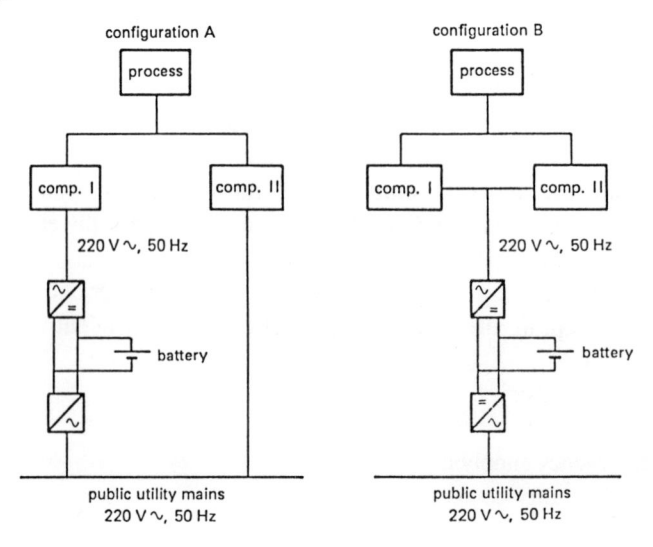

[Diagrams: configuration A and configuration B, each showing process connected to comp. I and comp. II with power supply circuits including batteries and public utility mains 220 V ~, 50 Hz]

- computer I fails (0.01);
- computer II fails (0.01);
- the batteries have failed (0.01);
- the public utility power fails (0.1).

It is further given that the probability that the public utility power line is not repaired within one hour is 0.1 and not within two hours is 0.01. With the respective fault trees, determine which of the two configurations results in the highest reliability.

8.9. We consider a smoke detection system in a given room. It uses a sensor which triggers an alarm when it detects a certain amount of smoke. However, this sensor can fail as follows (with the probability given):
- detects smoke if no smoke is present (0.1);
- does not detect smoke if there is smoke (0.15).

This system is obviously not reliably enough and for that reason it is decided to use three of these sensors.

a. At which state of the three sensors must we let the alarm go off for the detection probability to be maximal?
b. How large is then this detection probability?
c. How large is the likelihood of a false alarm?
d. The optimum decision strategy is that for which the product (detection probability × (1 − probability of false alarm)) most closely approaches unity. What is it in the case mentioned above and how large is this product then?

8.10. Safety regulations require lift cables to be fourfold redundant. Each cable is able to move the lift up and down on its own. Moreover, the lift must also be fitted with safety catches that clamp the car to the walls of the lift shaft as soon as a certain descending speed is exceeded. This safety system can fail with a failure rate λ_b, at which failure the system does not respond if the maximum allowed descending speed is exceeded.

If it is further given that the (identical) cables can break with failure rate λ_k, draw the Markov diagram corresponding to this lift system. Indicate clearly what the states represent and what the initial state(s) and the failure state(s) is (are).

9
Reliability of Computer Software

The use of computers has increased sharply over the years. Not only for scientific and administrative purposes (computer centres) but also for technical (automated measurement systems) and industrial purposes (process control engineering) computers are used at an ever larger scale. The use ranges from very critical applications (aircraft attitude control, navigation at sea, safety and monitoring systems) to playful, everyday applications (computer games). In Section 1.2 we have discussed the necessity of a high reliability for such systems interwoven with our everyday life.

Computer technology can be globally divided into the side concerned with the physical implementation (the logic systems, the communication paths, physical data storage, etc.) and the side concerned with the control of the (binary) processes taking place in the computer. This dichotomy is usually indicated as 'hardware' and 'software' computer technology. The treatment of the reliability of the hardware side of a computer is not different from that discussed so far. It is the software side that requires a special approach. This is primarily due to the fact that failures in software are caused by human design errors when writing the software. These more or less erratic human failures are more difficult to model mathematically than hardware failures, which are caused by all kinds of physical degeneration processes and can usually be reproduced in the laboratory. Software reliability engineering is still in its early stages. In the past few years software reliability has received much attention. This is because enormous costs can be involved in the development, verification, validation, maintenance and the documentation of software (see Figure 9.1).

However, this is not the only reason. Also the application of computers in critical areas where breakdown may bear a large price tag, where loss of production may cause inconvenience and irritation, or even cost human lives, has put the reliability of computer software in the lime light. In the following, we shall restrict ourselves to the treatment of software reliability and assume that the computer's hardware is perfect.

9.1 Introduction

A correctly functioning computer program may be considered as a mapping transformation that depicts the elements of an *input data set* onto elements of an *output data set*, as is illustrated in Figure 9.2.

Now the program may hang on only a certain subset of all possible input data, or only when this input data is presented to the program in a certain order. The program hangs in the sense that it reports a failure, produces a wrong result or never completes execution of

Introduction 197

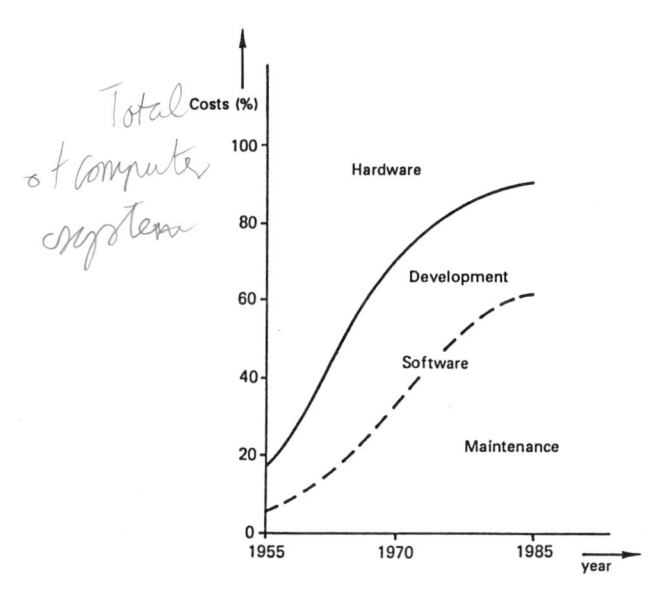

Figure 9.1 The relation of the total hardware costs and software costs for automatic test and measurement computer systems. This graph represents the total cost of ownership.

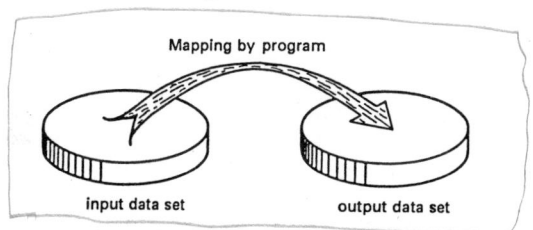

Figure 9.2 A computer program maps elements of an input (data) set onto an output (data) set.

the program. This does not mean that the failure actually occurs at the moment of execution of the instruction that causes the failure. This failure was already latently present in the program (*dormant failure*) and was activated by the accidental combination of input data (but valid combination of input data) which made the program follow an execution path during which the failure was activated that had so far been hidden. The remedy, in theory at least, is very simple: test the program (*all* instructions the program can execute) for all possible combinations of input data. It will be clear that this is hardly realistic with very complex programs (in some cases this would take several years).

Three activities should be distinguished in software reliability engineering: the writing and designing of reliable programs, the testing, verifying and repairing of programs, and the modelling of the program reliability. We shall discuss each of these activities in more detail in separate sections.

In general it can be said that one has to work strictly logically and consistently to obtain reliable programs. It should be kept in mind that there are a number of major differences from the hardware reliability which was discussed in the preceding chapters of this book:

- Software failures are exclusively made by the designer, assuming that a program is also updated or repaired by the designer. During the 'mass production' (copying) of software no further failures are introduced.

198 *Reliability of Computer Software*

- Software when left alone, i.e. when not updated or repaired, does not exhibit wear behaviour, nor drift failures.
- Software failures do not give precursors of an impending failure. They occur suddenly. After having been dormant for years they may become manifest. A certain combination of instructions and input data which make the program execution for the first time take a path in which one (or more) pitfalls occur excites these failures.
- Repairs or updates, i.e. changes in the program, may introduce new, unexpected failures, especially if they are done carelessly (having the program perform a 'new' function or fixing a bug in the program hastily).
- Older and longer-used software becomes more and more reliable (if it is regularly improved). The failure behaviour in time of maintained software packages is analogous to that of hardware with early failures. The failure rate of maintained software decreases in time. The associated reliability grows in time.
- Software reliability does not depend on the physical environment (vibrations, humidity, temperature, etc.). It does depend, however, on the machine environment (the type of computer, its configuration and the operating system).
- Executing the same program multiple times for creating redundancy makes no sense, possible failures are completely dependent and will therefore be exactly repeated. Executing the software developed by different programmers for the same task might be possible, but is not very efficient since the redundancy should be implemented at the lowest level of complexity to be effective.
- Flawless software does not fail anymore.
- In theory, software can be completely tested and if it passes the test it will never again produce failures.

9.2 Writing reliable software

As we have already seen, failures in computer software are made during the design (writing) of software packages and during upgrading or repair. To reduce the possibility of failure it is a prerequisite that the designer is familiar with both the 'task' the program is supposed to perform and the 'environment' in which the program is going to be used. Complex software is usually written by a team of programmers. A good coordination between the team members (*reliability management*) is of the greatest importance. Further, it is also important to develop software along well thought through lines, and to structure the developed software (*structured programming*). The common denominator of *structured programming* techniques is a *top-down design* composed of *modules* that can be tested separately as well as in their mutual relationship.

In a top-down design one starts with the software specification: What is the software supposed to do? Subsequently the architecture is determined, the modular structure, the size of the modules, the use of fixed subroutines, and so on. The core of the program is written first, temporarily with dummy subprograms at the next, lower level. Subsequently, these subprograms are further developed in a similar way. In this process, unstructured leap instructions such as the infamous 'goto' instruction are avoided, not because they would be inherently unreliable, but because they might change the order of execution, due to which the order of instruction execution by the program becomes

difficult to trace. As a result the program becomes difficult to read and it will therefore become difficult to check for failures. As part of reliable programming the entire program design should be well documented. This involves, for example, a *functional description* of the entirety and of every module (subroutine, procedure) individually. This description should consist of a:
- the name of the relevant module,
- the date, the version number and the name of the designer,
- the function of the relevant module,
- the required input parameters and their ranges,
- the resulting output parameters and their ranges,
- all modules or routines that can address the relevant module,
- all modules or routines the relevant module itself can address.

In addition the structure of every module has to be made clear, for example with diagrams as shown in Figure 9.3.

It is easy to see that transparent programs will result if the order in which the instructions can be read from the program listing corresponds with the order of execution of that program. This can be realized by composing the program of logic structures that have only one input, only one output, and that perform only one function. One should also restrict the use of the instructions: 'sequence', 'IF THEN ELSE', 'DO WHILE', 'DO UNTIL', and 'CASE'.

The use of 'GOTO' or 'JUMP' statements that produce 'control transfer' to a different part of the program depending on the data do not belong here either. From Figure 9.3 it can be seen that it is relatively simple to write 'goto'-free programs.

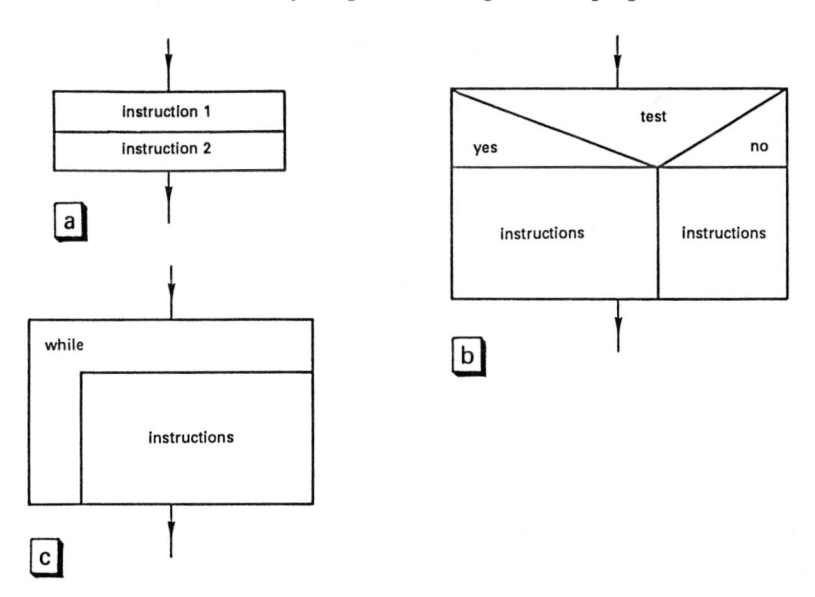

Figure 9.3 *Program structures with one input and one output. (a) Sequentially executing instructions. (b) The logic structure of the 'if then else' instruction. (c) With the 'while-wend' instruction the instructions within the block are executed time and again as long as the condition is met.*

200 Reliability of Computer Software

With the use of instructions like these the program can be completed purely sequentially. In Figure 9.4 it is demonstrated that such a program is considerably easier for the reader to examine from the listed program than a non-sequential program.

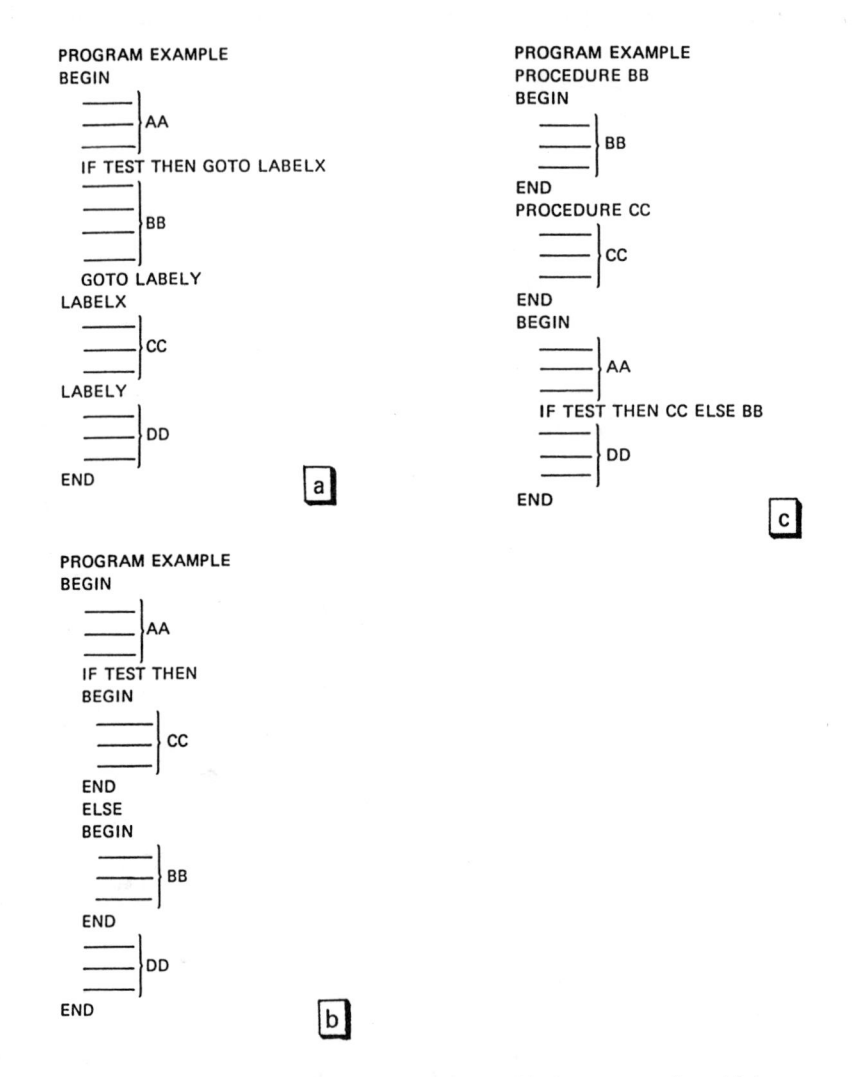

```
PROGRAM EXAMPLE
BEGIN
    ─── } AA
    ───
    IF TEST THEN GOTO LABELX
    ───
    ─── } BB
    ───
    GOTO LABELY
LABELX
    ───
    ─── } CC
    ───
LABELY
    ─── } DD
    ───
END
                                    (a)

PROGRAM EXAMPLE
BEGIN
    ───
    ─── } AA
    ───
    IF TEST THEN
    BEGIN
        ───
        ─── } CC
        ───
    END
    ELSE
    BEGIN
        ───
        ─── } BB
        ───
    END
    ─── } DD
    ───
END
                                    (b)

PROGRAM EXAMPLE
PROCEDURE BB
BEGIN
    ───
    ─── } BB
    ───
END
PROCEDURE CC
    ───
    ─── } CC
    ───
END
BEGIN
    ───
    ─── } AA
    ───
    IF TEST THEN CC ELSE BB
    ───
    ─── } DD
    ───
END
                                    (c)
```

Figure 9.4 *(a) A program with 'goto' instructions; (b) A program in which sequential execution is not perturbed; (c) Thanks to the use of so-called procedures (or subroutines) the structure of the program becomes beautifully clear.*

9.3 Reliability testing

The purpose of testing software is to make the program function as specified (validation) and to guarantee that it is free of failures (certification). Testing can be done in a variety of ways. It is to be recommended to have the testing done by people other than those involved in the design of the software. A first step is to check the written 'code' thoroughly (*manual code walk-throughs*). A second step is the use of special (software) tools to check the written software. These testing aids not only indicate failures but they also produce a numerical measure for the thoroughness of the testing.

The program module to be tested in its original code (the so-called 'source code') is converted into a so-called 'instrumented module', in which sensor and counter instructions have been added to the module to be tested without changing the function of the module. The addition of these diagnostic instructions is called 'instrumentation'. Subsequently, the instrumented module is prepared to be executed with test input data. The execution produces a set of instrumentation data in addition to the normal required output. The instrumentation data and the original program of the instrumented model are now added to a so-called 'analysis module'. This analysis module produces a report about the behaviour of the tested module during the processing of the the input test data. This reported information gives confidence in the structure and the instructions used by assuring that every branch in the program is passed at least once. A testing method that is also practically used is to realise the software in a completely different way, and then to run both programs in parallel and compare the results for a certain set of input test data.

9.4 Failure models for software

We have already seen that software fails through human errors. This causes a deterministic modelling of software failures to be impossible: Human beings are difficult to comprise in a deterministic model. Nevertheless, if complex software packages are involved on which a lot of programmers have collaborated, stochastic modelling is feasible.

The primary purpose of a software reliability model is to predict the behaviour of the software as soon as it becomes operational. The reliability of maintained system software grows drastically with time, since the program is continually tested and failures are repaired. This can be done in the postrelease phase by user feedback. The reliability and the MTTF generally increase as the accumulated computer time increases. For that reason, models are usually based on the accumulated execution time rather than on calender time. However, the actual time a program spends executing statements is often obscured by factors such as the time allocated to the respective applications in a multi-tasking environment. The central processor time which is actually used is a good practical measure for the failure-inducing stress to which the program is exposed. In the following *computer time model* we shall therefore determine the software program reliability R as a function of the totally accumulated processor time for that program. It should be noted that this model only supplies information about the period during which the program is tested and is being debugged (repaired). If the program were no longer maintained after the test period, its reliability will only depend on the period the program was tested; after that the model parameters do not change anymore.

The number of failures N *detected* during testing, and successfully *corrected* during debugging is an exponential function of the total CPU-time, accumulated during the test. This CPU-time shall be denoted by τ.

We then get for this number of detected and successfully corrected failures for a test of τ CPU seconds:

202   Reliability of Computer Software

$$N = N_0(1 - e^{-\tau C/M_0 T_0}).$$

Here N_0 is the initial number of failures in the software to be tested after completion of the system integration. C is an acceleration factor indicating how heavily the program is stressed by the testing with regard to the use of the program under normal circumstances. M_0 is the total number of failures possible for the duration of the test. It is equal to the number of failures found when the program would be tested until all failures would have been removed. T_0, finally, is the MTTF at the beginning of the test procedure.

Not every failure detected will be corrected successfully; a wrong correction may even introduce other, extra failures. If we assume that, on average, B failures per detected failure are removed from the program ($0 < B < 1$), we may write:

$$N_0 = B \cdot M_0$$

and, of course, also:

$$N = B \cdot M,$$

in which M is the number of failures detected during the test. The number of detected failures is now also a negative-exponential function of the accumulated computer time:

$$M = M_0(1 - e^{-\tau C/M_0 T_0}).$$

It now simply follows that the MTTF at the moment τ under normal user conditions of this program obeys the following expression:

$$\text{MTTF}(\tau) = \frac{C}{dM(\tau)/d\tau} = T_0 e^{+\tau C/M_0 T_0}.$$

The reliability for an operational period $(\tau, \tau + \Delta\tau)$ of normal use is then:

$$R(\tau, \Delta\tau) = e^{-\Delta\tau/\text{MTTF}(\tau)}.$$

If the test goal is to obtain a minimum MTTF of $T_{\min}$, one has to continue testing until ΔM more failures are detected. This will require the investment of an additional testing time $\Delta\tau$ if the current MTTF $= T$ (where $T \leq T_{\min}$) given by:

$$\Delta M = M_0 T_0 \left(\frac{1}{T} - \frac{1}{T_{\min}}\right),$$

$$\Delta\tau = \frac{M_0 T_0}{C} \ln\left(\frac{T_{\min}}{T}\right).$$

So far the time was expressed in CPU time. If it is known which fraction of the calendar time is used for actually running the program, the above model can be expressed in calendar time. Generally speaking one can state that in the beginning of the testing of computer software the average time between detected failures is so short that from time to time the testing has to be stopped to enable the programmers to correct observed failures. After further testing the interval between two consecutive failures becomes increasingly longer: Failure correction is no longer limited by the speed at which the programmers are able to repair failures, but by the speed at which the test team can execute the test procedure and analyse the failure data. Finally, for even more sparsely occurring failures the capacity of the computing facility becomes the restricting factor.

Three model parameters are needed to determine this execution-time model: C, M_0 and T_0. The acceleration factor C is determined by the test environment, whereas M_0 and T_0 are determined by the program complexity (and by the skills of its designers). The latter parameters may initially be estimated and can be determined more accurately as more data becomes available. The initial estimate is done as follows:

N_0 is predominantly determined by the complexity of the program and is about 5 failures per 1000 instructions for programs written in so-called 'assembler language'.

The failure reduction factor B depends on the thoroughness with which the program is repaired ($0 < B < 1$). M_0 can now be calculated. The parameter T_0 can be predicted from:

$$T_0 = \frac{1}{fkN_0},$$

in which the execution frequency f of the program is determined by the number of instructions the program presents to the CPU and the average execution time per instruction. The failure exposure factor k is the quotient of the average execution time of the program and the average accumulated execution time between two manifestations of the same failure. Or, in other words, k gives the average fraction of the total number of program executions in which a certain failure occurs. Finally, Figure 9.5 shows an example of how the estimated MTTF (with continually adapted values for C, M_0 and T_0), based on this execution time model, grows as a function of the calender time which passed during testing. Both outer curves mark the 75 % confidence interval around the estimated mean value (centre curve).

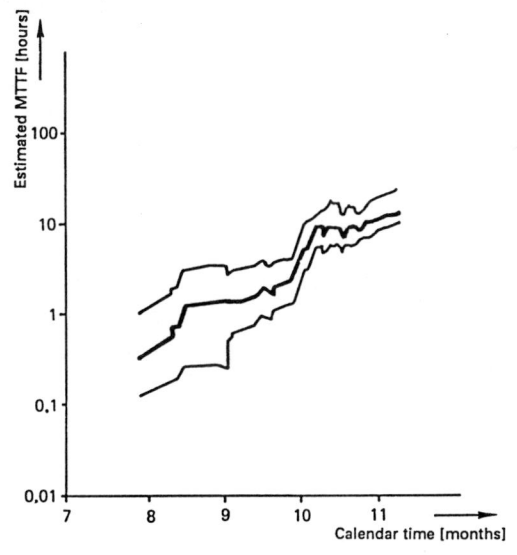

Figure 9.5 *The estimated 'present' MTTF of a software package during the test period. Estimated based on the execution time model.*

Problems

9.1. In a computer program written in BASIC the following sequence of instructions occurs:
```
400 IF A > 0 THEN GOTO 420
410 GOTO 430
420 IF B > 0 THEN GOTO 450
430 LET D = –C
440 GOTO 460
450 LET D = C
460 PRINT D
```
Rewrite this part of the program without using GOTO instructions.

9.2. Prove that the MTTF of a program after an accumulated test time τ is equal to:

$$T_0 \, e^{\tau C/M_0 T_0},$$

in which T_0 is the initial MTTF and M_0 the initial total number of failures that may occur during testing and correcting. C is the stress factor for the accelerated occurrence of failures during the test procedure relative to the occurrence of failures during a normal application of the program.

9.3. A program used to control a chemical process is required to have an MTTF of 10^6 hour. The program consists of 100,000 instructions, which may be assumed to have 600 failures prior to testing. The reduction factor B is 0.3; the execution frequency f is 5 per second; the exposure factor k is 10^{-2} and the acceleration factor C is 2.
   a. How much computer test time is needed to meet the above requirement?
   b. How many of the initial failures will still be present in the program (on average) after completion of the testing?

9.4. As Figure 9.1 shows, the costs of software development and maintenance become an increasingly larger share in the total costs of the computer system. Give a number of reasons.

9.5. The main purpose in the effort for reliable software is the prevention of software failures in as early a phase of its development as possible. Find a number of causes for the development of software failures:
   a. during the *specification phase*;
   b. during the *realisation phase*;
   c. in the *testing* and *implementation phase*.

9.6. Finding and correcting software failures becomes considerably more difficult as the program gets in a later stage of its development. Give a number of guide-lines to prevent software failures.

Solutions to Problems

1 Introduction

1.1. Reliability is the *probability* that a certain *system* uninterruptedly performs the *specified function* during a certain interval of a *life variable* provided that the system is used within a certain *specified environment*. (Misuse prevention clause.)

1.2. A uniform evaporation of a filament causes the resistance R of the filament to increase uniformly as well. With power supplied by a constant voltage source E the dissipation ($P = E^2/R$) will monotonically decrease with age, due to which the temperature of the filament will drop and the evaporation will decrease more and more.

1.3. If the power to an incandescent lamp were supplied by a constant current source I and the filament evaporates strictly uniformly, one would face the following situation: The evaporating wire will exhibit a higher and higher resistance R. The power dissipation $P = I^2 R$ will now rise monotonically with age. Since the evaporation rate is a very sensitive function of the filament temperature (for higher than normal operating temperatures) and therefore also is a sensitive function of the power dissipation, the lamp's filament will evaporate faster and faster. Finally the life would end in a bright flash (if the inside of the bulb would not be coated completely with the evaporated tungsten, making it nontransparant).

1.4. For temperatures below 0° C (32° F) and over 50° C (122° F) the amplifier apparently falls outside the specifications on gain, noise, etc. This may be due to frost formation below 0° C and the fairly large internal power dissipation (42 watt) which does not allow an ambient temperature over 50° C. Storage of this amplifier is still possible at lower and at higher temperatures without causing permanent damage, because the amplifier is not on while frosting occurs and has no temperature elevating internal power dissipation.

1.5. Deterministic reliability engineering, often called 'physics of failure', is concerned with the investigation of the physical deterioration processes (e.g. corrosion, fatigue, etc.) that lead to failure.

1.6. To manufacture a reliable product all aspects of the manufacturing process are of major importance: the design, the choice of components to be used, the testing, the compilation of mortality data, the assessment of the reliability to be expected, the analysis of the physical deterioration processes, and the (constant) quality of the production line. Usually it is impossible for one person to supervise and coordinate all these tasks. For that reason a *reliability group* (quality circles, etc.) has to be formed, each member of which is delegated a subtask. Without a proper division of tasks and an established feedback path for failure data, such a group will never be able to function optimally; an organisation geared towards these goals is very much essential to a high reliability product.

1.7. Examples of environmental factors which may have an influence on the ageing processes in integrated circuits are: humidity, vapours, temperature, aggressive chemical environment (acid, basic, solvent), mechanical vibrations and other mechanical stresses, and electric stress (supply voltage, input/output power, etc.).

2 Deterministic Reliability

2.1. For the Mean Time To Failure (MTTF) at the three different temperatures Arrhenius' model holds (MTTF inversely proportional to reaction rate):

$\text{MTTF}(T_1) = t_0 \exp[E_A/kT_1] = 6.5 \times 10^3$ hour with $T_1 = 373$ K,
$\text{MTTF}(T_2) = t_0 \exp[E_A/kT_2] = 2.4 \times 10^4$ hour with $T_2 = 358$ K,
$\text{MTTF}(T_3) = t_0 \exp[E_A/kT_3]$ with $T_3 = 298$ K.

From the first two equations the activation energy E_A can be determined:

$$E_A = \frac{k}{(1/T_2 - 1/T_1)} \ln\left[\frac{\text{MTTF}(T_2)}{\text{MTTF}(T_1)}\right] = 1 \text{ eV}.$$

MTTF(T_3) is then found from:

$$\text{MTTF}(T_3) = \text{MTTF}(T_1) \exp\left[\frac{E_A}{k}\left(\frac{1}{T_3} - \frac{1}{T_1}\right)\right] = 1.66 \times 10^7 \text{ hour}.$$

2.2. Heavy current conductors as used in power engineering have a much smaller current density (A/m^2) than the extremely small metallised traces in semiconductors (which are only a few μm wide and as little as 0.1 μm high). Owing to this, electromigration does not occur. The conductors in heavy current equipment are operated at these much lower current densities because the power losses have to be kept small.

2.3. Cathodic protection is used here, which utilizes a supply of metal attached to the hull near the expensive propeller. This supply of metal (electrode) is in electrical contact with the hull, the shaft and the propeller. It consists of a metal or alloy with a much higher electrogalvanic potential which dissolves more easily, thereby protecting the propeller from electrogalvanic corrosion (sacrificing electrode).

2.4. The dominant failure mechanism of a car of which the oil is never changed is caused by severe wear of the parts that must be lubricated. One of the possible consequences (one of the possible failure modes) is the crankshaft bearings breaking down or the piston rings fusing to the cylinder lining.

2.5. Testing *all* products prior to delivery in order to remove weak and failed components is called 'screening'. When the remaining products have not noticeably weakened from this test, the products shipped will have a higher reliability than the original group.

2.6. Early failures occur shortly after a product is taken into operation. They are caused by a production process which supplies products that are not all equal. Some products have weak spots which cause them to fail quickly. If a product has survived the early failure period (which should be approximately equal to the warranty period) without repair it may be assumed to have been manufactured well.

3 Statistical Reliability

3.1. The hazard rate is the conditional failure probability density and is defined as:

$$z(t) = \frac{f(t)}{R(t)} = -\frac{1}{R(t)}\frac{dR(t)}{dt}.$$

It can also be viewed as the *fractional* decrease in the number of working components in the next unit interval of time.

3.2. The definition of maintainability is given in Section 3.2

3.3. The various availability definitions are given in Section 3.2.

3.4. The reliability $R(t)$ is given by:

$$R(t) = \exp\{-\int_0^t z(t)dt\}.$$

Since it must also hold that $\lim_{t\to\infty} R(t) = 0$, it follows directly that:

$$\lim_{t\to\infty} \int_0^t z(t)dt = 0. \to \infty$$

3.5. The condition is (see Section 3.2.1):

$$\lim_{t\to\infty} tR(t) = 0.$$

3.6. The hazard rate is defined as:

$$z(t) = \frac{f(t)}{R(t)}.$$

It further holds that:

$$f(t) = -\frac{\Delta R}{\Delta t}.$$

So $z(t)$ may be written as:

$$z(t) = -\left(\frac{\Delta R(t)}{R(t)}\right)\frac{1}{\Delta t}.$$

Therefore $z(t)$ is equal to the relative *decrease* of the reliability function per unit of time.

3.7. The reliability $R(t)$ is given by:

$$R(t) = \exp\{-\int_0^t z(t)dt\},$$

assume $R(0) = 2$

and the hazard rate $z(t)$ by:

$$z(t) = \frac{f(t)}{R(t)}.$$

Consequently:

$$f(t) = z(t)\exp\{-\int_0^t z(t)dt\}.$$

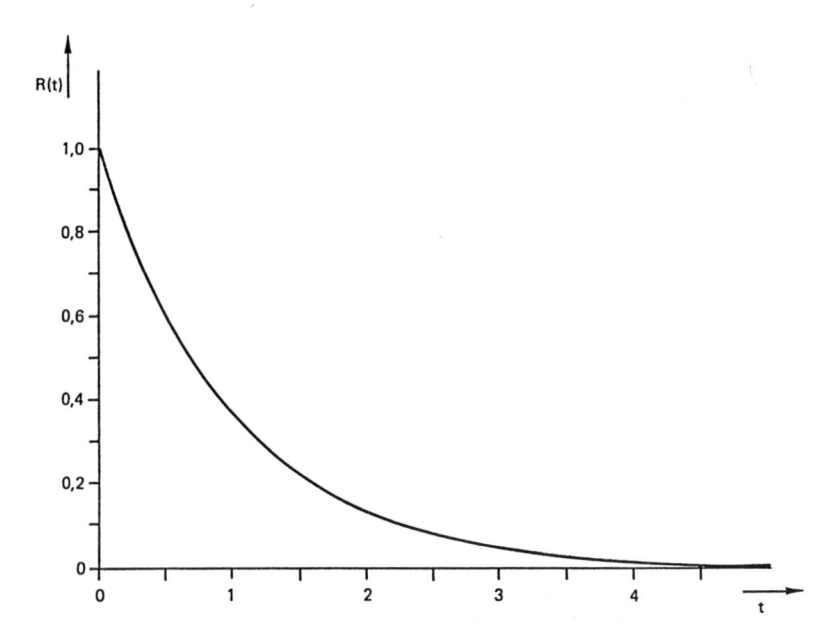

3.8. a. See figure above. The reliability function must meet the following requirements:
- $0 \leq R(0) \leq 1$;
- monotonic and not increasing over the interval $[0,\infty)$;
- $\lim_{t \to \infty} R(t) = 0$.

b. The reliability is given by:

$$R(t) = R(0) \exp\{-\int_0^t z(t)dt\}.$$

$R(t)$ will only be monotonically not increasing if the integral in the argument of the exponential function is monotonic and not decreasing with t. This means that for $z(t)$ must hold: $z(t) \geq 0$ in $[0,\infty)$. The fact that $R(t)$ must go to zero if t goes to infinity, requires from $z(t)$ that:

$$\lim_{t \to \infty} \int_0^t z(t)dt \to \infty.$$

$R(t)$ has to be equal to $R(0)$ for $t = 0$. So for $z(t)$ it must also hold that:

$$\lim_{t \downarrow 0} \int_0^t z(t)dt = 0, \text{ or } \lim_{t \downarrow 0} t z(t) = 0.$$

3.9. In Problem 3.8b we have found the requirements for a $z(t)$ function. These are met if $A \geq 0$ and $B \geq 0$. Furthermore, the physical dimension of A has to be: time^{-1} and that of B: time^{-2}.
In Problem 3.7 we found for $f(t)$:

$$f(t) = z(t) \exp\{-\int_0^t z(t)dt\}.$$

Substitution of $z(t) = A + Bt$ yields:

$$f(t) = (A + Bt) \exp\{-(At + \frac{Bt^2}{2})\}.$$

3.10. $S_t^R = \lim_{\Delta t \to 0} \frac{\Delta R(t)}{R(t)} \frac{1}{\Delta t} = \frac{1}{R(t)} \frac{dR(t)}{dt} = -\frac{1}{R(t)} \frac{dF(t)}{dt} = -\frac{f(t)}{R(t)} = -z(t).$

$-S_t^R$ is therefore equal to the hazard rate $z(t)$.

3.11. a. R (1000 hours) $= \exp(-10^{-3}) \approx 0.999$.
b. The failure probability is $1 - \exp(-10^{-3}) \approx 0.001$. If there are 10,000 of these pieces of equipment in use the number expected to fail during these 1000 hours is equal to 0.001 × 10,000 = 10.
c. The mean life is equal to $1/\lambda = 10^6$ hour. The reliability is equal to $e^{-1} \approx 0.37$ at that point in time.
d. The probability of surviving another 1000 hours, if it is given that the relevant equipment has survived the first 1000 hours, is given by:

$$P(2000 \text{ hours} \mid 1000 \text{ hours}) = \frac{P(2000 \text{ hours})}{P(1000 \text{ hours})},$$

and for the negative exponential distribution this is equal to:

$$P(1000 \text{ hours}) = \exp(-10^{-3}) \approx 0.999.$$

So, this survival probability is independent of age: any correctly operating piece of equipment is 'as new'.

3.12. $R(t) = R(0) \exp\{-\int_0^t z(t)dt\} = R(0) \exp\{-\int_0^{t_1} z(t)dt - \int_{t_1}^t \lambda dt\} =$

$= R(t_1) \exp\{-\lambda(t - t_1)\}.$

3.13. $R(1000) = \exp(-10^{-4} \times 1000) = 0.905.$

3.14. a. $R(t) = \exp(-\frac{a}{2}t^2)$; $z(t) = at.$
b. $T_1 = 10$ hours; $\exp(-\frac{a^2}{2}T_1^2) = \frac{4700}{5000}$ (300 failed);

$T_2 = 20$ hours; $\exp(-\frac{a^2}{2}T_2^2) = \{\exp -\frac{a^2}{2}T_1^2\}^4 = (\frac{47}{50})^4$. Thus, approximately 1096

systems have failed. Therefore, in the time interval from 10 to 20 hours, 796 are expected to fail.

4 Statistically Failing System Components

4.1. a. $R(t) = \sum_{i=0}^{1} \frac{(\lambda t)^i}{i!} e^{-\lambda t} = (1 + \lambda t)e^{-\lambda t}.$

From this the hazard rate can be calculated to be:

210 Solutions to Problems

$$z(t) = -\frac{1}{R(t)}\frac{dR(t)}{dt} = \frac{-1}{(1+\lambda t)e^{-\lambda t}}(-\lambda^2 t e^{-\lambda t}) = \frac{\lambda^2 t}{1+\lambda t}.$$

b. The $z(t)$ function found is monotonically increasing. For $t \to \infty$ it asymptotically approaches $z(\infty) = \lambda$. It is very improbable that no units would have failed here. If the system still functions, the last unit is engaged and the hazard rate of the system will be equal to the failure rate of this last single unit: $z_{max} = z(\infty) = \lambda$.

4.2. The components have a constant failure rate. This means that these components fail randomly in time and that these failure events are therefore distributed in accordance with a Poisson distribution. So, if on average a certain random event occurs m times in a time interval T, the probability of this event not occurring in this time interval T is equal to e^{-m}. In our case, the random event is the failure of a transistor, where the number of expected failures in the accumulated testing hours T (the so-called component hours) is λT, λ being the failure rate to be determined. We now want a solution for λ in which the probability that λ is estimated too low is only 10 %. From $e^{-m} = 0.1$ it follows that $m = 2.3$, so that the probability of λT being between 2.3 and ∞ is equal to 10 %. The probability that λT is between 0 and 2.3 is therefore 90 %. Consequently, with $\lambda T = 2.3$ and $T = 10^4$ component hours, we find $\lambda = 2.3 \times 10^{-4}$/hours. The failure rate is therefore between 0 and 2.3×10^{-4}/hour, with a confidence of 90 %.

4.3. A decreasing hazard rate depicts a systems still in its early failure phase (infancy region). Mainly the components that are already weak fail here.

4.4. $\lim_{t \to \infty} z(t) = \lim_{t \to \infty} \frac{f(t)}{R(t)} \stackrel{\text{L'Hopital}}{=} \lim_{t \to \infty} -\frac{f'(t)}{f(t)} = \lim_{t \to \infty} \frac{t-\theta}{\sigma^2} \to \infty.$

4.5. In accelerated life experiments the systems to be tested are exposed to an increased stress which causes particular failure mechanisms to speed up. In this way one hopes to collect more quickly sufficient mortality data to assess the reliability under normal conditions. The following conditions have to be met:
- the acceleration factor has to be known;
- the higher testing stress is not allowed to introduce *new* failure mechanisms;
- the higher testing stress is not allowed to mask normally occurring failure mechanisms.

5 Reliability Models

5.1. *Circuit A.*

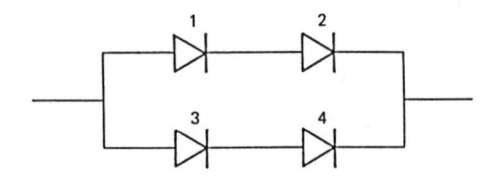

The probability of an open failure is: $\quad a_o = 4q_o^2 - 4q_o^3 + q_o^4.$

The probability of a short-circuit failure is: $a_s = 2q_s^2 - q_s^4$.
So, the total failure probability is: $a = a_o + a_s$.

Circuit B.

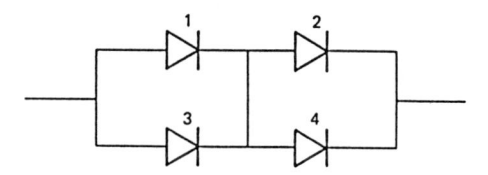

The probability of an open failure is: $b_o = 2q_o^2 - q_o^4$.

The probability of a short circuit failure is: $b_s = 4q_s^2 - 4q_s^3 + q_s^4$.
The total failure probability is: $b = b_o + b_s$
The difference between the two failure probabilities is:

$$a - b = q_o^2[2 - 4q_o + 2q_o^2] - q_s^2[2 - 4q_s + 2q_s^2].$$

We can now distinguish three regions:

- $q_o < q_s \Rightarrow a < b$: here we would rather not have a centre connection;
- $q_o = q_s \Rightarrow a = b$: no preference;
- $q_o > q_s \Rightarrow a > b$: we would advise to use a centre connection now.

5.2. Two disjunct events, each with non-zero probability, cannot be independent: the fact that the occurrence of one event excludes the occurrence of the other event makes the events dependent!

5.3.
$$R(t) = 1 - F(t) = 1 - \frac{t}{L}$$

$$F(t) = \int_0^t \frac{1}{L} dt = \frac{t}{L} \qquad 0 \le t \le L$$

$$z(t) = \frac{f(t)}{R(t)} = \frac{1}{L - t}$$

$$t > L \begin{cases} R(t) = 0 \\ F(t) = 1 \\ z(t) \text{ is undetermined} \end{cases}$$

5.4. a. With a constant failure rate the reliability is given by:

$$R(t) = e^{-t/\text{MTTF}}.$$

The failure probability during a 4-hour flight is therefore:

$$F(4) = 1 - e^{-4/1140} = 3.5 \times 10^{-3}.$$

This failure probability is independent of the age of the system.
b. Let us assume that such a flight lasts T hours: $e^{-T/\text{MTTF}} \ge 0.99$.

$$T \le -\text{MTTF} \ln 0.99 = 11.5 \text{ hours}.$$

So the maximum allowable duration of the flight is 11.5 hours.

212 Solutions to Problems

5.5. a. $R(t) = 1 - \int_0^t \{at \exp(-\frac{a}{2}t^2)\} dt = \exp(-\frac{a}{2}t^2)$,

and:

$$z(t) = f(t)/R(t) = at.$$

b. After 10 hours the reliability is:

$$\frac{4700}{5000} = \exp(-\frac{a}{2}T_1^2), \text{ with } T_1 = 10 \text{ hours.}$$

After 20 hours ($T_2 = 2T_1$) the reliability is:

$$\exp(-\frac{a}{2}T_2^2) = \{\exp(-\frac{a}{2}T_1^2)\}^4 = (\frac{47}{50})^4.$$

Therefore, $\{1 - (\frac{47}{50})^4\} \times 5000 = 1096$ units are expected to have failed then.

In the time interval from 10 to 20 hours 796 units are expected to fail. (1096 − 300)

5.6. See Figure 5.5: η is the ratio of the mean strength y to the mean stress x.

5.7.

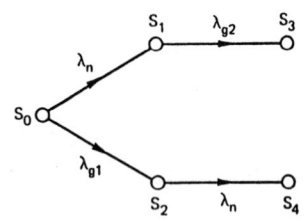

S_0 = grid and generator function;
S_1 = grid failed and generator functions;
S_2 = grid functions and generator failed (and can therefore no longer start up);
S_3 = grid failed and generator failed;
S_4 = grid failed and generator failed;
S_3 and S_4 are the absorbing 'down' states for this system.

6 Non-Maintained Systems

6.1. 'Dependent' is the stochastic phenomenon in which the occurrence of a certain event makes the occurrence of other events more probable or less probable. Examples are common-cause failures (e.g. power grid failure) and secondary failures (such as a bearing overheating because of an oil leak).

6.2. The Markov diagram for an active m-out-of-n homogeneous system is given in the figure below. The mean time for the transition from S_0 to S_1 is $1/n\lambda$; for the transition from S_1 to S_2 $1/(n-1)\lambda$; and so on. The MTTF is the mean time for the transition from S_0 to S_{n-m+1} and is therefore equal to:

$$\text{MTTF} = \frac{1}{n\lambda} + \frac{1}{(n-1)\lambda} + \ldots + \frac{1}{m\lambda} = \frac{1}{\lambda}\sum_{k=m}^{n}\frac{1}{k}.$$

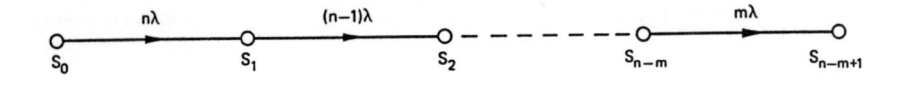

6.3. $\lambda_{eff} = (\frac{1}{4} \times 4 \times 10^{-8} + \frac{3}{4} \times 4 \times 10^{-9})/h = 1.3 \times 10^{-8}/h$.

6.4. a. The system fails if all n units have failed. The probability that a unit fails is $1 - e^{-\lambda t}$. So, the probability that the system fails is $\{1 - e^{-\lambda t}\}^n$. Consequently the reliability of this system is:

$$R(t) = 1 - \{1 - e^{-\lambda t}\}^n = \sum_{i=1}^{n}(-1)^{i+1}\binom{n}{i}e^{-\lambda i t}.$$

b. $\quad \theta = \int_0^\infty R(t)dt = \frac{1}{\lambda}\sum_{i=1}^{n}(-1)^{i+1}\binom{n}{i}\frac{1}{i}.$

The mean life can also be found with the Markov method. The Markov diagram is the same as that for Problem 6.2 with $m = 1$. For the mean life we find:

$$\theta = \frac{1}{\lambda}\sum_{i=1}^{n}\frac{1}{i}.$$

(Although the two solutions seem to be different this is not the case!)

6.5. a. $R_a = \{1 - (1 - R^n)^2\}^n = R^n(2 - R^n)^n;$
$\quad R_b = 1 - (1 - R^n)^2 = R^n(2 - R^n)$
$\quad \lim_{R \to 1} \frac{R_a}{R_b} = 1.$

b. $\quad \lim_{R \to 0} \frac{R_a}{R_b} = 2^{n-1}.$

c. See figure below.

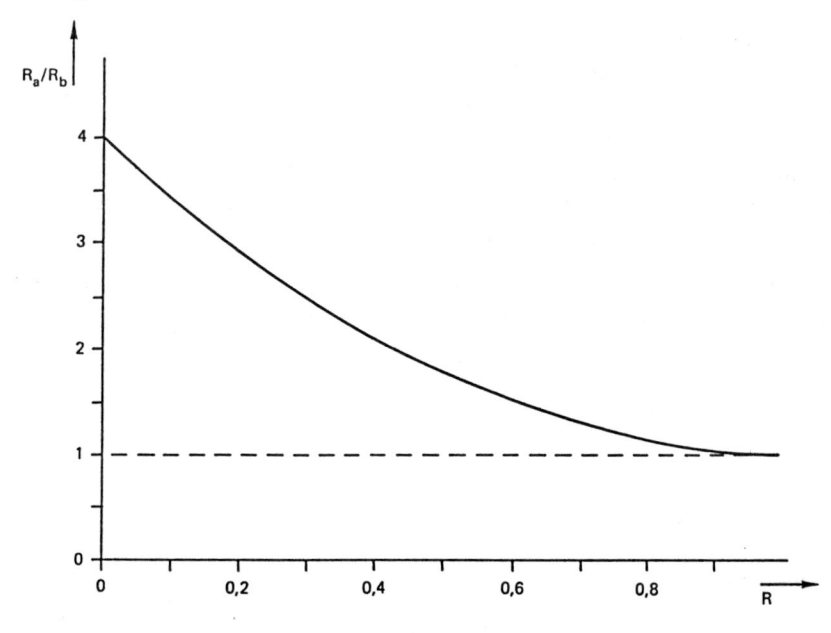

6.6. a. The failure probability of the system depicted can be determined with the tie-set, cut-set method (Section 6.9.2) or the decomposition method (Section 6.9.3). The solution with the first mentioned method is graphically represented in Figure 6.12. The reliability of the system is given by:

$$R = \sum_{i=0}^{n} (-1)^i R_i,$$

where n is the maximum number of loops and R_i is the sum of the reliabilities of all subgraphs with i loops. Denoting the central interconnecting branch as C we find:

$$\begin{aligned}R =\ & P(A)P(B) + P(A')P(B') + P(A)P(B')P(C) + P(A')P(B)P(C) + \\ & + P(A)P(B)P(B')P(C) - P(A)P(A')P(B)P(B') - P(A)P(A')P(B')P(C) + \\ & - P(A)P(A')P(B)P(C) - P(A')P(B)P(B')P(C) + 2P(A)P(A')P(B)P(B')P(C).\end{aligned}$$

With $P(A) = P(B') = 0.9$; $P(A') = P(B) = 0.8$ and $P(C) = 0.7$ we find $F = 1 - R = 0.05124$. It is obvious that the probability of making errors is considerable with this solution method. It is simpler to use the decomposition method. This is graphically demonstrated in Figure 6.13. The reliability is:

$$R = P(C)P(S/C) + P(\overline{C})P(S/\overline{C}),$$

with

$$R(S/C) = \{1 - P(A)P(A')\}\{1 - P(B)P(B')\} = 0.9604$$
$$R(S/\overline{C}) = 1 - \{1 - P(A)P(B)\}\{1 - P(A')P(B')\} = 0.9216$$
$$F = 1 - R = 0.05124.$$

b. The catastrophic failure model as described in this problem applies to a system comprising 1-out-of-2 active redundancy, where each channel consists of two different sub-units. If one sub-unit in one channel would have failed and the other, different sub-unit would have failed in the other channel, there is a switching unit (C) that is able to connect the two remaining unequal, still functioning sub-units. An example is a computer system consisting of two central processors and two memory units in an active redundant configuration. If one of the two processors and one of the two memories have failed the system is still able to function if the switching unit has not failed. So the central interconnection branch always results in an improvement with regard to the same system without such a branch.

6.7. $0 = R_a < R_e < R_b < R_f < R_c < R_d$. This can be seen as follows: Configuration a gives $R_a = 0$, since a single pump is not able to pump up the water 7 meters (23 feet). Configuration b results in $R_b = R_o^2$. For c we find:

$$R_c = 1 - (1 - R_o^2)^2 = R_o^2(2 - R_o^2)$$

and for R_d we find:

$$R_d = \{1 - (1 - R_o)^2\}^2 = R_o^2(2 - R_o)^2.$$

Configuration e results in:

$$R_e = R_o^2\{1 - (1 - R_o)^2\} = R_o^3(2 - R_o)$$

and configuration f in:

$$R_f = R_o^2(2 - R_o).$$

With $0 < R_o < 1$ we find the above order.

6.8. a. See the figure below.

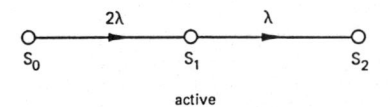

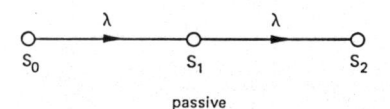

[Figure: active configuration $S_0 \xrightarrow{2\lambda} S_1 \xrightarrow{\lambda} S_2$; passive configuration $S_0 \xrightarrow{\lambda} S_1 \xrightarrow{\lambda} S_2$]

b. Active redundancy: the mean time for the transition from S_0 to S_1 is $1/(2\lambda)$. The mean time for the transition from S_1 to S_2 is $1/\lambda$. So the mean life of the system is therefore $3/(2\lambda)$.

Passive redundancy: with the same method as above we find a mean life of $2/\lambda$.

6.9. From the initial state S_0 three transitions are possible, viz. the network may fail, computer I may fail (transition to state S_1) or computer II may fail (transition to state S_2). If the power supplying network fails, the system is down (state S_3). In the other two states the system still functions. From state S_1 the system will fail if either the power or computer II fails. All this is illustrated in the Markov diagram below.

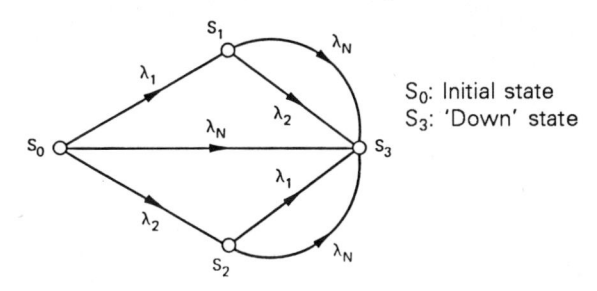

[Markov diagram with states S_0, S_1, S_2, S_3; S_0: Initial state, S_3: 'Down' state; transitions labelled λ_1, λ_2, λ_N]

6.10. The units are identical. This means that the reliability of the entire system is given by:

$$R_s = R_{\text{total}} = R^n.$$

The reliability of a series system is always lower than that of one single unit; the reliability of one unit is given by:

$$R = \sqrt[n]{R_s},$$

and the total costs by:

$$C_{\text{total}} = nCf(\sqrt[n]{R_s}).$$

6.11. In state S_0 three events may happen that cause a transition in the associated Markov diagram (see figure below):
- the active unit may fail (transition to state S_1);
- the stand-by unit may fail (transition to state S_2);
- The error detection and switching unit may fail (transition to state S_3).

216 Solutions to Problems

(Do not make the mistake here to think that this failure only occurs when the active unit fails and the stand-by unit must be switched in: That is indeed when one will notice that the error detection and switching unit has failed. The failing of the latter unit, however, has occurred *before* the failing of the active unit and must also be represented in the Markov diagram as such.)

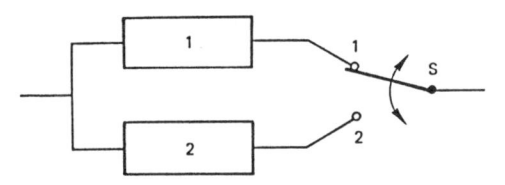

In both state S_1 and S_2 the switching unit might still fail, due to which the system is not able to switch over anymore. Since the redundant unit in S_1 has already been engaged and does not have to be switched anymore and since the redundant unit has already failed in state S_2 so that switching is useless in this state, a failure of the switching unit cannot influence the transition probability to 'system down' from these two states. This extra transition from these two states can be left out (is *de facto* incorporated in S_4). Similarly, also the failure of the redundant unit will have no influence on the 'system down' probability if the system is in state S_3.

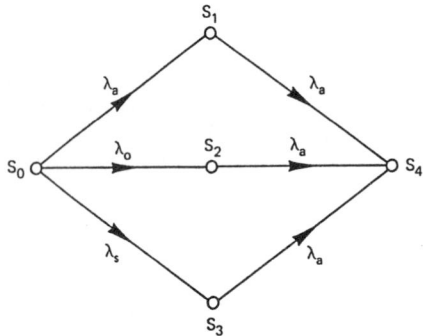

S_0: everything functions; S_1: only unit1 faulty
S_2: only unit 2 faulty (while S is in position 1)
S_3: only S faulty, stuck in position 1
S_4: system fails.

Since the transition probability of the states S_1, S_2 and S_3 changing into state S_4 is equal to (λ_a), these states can be taken together in one state (S_1'), with a transition rate λ_a leading into state S_4. The incoming branches of S_1, S_2 and S_3 are also taken together and the probability of a transition from S_0 to this combined state S_1' is the sum of the transition probabilities from S_0 to S_1, S_2 and S_3.

S_0: everything functions correctly;
S_1': only the unit that is engaged at the moment (1 or 2) can still be used; (the compound state joining S_1, S_2 and S_3);
S_3: system fails.

6.12. The Markov diagram has the same shape for both configurations. So, to determine which one has the highest reliability, we only have to look at the transition probabilities. From this it follows that if $\lambda_s > \lambda$ active redundancy has the highest reliability, while for $\lambda_s < \lambda$ the passive redundancy gives the highest reliability.

$$S_0 \xrightarrow{2\lambda} S_1 \xrightarrow{\lambda} S_2 \qquad\qquad S_0 \xrightarrow{\lambda+\lambda_s} S_1 \xrightarrow{\lambda} S_2$$
$$\text{active} \qquad\qquad\qquad\qquad \text{passive}$$

6.13. $\text{MTTF} = \dfrac{1}{\lambda} + \dfrac{1}{\lambda + \lambda_1 + \lambda_2}$.

This MTTF can be found by solving the differential equations describing the Markov diagram by means of Laplace-transformation. The MTTF is then given by:

$$\text{MTTF} = \sum_{i=1}^{4} \lim_{s \to 0} P_{S_i}(s).$$

It will be clear that this is rather time consuming for the Markov diagram given and might evoke mistakes. For that reason we should always first try to simplify the Markov diagram. There are two important rules for doing this:

- Different states having a transition to the same state all with the same transition probability can be taken together into one combined state, provided that there are no transitions possible from the states to be combined to still another state, (except for mutual transitions between these states). The transition probability of the combined state to the common state is equal to the original transition probability.

The above rule can be easily understood from the given Markov diagram. If we restrict ourselves to state S_5, state S_1 and state S_4, then, if it is given that the system is in either state S_1 or state S_4, the transition probability of the combined state $S_{1,4}$ to state S_5 is, of course, still $\lambda \Delta t$.

Repeating this process state $S_{1,4}$ can be combined with state S_2 and with state S_3 to form a new state $S_{1,2,3,4}$.

- If between two states more than one transition is possible, these transitions can be combined into one transition with a transition rate equal to the sum of the separate transition rates.

The latter rule is obvious. If we apply it to the Markov diagram resulting from the simplification by rule 1, the diagram becomes as shown in the figure below. The MTTF is now simple to calculate.

$$S_0 \xrightarrow{\lambda+\lambda_1+\lambda_2} S_1 \xrightarrow{\lambda} S_2$$

Note: The Markov diagram given represents a system as given in Problem 6.11. The MTTF of this relatively intricate system can be found virtually without any calculations!

218 Solutions to Problems

6.14. For a 2-out-of-3 active redundant system holds (see Section 6.5):
$$R(t) = (3 - 2e^{-\lambda t})e^{-2\lambda t}$$
For one unit holds: $R(t) = e^{-\lambda t}$.
For two units in series holds: $R(t) = e^{-2\lambda t}$.
These reliability functions are illustrated in the figure below.

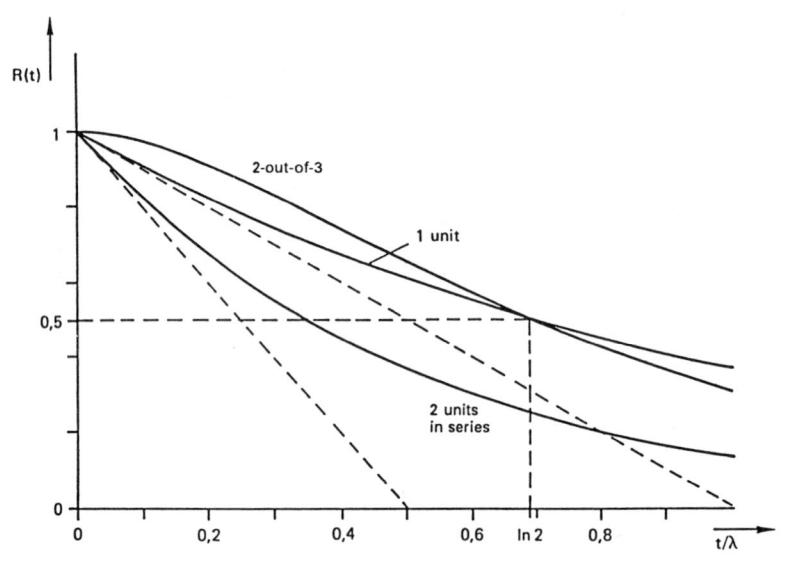

We see directly that for $t < (1/\lambda)\ln 2$ the redundant system has a higher reliability than a one-unit system. Here we once more see that information about the mean life does not provide sufficient information on the reliability of a system to determine the usefulness of various configurations; in this aspect a redundant system can become worse than a non-redundant system!

6.15. a. From the Markov diagram below it follows directly that:
$$\text{MTTF} = \frac{1}{2\lambda_n} + \frac{1}{\lambda_m} = 750 \text{ hours.}$$

(a)

b. The following differential equations can be inferred from the Markov diagram:
$$\frac{dP_0}{dt} = -2\lambda_n P_0,$$
$$\frac{dP_1}{dt} = 2\lambda_n P_0 - \lambda_m P_1,$$
$$\frac{dP_2}{dt} = \lambda_m P_1.$$
Laplace transformation results in:
$$P_0 = \frac{1}{s + 2\lambda_n},$$

$$P_1 = \frac{2\lambda_n P_0}{s + \lambda_m} = \frac{2\lambda_n}{(s + \lambda_m)(s + 2\lambda_n)},$$

$$R = P_0 + P_1 = \frac{s + \lambda_m + 2\lambda_n}{(s + \lambda_m)(s + 2\lambda_n)}.$$

Inverse transformation gives:

$$R(t) = \frac{\lambda_m}{\lambda_m - 2\lambda_n} e^{-2\lambda_n t} - \frac{2\lambda_n}{\lambda_m - 2\lambda_n} e^{-\lambda_m t} =$$

$$= 2e^{-2 \times 10^{-3} t} - e^{-4 \times 10^{-3} t}.$$

For one unit we find:

$$R(t) = e^{-\lambda_m t} = e^{-4 \times 10^{-3} t}.$$

The requested functions are sketched in the figure below.

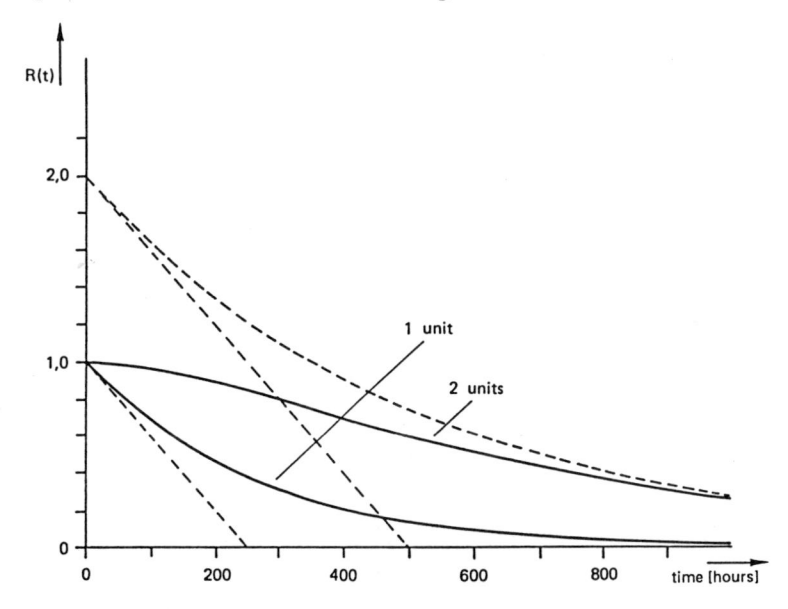

6.16. To determine the optimum strategy we will have to calculate the average costs per period for the three cases. For strategy a we find:

$$(1 - 0.95^3) \times \$ \, 10{,}000 + 0.01^3 \times \$ \, 25{,}000 = \$ \, 1426.28.$$

For strategy b:

$$(3 \times 0.05^2 \times 0.95 \times 0.05^3) \times \$ \, 10{,}000 +$$
$$(3 \times 0.01^2 \times 0.99 + 0.01^3) \times \$ \, 25{,}000 = \$ \, 79.95.$$

For strategy c:

$$(0.05^3 \times \$ \, 10{,}000 + (1 - 0.99^3) \times \$ \, 25{,}000 = \$ \, 743.78.$$

So, the optimal strategy is b.

220 Solutions to Problems

6.17. a. At least two pumps have to be active simultaneously. If one of them fails the passive pump is switched on. If, subsequently, yet another pump would fail the capacity required is not reached and the system is therefore 'down'. This is illustrated in the Markov diagram shown below under a. So the MTTF of the capacity required is:

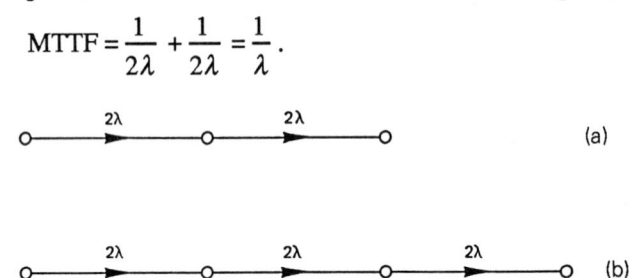

$$\text{MTTF} = \frac{1}{2\lambda} + \frac{1}{2\lambda} = \frac{1}{\lambda}.$$

$$\underset{\circ}{\xrightarrow{2\lambda}} \underset{\circ}{\xrightarrow{2\lambda}} \circ \qquad\qquad \text{(a)}$$

$$\underset{\circ}{\xrightarrow{2\lambda}} \underset{\circ}{\xrightarrow{2\lambda}} \underset{\circ}{\xrightarrow{2\lambda}} \circ \qquad\qquad \text{(b)}$$

b. If two pumps have failed the last pump will keep the mixture in motion to prevent clogging. The failure rate of this pump is 2λ. If this last pump fails the system is down. This is illustrated in Markov diagram b above. The MTTF of this clogging prevention system is:

$$\text{MTTF} = \frac{1}{2\lambda} + \frac{1}{2\lambda} + \frac{1}{2\lambda} = \frac{3}{2\lambda}.$$

6.18. First reduce

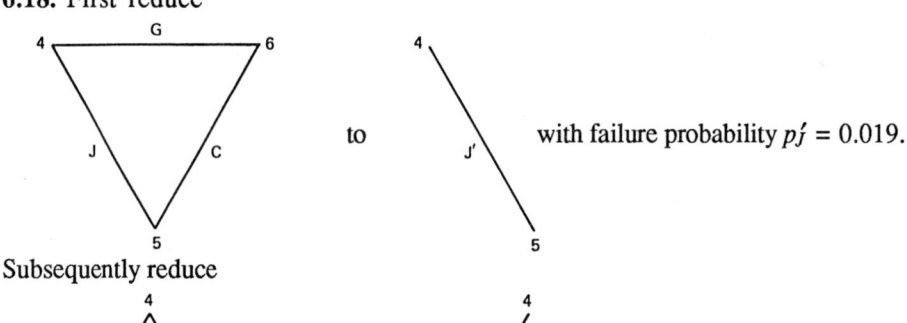

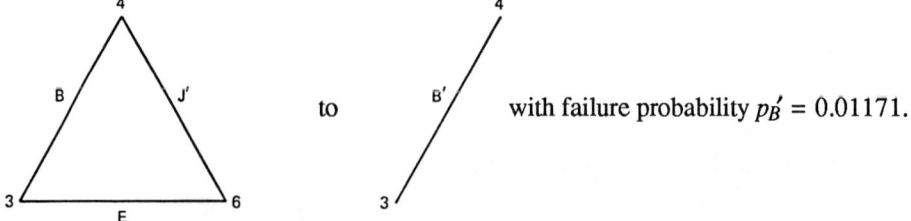

with failure probability $p'_J = 0.019$.

Subsequently reduce

with failure probability $p'_B = 0.01171$.

The following bridge structure is left now:

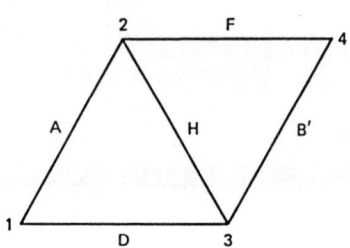

Using the decomposition method this can be solved as follows ($S \equiv$ system):
$$R_S = P(H)P(S|H) + P(\overline{H})P(S|\overline{H})$$
$$= R_h[(1 - F_aF_d)(1 - F_fF_{b'})] + F_h[1 - (1 - R_aR_f)(1 - R_{b'}R_d)] = 0.988.$$

6.19. Solving this problem we must keep in mind that a series circuit of diodes is redundant for short-circuit failures, whereas a parallel circuit of diodes is redundant for open failures. We can thus draw up the following table of failure probabilities:

circuit	open failure	short circuit failure	total probability
1	$p_o = 0.02$	$p_s = 0.01$	0.03
2	$\approx 2p_o = 0.04$	$p_s^2 = 10^{-4}$	0.0401
3	$p_o^2 = 4 \times 10^{-4}$	$\approx 2p_s = 0.02$	0.0204
4	$\approx 3p_o = 0.06$	$p_s^3 = 10^{-6}$	0.060001
5	$\approx p_o + p_o^2 = 0.0204$	$\approx 2p_s^2 = 2 \times 10^{-4}$	0.0206
6	$\approx 2p_o^2 = 8 \times 10^{-4}$	$\approx p_s^2 + p_s = 0.0101$	0.0109
7	$p_o^3 = 8 \times 10^{-6}$	$\approx 3p_s = 0.03$	0.03

We see that with the probabilities given for the diode failures ($p_o = 0.02$ and $p_s = 0.01$) circuit 6 is optimal.

6.20. a. Markov diagram:

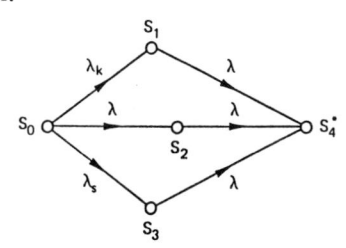

S_0: all units and the switch are functioning correctly.
S_1: the switch sticks to unit 1.
S_2: one unit failed; the second unit is on.
S_3: the switch is incorrectly in position 2 and so unit 2 is active.
S_4: the system fails.

b. The above Markov diagram can be reduced to:

(After all, it does not make a difference whether the system is in state S_1, S_2 or S_3. For all these states the system fails with failure rate λ). The mean life then is the mean time needed to get from S_0 to S_1' plus the mean time for getting from S_1' to S_2':

$$\text{MTTF} = \frac{1}{\lambda + \lambda_s + \lambda_k} + \frac{1}{\lambda}.$$

222 Solutions to Problems

6.21. The hydraulic pressure failure is best displayed if before the failure happened less than a majority indicates alarm and after the failure a majority signals alarm. The system configurations and the probability of a correct alarm are given below:

1. single: $P_{sign} = 0.8\{1c\}$;
2. double: $P_{sign} = 0.8^2\{2c\} + 2\times0.89\times0.1\{1c,1n\} = 0.8$;
3. triple: $P_{sign} = 0.8^3\{3c\} + 3\times0.8^2\times0.1\{2c,1f\} + 3\times0.8^2\times0.1\{2c,1n\} +$
   $+ 6\times0.8\times0.1\times0.1\{1c,1f,1n\} = 0.944$;
4. fourfold: $P_{sign} = 0.8^4\{4c\} + 4\times0.8^3\times0.1\{3c,1f\} + 4\times0.8^3\times0.1\{3c,1n\} +$
   $+ 6\times0.8^2\times0.1^2\{2c,2n\} + 12\times0.8^2\times0.1\times0.1\{2c,1n,1f\} +$
   $+ 12\times0.8\times0.1^2\times0.1\{1c,2n,1f\} = 0.934.$

In the above we have indicated in brackets a correctly functioning alarm by a c, a nuisance alarm by an n, and a unit failing to indicate alarm by an f. We see that the triple system is the best choice.

6.22. In the table below the reliability of the three system partitions (blocks) is given for different numbers of components in parallel.

Partition:	comp$_1$	comp$_2$	comp$_3$
Redundancy			
×1	0.9	0.8	0.5
×2	0.99	0.96	0.75
×3	0.999	0.992	0.875
×4	0.9999	0.9984	0.9375

The first step is clear: to reach a total reliability of 0.8 some partitions must be allotted a higher reliability. We therefore start with the configuration (in units U_i):

$$U_1(\times 1), U_2(\times 2), U_3(\times 3),$$

which has a reliability of:

$$R_s = R_1(1 - F_2^2)(1 - F_3^3) = 0.756,$$

which is not sufficient yet.

Adding an extra component to a partition results in the following improvement factors for the system reliability (R_p denotes the redundant partition reliability before the addition of the extra unit, R_p' that of the partition after adding the extra unit):

$$\frac{R_{p1}'}{R_{p1}} = \frac{0.99}{0.9} = 1.1 \text{ for partition 1 } (R_s' = 0.832);$$

$$\frac{R_{p2}'}{R_{p2}} = \frac{0.992}{0.96} = 1.0333 \text{ for partition 2 } (R_s' = 0{,}7812);$$

$$\frac{R_{p3}'}{R_{p3}} = \frac{0.9375}{0.875} = 1.0714 \text{ for partition 3 } (R_s' = 0.81).$$

Consequently the optimum configuration becomes: $U_1(\times 2), U_2(\times 2), U_3(\times 3)$.

7 Maintained Systems

7.1. A maintainable system is a system which is suitable to maintain, i.e. after it has failed can be restored to a working order by human intervention. A maintained system is a maintainable system that is actually maintained. Maintenance is also defined in Section 3.2.

7.2. Preventive maintenance is performed *before* a system has failed. It intends to protect the system from failing in the future. The hazard rate $z(t)$ of a system has to be a monotonically increasing function for preventive maintenance to be useful for that system. After such preventive maintenance the system is restored from a high hazard rate to a state with a lower hazard rate. If the hazard rate is a decreasing function of time the system would, by renewing (part of) it, be brought back to a state with a higher hazard rate immediately after preventive maintenance. Here preventive intervention clearly does not work; it is even counterproductive.

7.3. Optimisation of the 'life cycle cost' of a system encompasses the following costs:
- the initial investment costs;
- the maintenance and exploitation costs;
- the costs of discarding the system, if any.

7.4. a. MTTF = $3/2\lambda$ (see also Problem 6.8).
b. It is easy to deduce from the Markov diagram given in the solution of Problem 6.8 that for the system without scheduled maintenance the reliability $R(t)$ is given by:

$$R(t) = 2e^{-\lambda t} - e^{-2\lambda t}.$$

For a system with scheduled maintenance at time interval T holds:

$$\text{MTTFF} = \frac{\int_0^t R(t)dt}{1 - R(T)} = \frac{3}{2\lambda} \frac{1 - \frac{1}{3}e^{-\lambda T}}{1 - e^{-\lambda T}}.$$

Note: If we choose T equal to the mean life of one unit (which is $1/\lambda$) the MTTFF becomes equal to $1.39 \times 3/2\lambda$. This shows that scheduled maintenance conducted at these large intervals results in an improvement with regard to the same system without scheduled maintenance of only 1.39. So, scheduled maintenance is only efficient if T is chosen sufficiently small with regard to the mean life of the respective units.
For very small T the MTTF approaches $1/\lambda^2 T$; it becomes very large with respect to $1/\lambda$.

7.5. a. The reliability of a system with scheduled maintenance is given by:

$$R_s(t) = R(T)^i R(t - iT) \qquad (iT \le t < (i+1)T).$$

With $R(t) = 1 - \frac{1}{2}t$ this becomes:

$$R_s(t) = (\tfrac{1}{2})^i \{1 - \tfrac{1}{2}(t - i)\} \qquad (i \le t < i+1).$$

This $R_s(t)$ is plotted in the figure below.

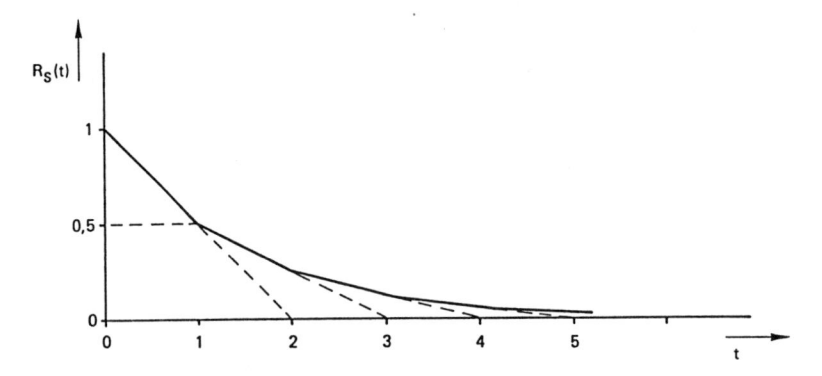

b. This mean life is the area under the $R_s(t)$-function:
$$\text{MTTFF} = \frac{3}{4} + \frac{3}{8} + \frac{3}{16} + \ldots = \frac{3}{2}.$$
The mean life can also be found with:
$$\text{MTTFF} = \frac{\int_0^t R(t)dt}{1 - R(T)} = \frac{3}{2}.$$

7.6. a. The Markov diagram of the system described is shown in the figure below. Only in state S_0 does the system function correctly: All other states are 'system down'-states.

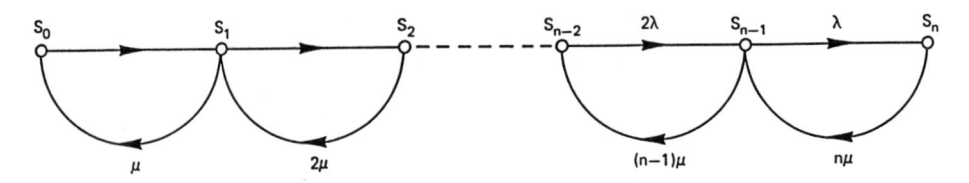

b. For the steady-state availability the occupation of all states in the diagram has reached equilibrium: the probability of a transition into a state is just as large as the probability of a transition out of that state. So for state S_i it must hold that:

S_n: $n\mu P_n = \lambda P_{n-1}$
S_{n-1}: $(n-1)\mu P_{n-1} = 2\lambda P_{n-2}$
$\vdots$
S_2: $2\mu P_2 = (n-1)\lambda P_1$
S_1: $\mu P_1 = n\lambda P_0.$

From this we see directly that:
$$P_i = \binom{n}{i}\left(\frac{\lambda}{\mu}\right)^i P_0.$$

With $\sum_{i=0}^n P_i = 1$ we then find:
$$P_0\left\{\sum_{i=0}^n \binom{n}{i}\left(\frac{\lambda}{\mu}\right)^i\right\} = P_0\left(1 + \frac{\lambda}{\mu}\right)^n = 1.$$

So, for the steady-state availability holds:

$$A_\infty = P_0 = (1 + \frac{\lambda}{\mu})^{-n}.$$

7.7. The Markov diagram of a homogenous 2-out-of-3 passive redundant system with two collaborating repairmen with efficiency α is given in the figure below. S_2 and S_3 are the failed states.

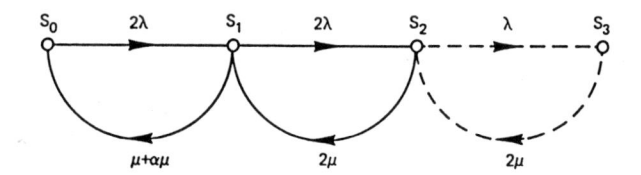

7.8. The Markov diagram is as follows:

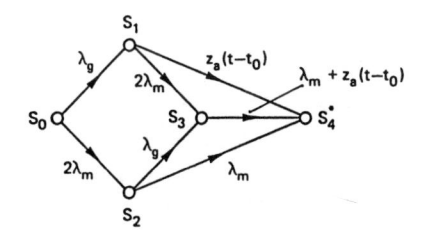

Inventory of states

S_0: This is the initial state; the diesel gen-set and both engines are switched on and may therefore fail.

S_1: The diesel gen-set has failed and the power is supplied by the battery. Both the battery and the two engines may fail in this state.

S_2: One of the engines has failed. In this state the remaining engine and the diesel gen-set may fail.

S_3: The diesel gen-set and one engine have failed. The remaining engine is powered from the battery. Both the battery and this engine can still fail.

S_4: This is the final state of the system. The propulsion system has failed.

7.9. In state S_1 of the accompanying Markov diagram only one unit has failed, so, the repairmen will not come into action yet. In state S_2 two units have failed and both repairmen are now working (branch to S_3). In state S_3 one unit has been repaired and, since the repairmen do not collaborate, only one repairman remains busy (branch to S_0). Since in state S_3 two units may fail, there also is a branch from S_3 back to S_2 which represents both repairmen working again. In state S_2 two units may, of course, fail which results in a transition to S_4. In state S_4 only one unit may still fail. From S_4 and S_5 both repairmen are active, which creates the transitions from S_5 to S_4 and from S_4 to S_2. Since three units have failed in state S_4, S_4 and S_5 are in 'down' states.

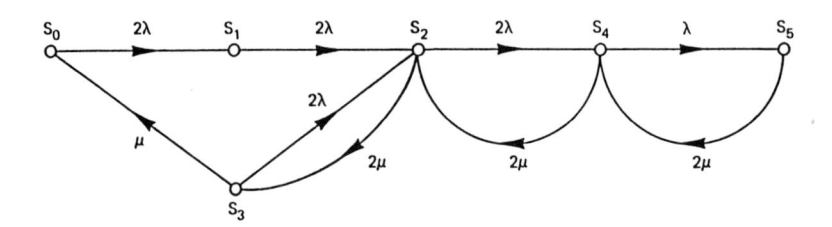

7.10. In state S_0 three lines are free. The probability that a call is placed, i.e. the probability of a transition from state S_0 to S_1 is $\lambda \Delta t$. In state S_1 one line is occupied and the probability of a transition back to S_0 is equal to $\mu \Delta t$. However, there is also the probability that still another call is placed, i.e. the probability of a transition to S_2. The transition probability from S_2 (two lines occupied) to S_1 (one line occupied) now is $2\mu\Delta t$. The rest of the Markov diagram is self-explanatory. In state S_3 no free lines are available.

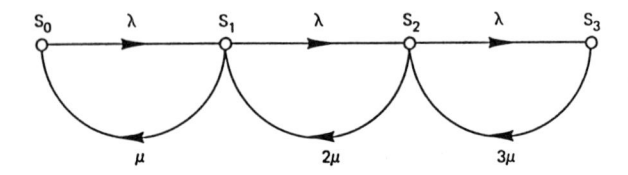

With the method given in the solution to Problem 7.6 we find:

$$\lambda P_2 = 3\mu P_3,$$

$$\lambda P_1 = 2\mu P_2 = \frac{6\mu^2}{\lambda} P_3,$$

$$\lambda P_0 = \mu P_1 = \frac{6\mu^3}{\lambda} P_3,$$

$$P_0 + P_1 + P_2 + P_3 = 1,$$

$$\overline{A}_\infty = P_3 = \frac{1}{1 + \frac{3\mu}{\lambda} + \frac{6\mu^2}{\lambda^2} + \frac{6\mu^3}{\lambda^3}}.$$

So for the steady-state availability we find:

$$A_\infty = \frac{\frac{3\mu}{\lambda} + \frac{6\mu^2}{\lambda^2} + \frac{6\mu^3}{\lambda^3}}{1 + \frac{3\mu}{\lambda} + \frac{6\mu^2}{\lambda^2} + \frac{6\mu^3}{\lambda^3}}.$$

7.11. a. The mean life (MTTFF) of the system is given by:

$$\text{MTTFF} = \frac{\int_0^T R(t)dt}{1 - R(T)} = \frac{T + \tau}{T} \tau \ln\left(\frac{T + \tau}{\tau}\right).$$

Scheduled maintenance with period $T = \tau$ yields:

$$\text{MTTFF} = 2\tau \ln 2,$$

and for $T = 3\tau$:

$$\text{MTTFF} = \frac{8}{3}\tau \ln 2.$$

b. For $T = 3\tau$ we find the longest mean life; so for scheduled maintenance with the longest maintenance interval T! The explanation can be easily found by determining the hazard rate $z(t)$:

$$z(t) = \frac{1}{t+\tau}.$$

This is a monotonically *decreasing* function with time. Here scheduled maintenance keeps bringing the system back from a low hazard rate to the higher hazard rate which is brought about by the new components introducing infancy failures all over again. This shortens the mean life and the decrease in mean life increases with the frequency of the maintenance. Therefore, the conclusion has to be that this system is not suitable for scheduled maintenance.

7.12. a. If no corrective maintenance is performed we find Markov diagram a. The associated MTTFF is easy to determine:

$$\text{MTTFF} = \frac{1}{2\lambda} + \frac{1}{\lambda} = \frac{3}{2\lambda} = 1500 \text{ hours}.$$

b. For determining the MTTFF if repairs are performed at the unit-level, we have to start with a Markov diagram that represents that no repairs are performed anymore after the system is down. This is the case in Markov diagram b. We can now calculate the MTTFF by first determining $R_s(t)$ from the differential equations and a Laplace transformation. Subsequently, we compute the time integral from 0 to infinity of $R_s(t)$. However, this is a rather circuitous method. We shall demonstrate a simpler method here.
The MTTFF is defined as:

$$\text{MTTFF} = \int_0^\infty R_s(t)\,dt = \int_0^\infty P_0(t)\,dt + \int_0^\infty P_1(t)\,dt.$$

The differential equations arrived at from the Markov diagram are:

$$\frac{dP_0}{dt} = -2\lambda P_0 + \mu P_1,$$

$$\frac{dP_1}{dt} = 2\lambda P_0 - (\mu + \lambda)P_1,$$

$$\frac{dP_2}{dt} = \lambda P_1.$$

Integrating over the time from zero to infinity gives:

$$-1 = -2\lambda \int_0^\infty P_0 dt + \mu \int_0^\infty P_1 dt,$$

$$0 = 2\lambda \int_0^\infty P_0 dt - (\mu + \lambda) \int_0^\infty P_1 dt,$$

$$1 = \lambda \int_0^\infty P_1 dt.$$

Consequently:

$$\int_0^\infty P_1 dt = \frac{1}{\lambda}; \quad \int_0^\infty P_0 dt = \frac{\mu + \lambda}{2\lambda^2}.$$

So the MTTFF becomes:

$$\text{MTTFF} = \frac{\mu + 3\lambda}{2\lambda^2} \approx 50{,}000 \text{ hours}.$$

Note: For the determination of the MTTFF the differential equations can (in principle) be skipped. To this end, for the initial state, dP_0/dt is replaced by -1. (For $t = 0$ the probability of functioning well is assumed equal to 1). For the final state ($t \to \infty$) we replace dP_n/dt by $+1$. (For $t \to \infty$ the units have a failure probability 1). For all other states dP_i/dt is made equal to 0. All probabilities P_i are replaced by the time integral of that probability.

From the so created equations the time integrals of all probabilities are easy to determine. The MTTFF is now simply equal to the sum of the time integrals of all probabilities of which the corresponding state is a good state (in which the system still functions correctly).

c. If it holds that $\lambda \ll \mu$ we only have to consider the first loop of the Markov diagram. The mean time to get from state S_0 to state S_1 is $1/2\lambda$. The mean time to get back from S_1 to S_0 is $1/\mu$. The mean time to complete the loop is therefore:

$$\frac{1}{2\lambda} + \frac{1}{\mu}.$$

The addressing frequency of the repair channel thus becomes:

$$f = \frac{1}{\frac{1}{2\lambda} + \frac{1}{\mu}} \approx 2\lambda = 2 \times 10^{-3}/\text{hour}.$$

d. If we do not perform repairs at the unit level the costs per unit of time become:
$ 5000/1500 hours = $ 3.33 / hour.
If we do perform repair at the unit level, the costs per unit of time are given by:

$$\frac{2 \times 10^{-3} \times 50{,}000 \times \$\,500 + \$\,5000}{50{,}000} = \$\,1.10\,/\,\text{hour}.$$

We may conclude that repairs at the unit level are economically justifiable.

7.13. The Markov diagram of this problem is given in the accompanying figure below.

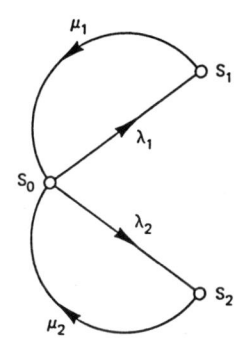

For the states S_1 and S_2 holds:

$$\mu_1 P_1 = \lambda_1 P_0,$$

$$\mu_2 P_2 = \lambda_2 P_0.$$

It furthermore holds that:

$$P_0 + P_1 + P_2 = 1.$$

Subsequently:

$$A_\infty = P_0 = \frac{1}{1 + \dfrac{\lambda_1}{\mu_1} + \dfrac{\lambda_2}{\mu_2}}.$$

7.14. In figure (a) below the configuration is shown schematically. We assume that we have to take along M spare transmitters and M spare receivers. (Because, statistically speaking, the transmitters and receivers fail equally often it is obvious to choose equal numbers). The reliability is equal to the product of the reliabilities of the transmitter and receiver system. We can therefore restrict ourselves to the determination of the reliability of either one. The Markov diagram (b) below for the transmitter system is easy to determine. With $n = M + 1$ we find:

$$\frac{dP_0}{dt} = -\lambda P_0,$$

$$\frac{dP_1}{dt} = \lambda P_0 - \lambda P_1,$$

$$\vdots$$

$$\frac{dP_{n+1}}{dt} = \lambda P_M.$$

230 Solutions to Problems

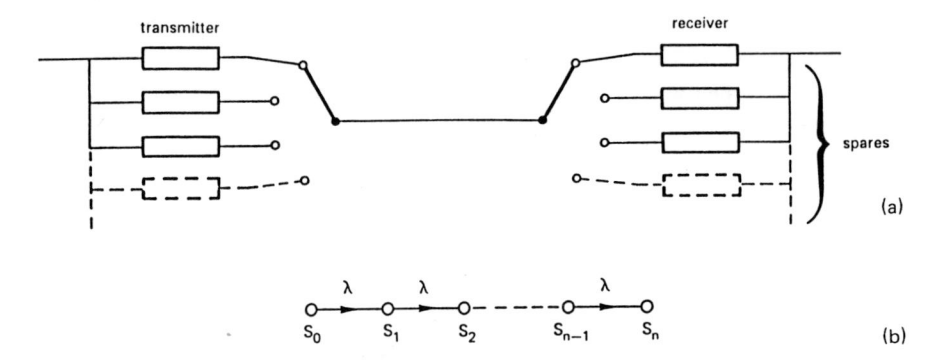

The Laplace transforms are:

$$P_0 = \frac{1}{s + \lambda},$$

$$P_1 = \frac{\lambda}{(s + \lambda)^2},$$

$$P_2 = \frac{\lambda^2}{(s + \lambda)^3},$$

$$\vdots$$

$$P_M = \frac{\lambda^M}{(s + \lambda)^{M+1}}.$$

For the reliability of the transmitter system we now find:

$$R_T(s) = \sum_{i=0}^{M} \frac{\lambda^i}{(s+\lambda)^{i+1}}.$$

Inverse transformation results in:

$$R_T(t) = \sum_{i=0}^{M} \frac{(\lambda t)^i}{i!} e^{-\lambda t}.$$

(This is the Erlang-distribution, see Section 4.1.5). For the receiver system we, of course, find the same reliability. So the reliability of the entire configuration after time T is:

$$R = e^{-2\lambda T} \left\{ \sum_{i=0}^{M} \frac{(\lambda t)^i}{i!} \right\}^2.$$

With $R \geq 98\%$, $\lambda = 10^{-3}$/h and $T = 1000$ hours we find for the number of required spare transmitters and receivers $M = 4$. R is then equal to 0.9927. A number of 3 spare transmitters and 5 receivers (or the other way around) would meet the reliability requirement, but this choice is not is not optimal (R is then equal to 0.980) because the failure probabilities are equal for both subsystems.

7.15. a. The system fails if the mechanical part fails or if both electronic units fail. Its catastrophic failure model is given below. The reliability is equal to the product of the reliability of the mechanical part and the two electronic parts in parallel:

$$R(t) = e^{-\lambda_m t}[1 - \{1 - e^{-\lambda_e t}\}^2] =$$

$$= \{2 - e^{-\lambda_e t}\} \, e^{-(\lambda_m + \lambda_e)t}.$$

(a)

b. To construct the Markov diagram we first assume there are no repairs. In the initial state S_0 two events may happen: an electronic unit fails (transition to S_1) or the mechanical part fails (transition to S_2). In state S_1 also two events may happen: the second electronic unit fails (transition to S_4) or the mechanical part fails (transition to state S_3). In state S_4 only the mechanical part may fail and in state S_3 only the second electronic unit (transition to S_5) may fail. It will be clear that the states S_2, S_3, S_4 and S_5 are 'system down' states.

Let us now introduce repair. Since only one repairman can be working at a time, only one branch will lead back from each state (except the initial state S_0). Where it leads to is determined by the repair strategy: in state S_5, in which everything has failed, at first an electronic unit is repaired, i.e. a branch to S_3, in which state the mechanical part and the second electronic unit are still to be repaired. In state S_3 we shall first repair the mechanical part because this enables the system to function again (transition to S_1). In S_1 only one electronic unit is left to be repaired which results in a branch to S_0. In S_4 both electronic units are being worked on, which results in a branch to S_1. In S_2 the mechanical part is being repaired and this results in a branch to S_0. This makes the Markov diagram complete:

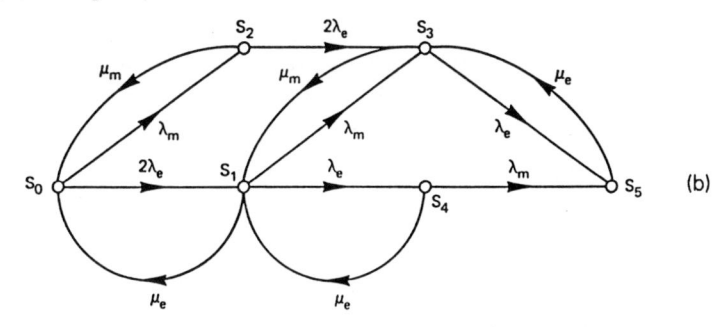

(b)

c. If the system is switched off when it is no longer able to function (the diagram has not been developed beyond the first 'system down' state), we find the Markov diagram below:

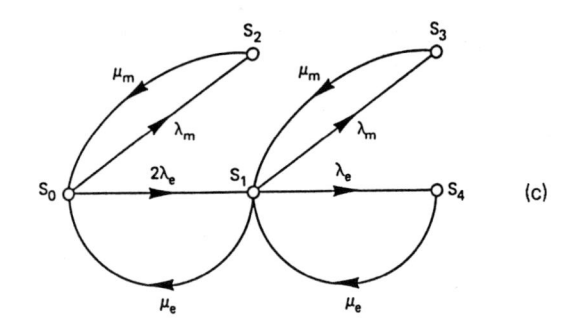

We are now able to determine the steady-state availability A_∞ with the familiar method:

$$\mu_e P_4 = \lambda_e P_1,$$

$$\mu_m P_3 = \lambda_m P_1,$$

$$\mu_m P_2 = \lambda_m P_0,$$

$$\mu_e P_1 = 2\lambda_e P_0,$$

$$P_0 + P_1 + P_2 + P_3 + P_4 = 1,$$

$$A_\infty = P_0 + P_1,$$

$$A_\infty = \frac{1}{1 + \beta_m + \dfrac{2\beta_e^2}{1 + 2\beta_e}} \approx 0.974,$$

$(\beta_m = \dfrac{\lambda_m}{\mu_m}$ and $\beta_e = \dfrac{\lambda_e}{\mu_e})$.

7.16. If the best unit (with $\lambda = \lambda_L$) is switched on first, the switch is likely to fail sooner than the unit because of its large hazard rate ($10\lambda_L$). The transition phenomenon in the $a(t)$-function will therefore have a calm and monotonic *decay* until it reaches the steady-state-value. If, however, one starts with one of the worse units there is a substantial likelihood that a good unit cannot be reached anymore because the switch got stuck. The mortality is then higher, in first instance, than the repair channel is able to cope with. Because further in time the mean mortality is considerably lower (the best unit will get a chance then) the backlog in repairs will be partly caught up. The availability $A(t)$ will then slowly *increase* to the steady-state value A_∞. This means that the relative decrease $a(t)$ per unit of time of the availability $A(t)$ becomes temporarily negative.

7.17. Because only scheduled maintenance is performed and the system is 'as new' again after each maintenance, the initial section of the reliability curve of the system without maintenance will be repeated time and again. So:

a. $A(t) = R(t - iT) = 1 - F(t - iT) = 1 - 50 \times 10^{-12}(t - iT)^2$ for $iT \le t < (i + 1)T$, with $i = 0, 1, 2, \ldots$ ($T = 50 \times 10^3$h). This is shown in the figure below:

Solutions to Problems of Chapter 7   233

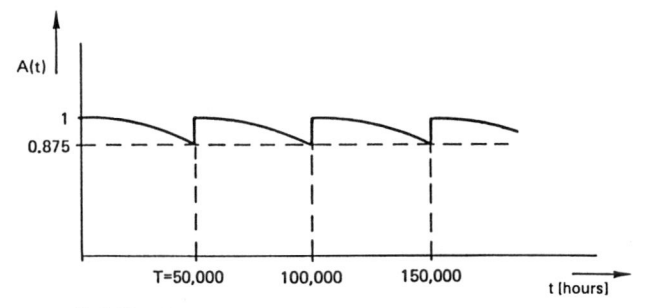

b. The mean availability is:

$$A_{\text{mean}} = \frac{1}{T}\int_0^T A(t)\,dt = 0.958.$$

7.18. The associated Markov diagram is:

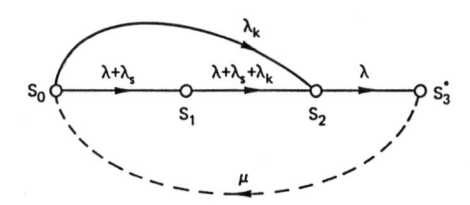

$$S_0 \xrightarrow{\lambda+\lambda_s} S_1 \xrightarrow{\lambda+\lambda_s+\lambda_k} S_2 \xrightarrow{\lambda} S_3^{\bullet}$$

(with λ_k loop from S_1 back, and μ return)

To calculate the MTTFF the branches in the Markov diagram starting at the down state(s) have to be neglected (after all, the system is only considered *until* it reaches the down state, when the user first notices that it can no longer be used). By integrating the differential equations on the left- and right-hand sides of the equality sign over the time interval $[0,\infty)$ one obtains the following equations:

$$\left.\begin{array}{rl}
-1 =& (\lambda + \lambda_s + \lambda_k)\theta_0 \\
0 =& (\lambda + \lambda_s)\theta_0 - (\lambda + \lambda_s + \lambda_k)\theta_1 \\
0 =& \lambda_h\theta_0 + (\lambda + \lambda_s + \lambda_k)\theta_1 - \lambda\theta_2 \\
1 =& \lambda\theta_2
\end{array}\right\} \text{ with } \theta_i = \int_0^\infty P_{S_1}(t)\,dt.$$

Consequently:

$$\text{MTTFF} = \theta_0 + \theta_1 + \theta_2 = \frac{1}{4} + \frac{1}{2} + \frac{3}{16} = \frac{15}{16} = \text{MTBF}.$$

The steady-state availability is then:

$$A_\infty = \frac{\text{MTBF}}{\text{MTBF} + \text{MTTR}} = \frac{15/16}{15/16 + 1/10} = 0.904.$$

7.19. The Markov diagram is as follows:

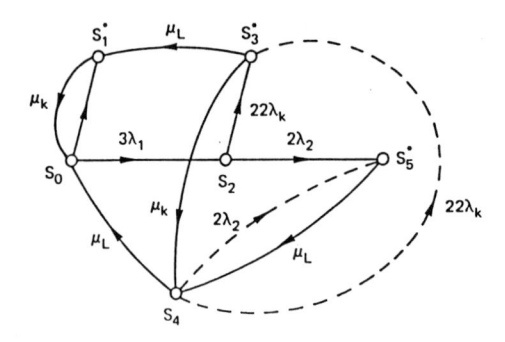

Inventory of states

S_0: Everything functions; the three locomotives each have a failure rate λ_1 and the couplings each have a failure rate λ_c.

S_1: One coupling has failed. The train stands still and the coupling is being repaired, while the other components cannot fail.

S_2: One locomotive has failed. Both the other locomotives now have a failure rate λ_2. Because the train is still able to ride no repairs are performed yet.

S_3: One locomotive and one coupling have failed. Both are being repaired.

S_4: This is an interim state in the repair process. Only one locomotive is still defective (and is being repaired). The train is able to ride again, all other components may fail again. The locomotive that is still broken is being repaired since the system has been down (in S_3 or S_5). If the train remains stopped until all components have been repaired, the dotted branches from S_4 to S_3 and to S_5 can be deleted.

S_5: Two locomotives have failed, due to which the system fails. One of the failed locomotives is being repaired. The other components cannot fail.

8 Evaluation Methods

8.1. Yes, of course it is possible to use *success trees* instead of *fault trees*. To build a success tree we may start with the expression for the top event of the fault tree given $T = A \cdot B + C$. For the non-occurrence of the top event:

$$\overline{T} = \overline{A \cdot B + C} = \overline{A \cdot B} \cdot \overline{C} = (\overline{A} + \overline{B}) \cdot \overline{C}.$$

This is illustrated in the success tree below.

So the top event T occurs if A and/or B is successful *and* if C is successful. The catastrophic failure model can also be determined simply and is indicated in the right-hand side of the figure below.

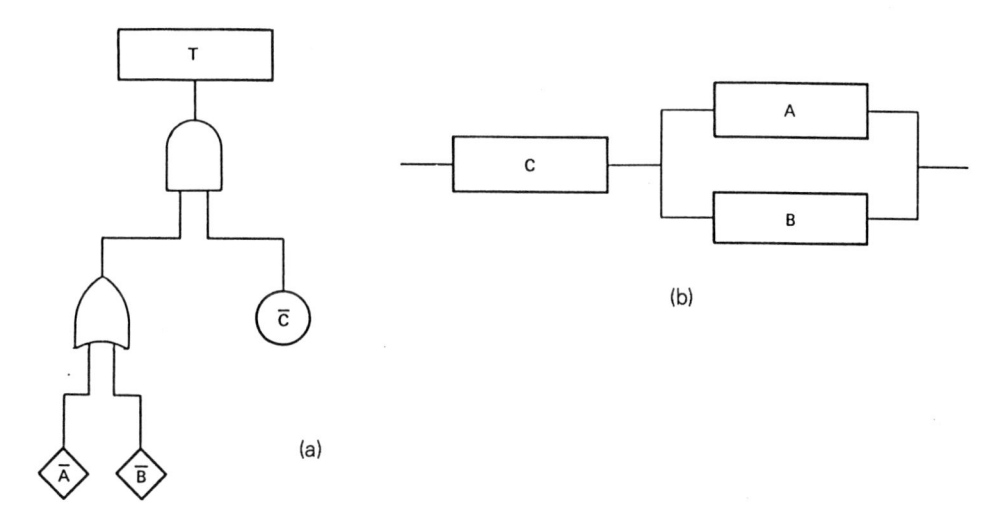

8.2. a. The figure below shows that the minimum cut sets are: A,E; B,C and B,D.

```
                    A,E
T —1,2 ── A,2 ── A,3
      ╲ 3,2 ── 3 ──── 4,5 ── A,5 ── A,6 ── A,B,C
            ╲ 3,E          ╲ B,5 ── A,7 ── A,B,D
                                  ╲ B,6 ── B,C
                                   ╲ 5,7 ── B,D
```

b. The reduced fault tree is given in the figure below.

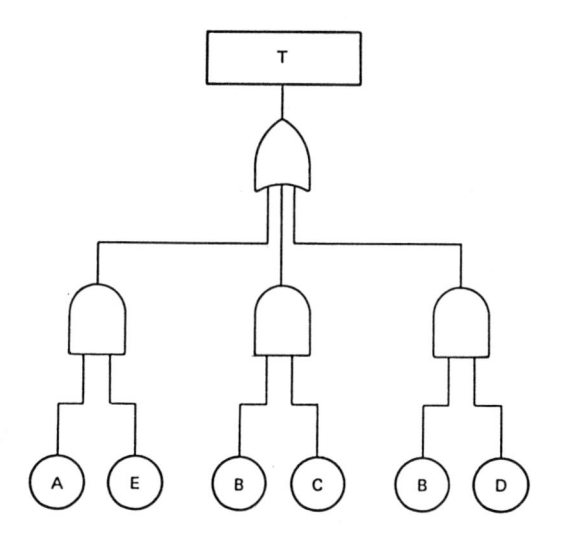

8.3. a. $T \;-\; 1,2 \;-\; 2,3,4 \;-\; 2,3,A,D \;-\; 3,A,D \;-\; A,B,D$
$ \searrow 3,A,C,D \searrow A,D$

So the minimum cut set is A,D. The reduced fault tree is given as below.

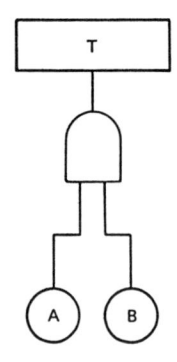

b. The system will fail if both events A and D occur. Consequently the failure probability becomes:

$$F(t) = \{1 - e^{-\lambda_A t}\}\{1 - e^{-\lambda_D t}\}.$$

8.4. a. The probability that all four engines operate correctly throughout the flight is equal to $\binom{4}{0}(0.9)^4$. The probability that only three engines do so is $\binom{4}{1}(0.9)^3(0.1)$, while the probability that only two engines give no problems is $\binom{4}{2}(0.9)^2(0.1)^2$. In each of these cases the airplane arrives safely. The probability of this happening is the sum of the above probabilities. It is equal to 0.9963.

b. The restriction that each wing should at least have one working engine makes that the airplane would not arrive safely if both starboard engines and/or both port engines fail. This is illustrated in the fault tree below.

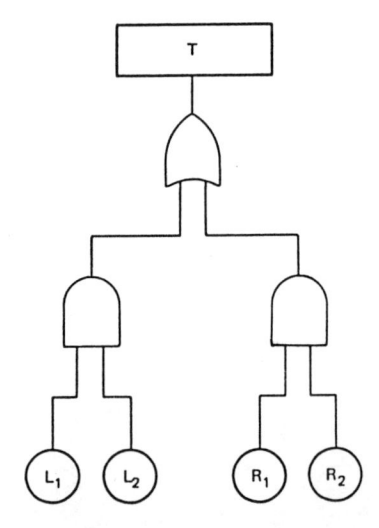

c. The probability that the airplane does not arrive safely is the probability of the top event T. From the above tree we see:

$$P(T) = P(L_1)P(L_2) + P(R_1)P(R_2) - P(L_1)P(L_2)P(R_1)P(R_2).$$

The probability that the airplane does arrive is:

$$1 - P(T) = 0.9801.$$

d. The reliability R of an engine for a flight time T is $\exp(-\lambda T)$. The failure probability is $1 - R$. If we substitute this in the expression under c, we find for the *mission reliability*:

$$e^{-2\lambda T}(2 - e^{-\lambda T})^2.$$

8.5. a. Let us first determine when the transceivers can no longer be used. The primary transmitter can no longer be used if either the transmitter or the 24 V board power supply has failed. The emergency transmitter can no longer be used if it has failed and/or if there is no more power supply available. There is no power supply available if both the 24 volt board supply and the battery pack have failed. This gives rise to the left-hand part of the fault tree below. The construction for the receiver part is similar. This results in the right-hand part of the tree. If both transmitters and/or both receivers are unserviceable the top event results: no communication is possible anymore.

b. With the Fussel-Vesely algorithm we are able to determine the minimum cut sets:

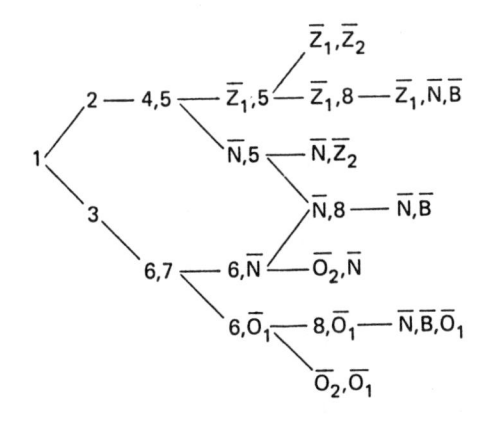

$$\begin{array}{c}
\bar{Z}_1,\bar{Z}_2 \\
2 - 4,5 - \bar{Z}_1,5 - \bar{Z}_1,8 - \bar{Z}_1,\bar{N},\bar{B} \\
1 \diagup \qquad \bar{N},5 - \bar{N},\bar{Z}_2 \\
3 \qquad\qquad \bar{N},8 - \bar{N},\bar{B} \\
6,7 - 6,\bar{N} - \bar{O}_2,\bar{N} \\
6,\bar{O}_1 - 8,\bar{O}_1 - \bar{N},\bar{B},\bar{O}_1 \\
\bar{O}_2,\bar{O}_1
\end{array}$$

The minimum cut sets are also indicated in the fault tree below.

The probability of the top event is given by:

$$\begin{aligned}
P_1 = &\, P(T_1)P(T_2) + P(R_1)P(R_2) + P(N)P(T_2) + P(N)P(R_2) + P(N)P(B) - \\
& - \{P(T_1)P(T_2)P(R_1)P(R_2) + \ldots \text{ all double intersections}\} + \\
& + \{\text{all triple intersections}\} - \text{etc.} \ldots = 0.04471.
\end{aligned}$$

Figure of Problem 8.5.b

minimum cuts

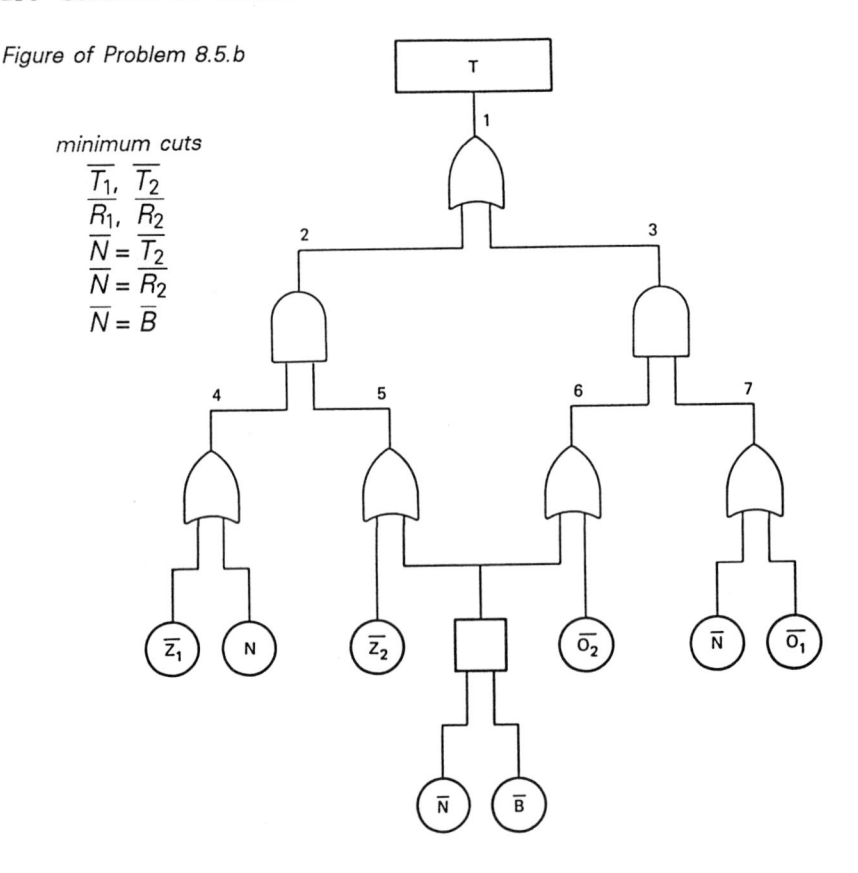

$\overline{T_1}, \overline{T_2}$
$\overline{R_1}, \overline{R_2}$
$\overline{N} = \overline{T_2}$
$\overline{N} = \overline{R_2}$
$\overline{N} = \overline{B}$

8.6. a. The reliability graph is given in the figure below. It will be clear that reliability-wise valve K_{a1} and pump A are in series. This also holds for valve K_{b1} and pump B. Since valve K is able to let the water flow in both directions, both branches have to be incorporated in the graph. In the reliability model the electrical mains E is in series with the rest of the system. The system will fail if the electricity supply stops.

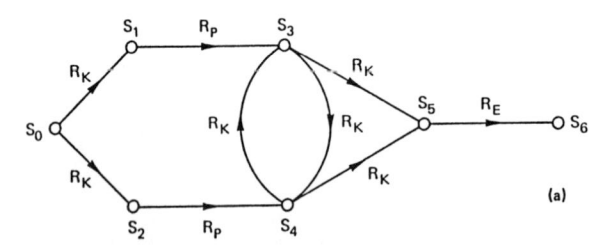

(a)

The reliability of the system without branch R_E can now be solved with the method given in the solution of Problem 6.6. To that end substitute: $R_A = R_{A'} = R_K P_P$ and $R_B = R_{B'} = R_C = R_K$.

To arrive at the overall reliability the reliability found above has to be multiplied by the reliability of the electricity supply (R_E). We then find:

$$R = R_E(R_K + R_P)\{2R_K(1 + R_K - R_K^2) + R_K^2(2R_K - 3)(R_K + R_P)\}.$$

b. The fault tree describing this system is easy to construct and is given below.

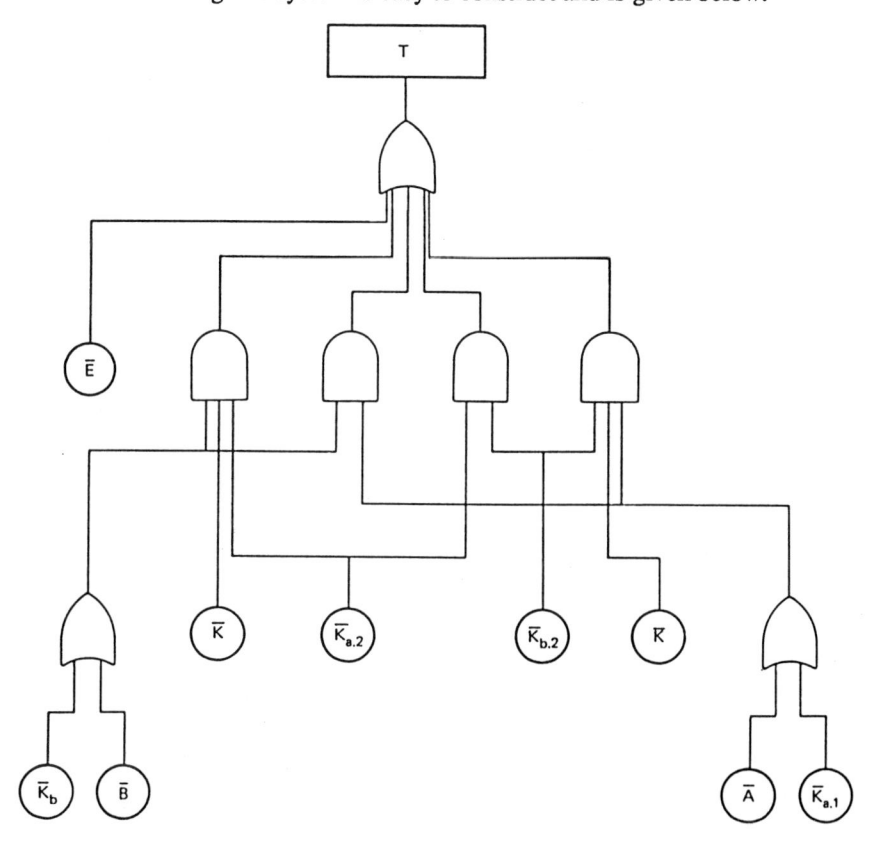

8.7. The fault tree associated with this problem is given on page 240.
The basic events in this fault tree are:
- A: computer I fails,
- B: computer II fails,
- C: the public utility power line fails,
- D: the batteries fail,
- E: restoration of the power lasts longer than 1 hour.

The resulting events are:
- F: batteries depleted,
- G: batteries fail,
- H: power supply to computers fails,
- I: redundancy channel I fails,
- J: redundancy channel II fails.

The top event is:
- T: process control stops.

It is, of course, also possible to find another configuration for the fault tree. All fault trees are correct if and only if the top event is equal to:
$$(A \cap B) \cup (C \cap D) \cup (C \cap E).$$

240 Solutions to Problems

Figure of Problem 8.7

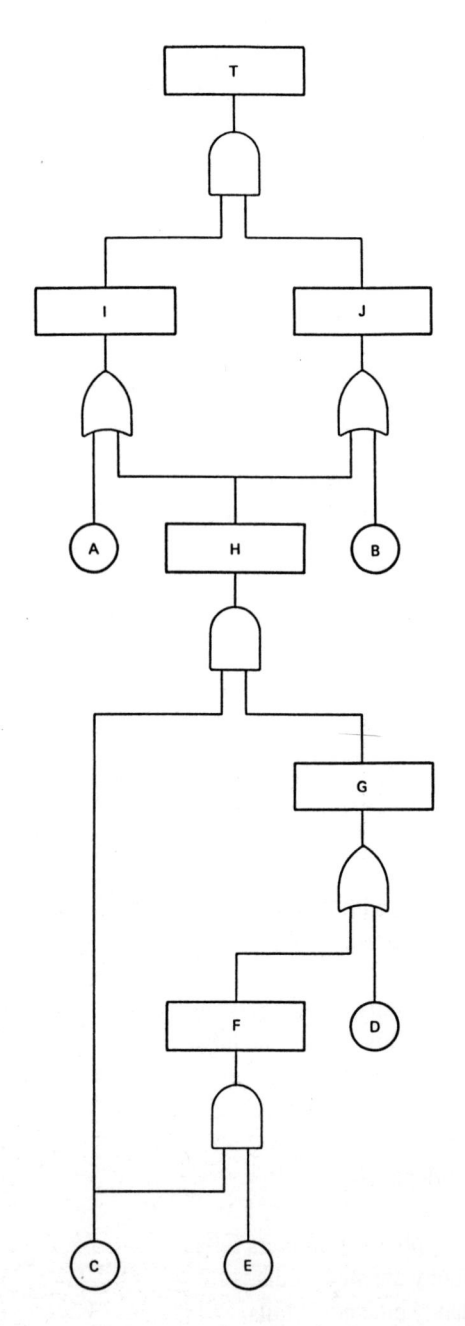

8.8. The reduced fault trees for both configurations are given below, together with the calculation of the probability of the top event.

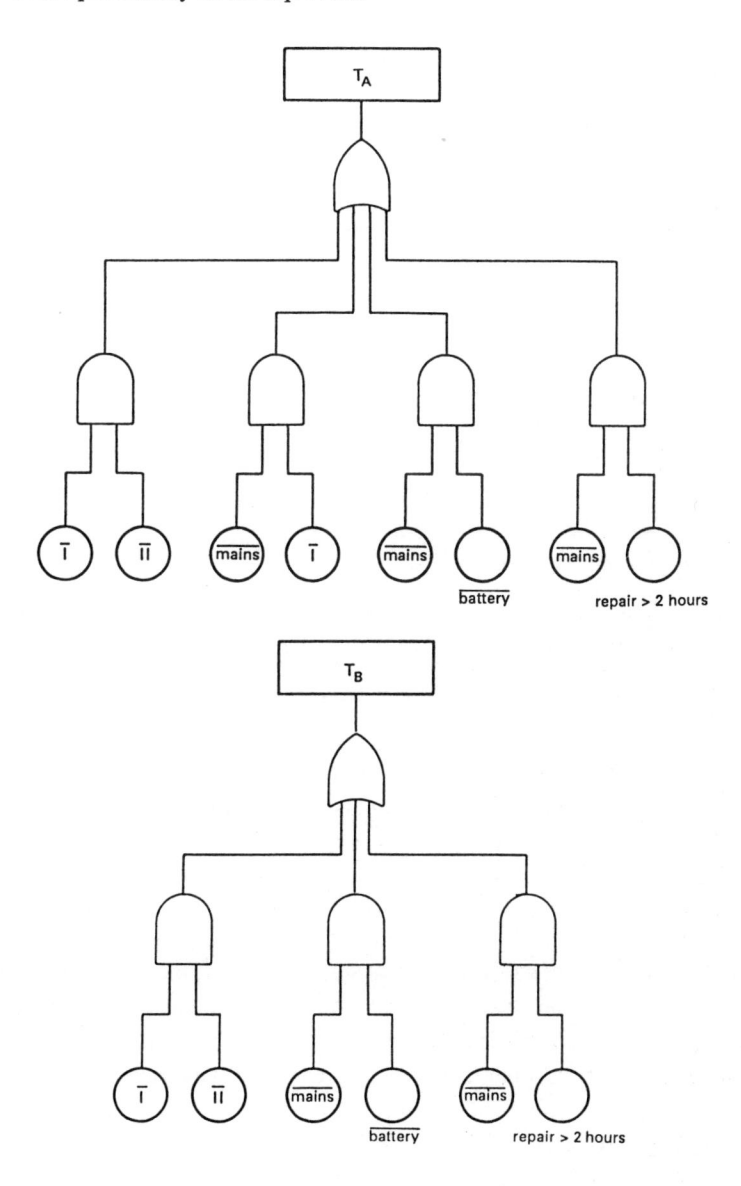

$$P(T_A) = I \cdot II + I \cdot N + N \cdot A + N \cdot R - I \cdot II \cdot N - I \cdot II \cdot N \cdot A - I \cdot II \cdot N \cdot R - I \cdot N \cdot A - I \cdot N \cdot A - I \cdot N \cdot R +$$
$$- N \cdot A \cdot R + I \cdot II \cdot N \cdot A + I \cdot II \cdot N \cdot R + I \cdot N \cdot A \cdot R + I \cdot II \cdot N \cdot A \cdot R - I \cdot II \cdot N \cdot A \cdot R = 0.0030601$$
$$P(T_B) = I \cdot II + N \cdot A + N \cdot R - I \cdot II \cdot N \cdot A - I \cdot II \cdot N \cdot R - N \cdot A \cdot R + I \cdot II \cdot N \cdot A \cdot R = 0.01099891$$

So, configuration A gives the highest reliability. This practical example clearly illustrates that with given units and components the reliability of a system is largely determined by the choice of the configuration of that system.

8.9. a. Of course, the detection probability of smoke is maximal if the alarm goes off when one or more sensors detect smoke.
b. No smoke is being detected even if there were smoke, if none of the three detectors would detect smoke. The probability of this happening is $(0.15)^3$. The probability that the smoke is detected, however, is $1 - (0.15)^3 = 0.997$.
c. The probability of a false alarm is $1 - $ (probability of no false alarm) $= 1 - (0.9)^3 = 0.27$.
d. If two or more sensors detect smoke the alarm must go off. The product {detection probability $\times$ (1 − probability of false alarm)} is then:

$$\{\binom{3}{1}(0.85)^2 0.15 + (0.85)^3\} \times \{1 - \binom{3}{1}(0.1)^2 0.9 - (0.1)^3\}(1 - 0.028) = 0.913.$$

For the strategy that the alarm goes off if at least one sensor detects smoke we find for the above product 0.723. And for the strategy that all three sensors must detect smoke before the alarm goes off we find the product to be $(0.85)^3(1 - (0.1)^3) = 0.614$.

8.10. The Markov diagram is sketched in the figure below.

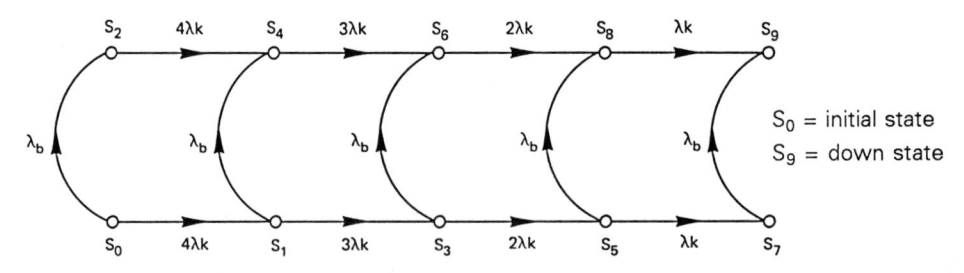

S_0 = initial state
S_9 = down state

From the initial state S_0 one of the four cables may break (transition to state S_1) or the safety system may fail, when the safety catches would not clamp the car to the walls if it starts plummeting (transition to S_2).
In state S_1 one of the three remaining cables may break (transition to S_3) or the safety system may fail (transition to S_4). In this manner one can trace the entire diagram.
If the system reaches state S_7 the system is unusable and hence 'down', since all four cables have broken. From a safety point of view, however, the system has not failed yet: in state S_7 the safety system has worked, after all. The lift car is stuck somewhere in the lift shaft. State S_9 is not just a down-state from a reliability point of view but also literally a down state: it is the fastest but fatal way to get to the ground floor!

9 Reliability of Computer Software

9.1. 400 IF A>0 AND B>0 THEN D=C ELSE D=-C
401 PRINT D

9.2. During the execution of the program, failures will occur at random. This means that the mean time to first failure (MTTFF) will be inversely proportional to the number of bugs present. This makes that the product of the MTTFF and the number of failures possible is a constant and equal to $M_0 T_0$. After a test time τ there still are $M_0 e^{-\tau C/M_0 T_0}$ possible failures, so that the MTTF(τ) is equal to $T_0 e^{\tau C/M_0 T_0}$.

9.3. a. The total number of failures possible during testing and correcting is the number of programming errors N_0 divided by the reduction factor B, so $M_0 = 2000$. The initial MTTF T_0 is given by $(fKN_0)^{-1} = 1/30$ s.
The test time τ is now given by:

$$\tau = \frac{M_0 T_0}{C} \ln \frac{\text{MTTF}(\tau)}{T_0} = 847 \text{ s}.$$

b. The number of initial failures detected is given by:

$$N = N_0 (1 - e^{-\tau C/M_0 T_0}).$$

So the remaining number of initial failures is:

$$N_0 \, e^{-\tau C/M_0 T_0} = 5.6 \times 10^{-9}.$$

The program has to be virtually without errors to meet the requirement stated in this problem.

9.4. The prices of basic components such as microprocessors and memories are falling drastically while their computing potential and reliability increase. The tendency therefore is to use as much as possible these basic components and to realise the required function under program control (in software). However, the programs to be used differ for each application and will, to a large extent, determine the system reliability. For a high reliability not only must the programs be constructed in a modular, structured and transparent form, but also extra money must be earmarked for testing and repairing the software and for writing thorough documentation.

9.5. a. Possible causes of failures in program development during the specification phase are:
- misunderstanding, and ambiguities in the formulation of the problem;
- insufficient attention for limiting conditions of the system such as processor capacity, memory space, calculation time and input and output capabilities;
- a vague definition of interfaces and communication protocols;
- overestimation of the knowledge or education of the user.

b. Possible causes of failures in the realisation phase are:
- incomplete or unclear detailed specifications;
- false interpretation of specifications;
- contradictions within the algorithms;
- incorrect initialisation;
- uncertainties about the program structure.

c. Failures in the test and user phase may be caused by:
- introducing new programming errors through carelessness in program improvement;
- vague or incomplete documentation;
- changes in the hardware or the software environment;
- lack of capacity (overflowing of buffers, lack of memory locations).

9.6. Some guidelines to prevent program failures are:
- laying down directives and working methods beforehand, which includes quality characteristics and the procedures to be followed;
- the detailed specifications should observe the program structure and be specified as accurately as possible;
- higher level, generally recognised programming languages are to be preferred;
- modular and structured programs, the modules having as great an independence as possible and having the possibility to be tested separately;
- keep in mind the communication restrictions imposed by the peripherals;
- realise testability (degree of error detection and of error location, duration of tests and correctness of the test results) as well as possible in the program;
- in order to make the program accessible for other users as well, add as many clarifying and meaningful comments as possible to the program;
- provide correct and clear documentation;
- rigidly apply a logical program structure, together with extensive testing capabilities, both for the modules individually as well as jointly.

Appendix

A.1 Applied Laplace Transforms

$f(t) = L^{-1}\{F(s)\}$	$F(s) = L\{F(t)\}$	
1	$\dfrac{1}{s}$	$(s > 0)$
e^{-at}	$\dfrac{1}{s+a}$	
$\dfrac{t^n}{n!}e^{-at}$	$\dfrac{1}{(s+a)^{n+1}}$	
$\lim\limits_{t\to\infty} f(t)$	$\lim\limits_{s\downarrow 0} s\,F(s)$	(final value theorem)
$\lim\limits_{t\downarrow 0} f(t)$	$\lim\limits_{s\to\infty} s\,F(s)$	(initial value theorem)

A.2 The Central Limit Theorem

The central limit theorem in probability theory states that the distribution of the sum of a large number of mutually independent stochastic variables x_i, each having a finite expectation and variance, converges to a normal distribution (under the Lindeberg condition which is usually satisfied for practical distributions for $\underline{x_i}$). Thus the variable y:

$$y = \lim_{n\to\infty} \sum_{i=1}^{n} x_i$$

has a normal or Gaussian distribution.

A.3 Most Commonly Used Symbols

symbol	name	dimension
A	availability	—
A_∞	steady-state availability	—
F	failure probability	—
R	reliability	—
f	failure probability density function	s^{-1}
z	hazard rate (conditional failure probability density function)	s^{-1}
λ	failure rate	s^{-1}
μ	repair rate	s^{-1}
α	efficiency	—
θ	mean life	s
θ_m	median life	s
MTTF	mean time to failure	s
MTTFF	mean time to first failure	s
MTBF	mean time between failures	s
σ	standard deviation	s
η	safety factor or degree of redundancy	—
$P(x)$	probability of event x	—
S_i	state i in Markov diagram	—
P_{S_i}	probability of state i	—
N	number	—
n	number	—
t	time	s
s	Laplace variable	—

Literature

Introductory

Cho, C.K., *Introduction to Software Quality Control*, Wiley, New York, 1980.
Becker, P.W., et al, *Design of Systems and Circuits for Maximum Reliability and Production Yield*, McGraw-Hill, New York, 1977.
Kapur, K.C., *Reliability in Engineering Design*, Wiley, New York, 1977.
Singh, C., et al., *System Reliability Modelling and Evaluation*, Hutchinson, London, 1977.
Tsokos, C.P., et al., *The Theory and Applications of Reliability*, Academic Press, New York, 1977.
Smith, C.O., *Introduction to Reliability in Design*, McGraw-Hill, 1976.
Mann, N.R., et al., *Methods for Statistical Analysis of Reliability and Life Data*, Wiley, New York, 1974.
Smith, D.J., *Reliability Engineering*, Pitman, New York, 1973.
Smith, D.J., et al., *Maintainability Engineering*, Pitman, New York, 1973.
Brook, R.H.W., *Reliability Concepts in Engineering Manufacture*, Butterworth, London, 1972.
Green, A.E., et al., *Reliability Technology*, Wiley Interscience, London, 1972.
Smith, D.J., *Reliability Engineering*, Barnes and Noble Books, New York, 1972.
Kozlov, B.A., et al., *Reliability Handbook*, Holt Rinehart, New York, 1970.
Störmer, H., *Mathematische Theorie der Zuverlässigkeit*, Akademie-Verlag, Berlin, 1970.
Dummer, G.W.A., et al, *An Elementary Guide to Reliability*, Pergamon, Elmsford, NY, 1968.
Hofman, W., *Zuverlässigkeit von Mess-, Steuer-, Regel-, und Sicherheits-Systemen*, Verlag Karl Thiemig, 1968.
Polovko, A.M., *Fundamentals of Reliability Theory*, Academic Press, New York, 1968.
Shooman, M.L., *Probabilistic Reliability*, McGraw-Hill, New York, 1968.
Dummer, G.W.A., et al., *Electronics Reliability, Calculation and Design*, Pergamon Press, Oxford, 1966.
Roberts, N.H., *Mathematical Methods in Reliability Engineering*, McGraw-Hill, New York, 1965.
Goldman, A.S., et al., *Maintainability*, Wiley, New York, 1964.
Pieruschka, E., *Principles of Reliability*, Prentice-Hall, Englewood Cliffs NJ, 1963.
Lloyd, D.K., et al., *Reliability*, Prentice-Hall, Englewood Cliffs NJ, 1962.
Bazovsky, I., *Reliability Theory and Practice*, Prentice-Hall, Englewood Cliffs NJ, 1961.

Handbooks, Standards

INSPEC: "*Electronic Reliability Data, a Guide to Selected Components*", The Institution of Electrical Engineers, London, 1981.
IEEE: "*Recommended Practice for Design of Reliable Industrial and Commercial Power Systems*", Wiley, New York, 1980
IEC publication 605-2: "*Equipment Reliability Testing, part 2: Guidance for the Design of Test cycles*", 56(sec) 124, IEC, Geneva, 1979.
Kozlov, B.A., et al., *Handbuch zur Berechnung der Zuverlässigkeit für Ingenieure*, Hanser Verlag, Munich, 1979 (598 pages, supplemented and revised edition of the preceding book).
MIL-HDBK-217C: "*Reliability Prediction of Electronic Equipment*", US Department of Defence, Washington, 1979.
IEC publication 605-1: "*Equipment Reliability Testing, part 1: General requirements*", IEC, Geneva, 1978.
IEEE: "*Nuclear reliability data manual*", IEEE Guide to the Collection and Presentation of Electrical, Electronic, and Sensing Components Reliability Data for Nuclear-power Generating Stations, IEEE, New York, 1977.
MIL-STD-781C: "*Reliability Design Qualification and Production Acceptance Tests: Exponential Distribution*", US Department of Defence, Washington, 1977.
NAVMAT-08E: "*Human Reliability Production*", US Department of the Navy, Washington, 1977
IEC publication 271: "*List of Basic Terms, Definitions, and Related Mathematics for Reliability*", IEC, Geneva, 1974.
Kozlov, B.A., et al., *Reliability Handbook*, Holt Rinehart and Winston, New York, 1970.
Ireson, W.G., editor, Reliability Handbook, McGraw-Hill, New York, 1966.
MIL-HDBK-472: "*Maintainability Prediction*", US Department of Defense, Washington, 1966.
NAVAIR-00-65-502/NAVORD OD-41146: "*Reliability Engineering Handbook*", June 1, 1964.

Juran, J.M., *Quality Control Handbook*, McGraw-Hill, New York, 1962.
British Standard BS4778: *"Glossary of Terms used in Quality Assurance"*, British Standards Institution, London.
British Standard BS5760: *"Reliability of Systems, Equipments and Components"*, British Standards Institution, London.
MIL-STD-721: *"Definitions of Effectiveness Terms for Reliability, Maintainability, Human Factors and Safety"*, Issue B, US Department of Defence, Washington.
MIL-STD-785B: *"Reliability Program for Systems and Equipment Development and Production"*, Naval Publications and Forms Center, 5801 Tabo Avenue, Philadelphia, PA, 19120 USA.
MIL-STD-2164: *"Environmental Stress Screening Process for Electronic Equipment"*, Naval Publications and Forms Center, 5801 Tabo Avenue, Philadelphia, PA, 19120 USA.

Journals
IEEE Transactions on Reliability (IEEE, New York).
Human Factors (Prentice-Hall, Englewood Cliffs).
Journal of Quality Technology (American Society for Quality Control).
Microelectronics and Reliability (Pergamon Press, Oxford).

Symposia
Annual Reliability and Maintainability Symposium (IEEE publishes the proceedings).
European Conference on Electrotechnics (Eurocon '82, proceedings available from North-Holland, Amsterdam).
Reliability Physics Symposium (annually, IEEE publishes the proceedings).
Reliability and Maintainability Symposium (annually, IEEE publishes proceedings).
Reliability Symposium SRE (organised biennially by Canadian Society of Reliability Engineers, proceedings in *Microelectronics and Reliability*).
SIAM '75 Philadelphia: *Reliability and Fault Tree Analysis*, Barlow, R.E. et al editors, 1975

Capita Selecta

Failure/Survival Models and Distributions
Lawless, J.F., *Statistical Models and Methods for Lifetime Data*, Wiley, New York, 1982.
Elandt-Johnson, R.C. and Johnson, N.L., *Survival Models and Data Analysis*, Wiley, New York, 1980.
Bury, K.V., *Statistical Models in Applied Science*, Wiley, New York, 1975.
Gross, AJ., et al, *Survival Distributions: Reliability Applications in the Biomedical Sciences*, Wiley, New York, 1975.
Ross, S.M., *Introduction to Probability Models*, Academic Press, New York, 1972.
Johnson, N.L., *Continuous Univariate Distributions*, Houghton Mifflin Company, Boston, 1970.
Ross, S.M., *Applied Probability Models with Optimization Applications*, Holden-Day, San Francisco, 1970.
Dixon, W.J., et al., *Introduction to Statistical Analysis*, McGraw-Hill, New York, 1969.
Gertsbakh, I.B., et al., *Models of Failure*, Springer Verlag, Munich, 1969.
Hahn, G.J. and Shapiro, S.S., *Statistical Models in Engineering*, Wiley, New York, 1967.
Cox, D.R. and Lewis, P.A., *The Statistical Analysis of Series of Events*, Wiley, New York, 1966.
Peck, D.S., *The Uses of Semiconductor Life Distributions*; In: Semiconductor Reliability, Vol. 2, W. Von Alren ed., Engineering Publishers, Elizabeth NJ, 1962.
Bowker, A.H. and Lieberman, G.J., *Engineering Statistics*, Prentice-Hall, Englewood Cliffs NJ, 1959.
Gumbel, E.J., *Statistics of Extremes*, Columbia University Press, New York, 1958.

Reliability Analysis, Theory
Martz, H.F., *Bayesian Reliability Analysis*, Wiley, New York, 1982.
Cooper, R.B., *Introduction to Queueing Theory*, North-Holland, Amsterdam, 1981.
Henley, E.J., *Reliability Engineering and Risk Assessment*, Prentice-Hall, Englewood Cliffs NJ, 1981.
McCormick, N.J., *Reliability and Risk Analysis*, Academic Press, 1981.
Miller, R., *Survival Analysis*, Wiley, New York, 1981.
Tillman, F.A., et al, *Optimization of Systems Reliability*, Marcel Dekker, New York, 1980.
Bain, L.J., *Statistical Analysis of Reliability*, Dekker, New York, 1978.
Kapur, K.C., et al., *Reliability in Engineering Design*, Wiley, New York, 1977.
Kaufmann, A., et al., *Mathematical Models for the Study of the Reliability of Systems*, Academic Press, New York, 1977.
Bompas Smith, J.H., *Mechanical Survival*, McGraw-Hill, London, 1973.
Jardine, A.K.S., *Maintenance, Replacement and Reliability*, Halsted Press-Wiley, New York, 1973.
Schneeweisz, W., *Zuverlässigkeitstheorie*, Springer Verlag, Berlin, 1973.

Literature 249

Amstadter, B.L., *Reliability Mathematics*, McGraw-Hill, New York, 1971.
Bitter, P., et al., *Technische Zuverlässigkeit*, Messerschmitt Bölkov, Springer Verlag, Berlin, 1971.
Krishaniah, P.R., *Multivariate Analysis III*, Academic Press, 1971.
Feller, W., *Probability Theory and its Applications*, Vol. I and II, Wiley, New York, 1970.
Jardine, A.K.S., *Operational Research in Maintenance*, Halsted Press-Wiley, New York, 1970.
Rau, J.J., *Optimization and Probability in Systems Engineering*, Van Nostrand-Reinhold, New York, 1970.
Gnedenko, B.V., et al., *Mathematical Models of Reliability Theory*, Academic Press, New York, 1969.
Grouchko, D., eds., *Operations Research and Reliability*, Gordon and Breach, New York, 1969.
Krishaniah, P.R., *Multivariate Analysis II*, Academic Press, 1969.
Polovko, A.M., *Fundamentals of Reliability Theory*, Academic Press, New York, 1968.
Jorgenson, D.W., et al., *Optimal replacement policy*, Rand McNally, Chicago, 1967.
Barlow, R.E., et al., *Mathematical Theory of Reliability*, Wiley, New York, 1965.
Hummitzsch, P., *Zuverlässigkeit von Systemen*, Vieweg und Sohn, Braunschweig, 1965.
Roberts, N., *Mathematical Methods in Reliability Engineering*, McGraw-Hill, New York, 1964.
Sandler, G.H., *System Reliability Engineering*, Prentice-Hall, Englewood Cliffs NJ, 1963.
Von Alven, W.H., editor, *Reliability Engineering*, Prentice Hall, Englewood Cliffs NJ, 1963.
Zelen, M., ed., *Statistical Theory of Reliability*, University of Wisconsin Press, Madison WI, 1963.
Cox, D.R., *Renewal Theory*, Methuen, London, 1962.
Khintchine, A., *Mathematical Methods in the Theory of Queueing*, Griffin, London, 1960.
Morse, P.M., Queues, *Inventories and Maintenance*, Wiley, New York, 1958.

Testing for Reliability
Lee, K.T., *Statistical Methods for Survival Data Analysis*, Wadsworth, Belmont CA, 1982.
Nelson, W., *Applied Life Data Analysis*, Wiley, New York, 1982.
Kalbfleisch, J.D., et al, *The Statistical Analysis of Failure Time Data*, Wiley, New York, 1980.
Lee, E., *Statistical Methods for Survival Data Analysis*, Lifetime Learning, Belmont CA, 1980.
Sinha, S.K., et al, *Life Testing and Rreliability Estimation*, Wiley Eastern, New Delhi, 1980.
Barlow, R.E., et al., *Statistical Theory of Reliability and Life Testing*, Holt Rinehart, New York, 1975.
Little, R.E., et al, *Statistical Design of Fatigue Experiments*, Halsted, New York, 1975.
Mann, N.R.R.E., et al., *Methods for Statistical Analysis of Reliability and Life Data*, Wiley, New York, 1974.
Lipson, C., et al, *Statistical Design and Analysis of Engineering Experiments*, McGraw-Hill, New York, 1973.
King, J.R., *Probability Plots for Decision Making*, Industrial Press, New York, 1971.
Johnson, L.G., *The Statistical Treatment of Fatigue Experiments*, American Elsevier Pub., New York, 1964.
Johnson, L.G., *Theory and Techniques of Variation Research*, American Elsevier Pub., New York, 1964.
Mace, A.E., *Sample-size Determination*, Reinhold Publishing Co., New York, 1964.
Myers, R.H., *Reliability Engineering for Electronic Systems*, Wiley, New York, 1964.
Kulldorff, G., *Estimation from Grouped and Partially Grouped Samples*, Wiley, New York, 1961.
Weibull, W., *Fatigue Testing and the Analysis of Results*, Pergamon, New York, 1961.

Maintenance, Organisation and Management, Production
Kletz, T. A., *What Went Wrong? Case Histories of Process Plant Disasters*, Gulf Publishing Company, Houston Texas, 1985.
Petak, W.J., et al, *Natural Hazard Risk Assessment and Public Policy*, Springer Verlag, München, 1982.
Fishhoff, B., et al, *Acceptable Risk*, Cambridge University Press, 1981.
Patton, J.D., *Maintainability and Maintenance Management*, Instrument Society of America, P.O. Box 12277, Research Triangle Park, NC 27709, USA, 1981.
Smith, D.J., *Reliability and Maintainability in Perspective*, Wiley, 1981.
Halpern, S., *The Assurance Sciences: An Introduction to Quality Control and Reliability*, Prentice-Hall, Englewood Cliffs NJ, 1978.
Welch, A., *Accidents Happen*, John Murray Ltd., London, 1978.
Enrick, N.L., *Quality Control and Reliability*, Industrial Press, New York, 1977.
Rowe, W.D., *An Anatomy of Risk*, Wiley, New York, 1977.
Lowrance, W.W., *Of Acceptable Risk: Science and the Determination of Safety*, William Kaufman, Los Altos CA, 1976.
Carrubba, E.R., et al., *Assuring Product Integrity*, Lexington Books, Massachusetts, 1975.
Locks, M.O., Reliability, *Maintainability and Availability Assessment*, Hayden, Rochelle Park NJ, 1973.
Smith, D.J., et al., *Maintainability Engineering*, Wiley, New York, 1973.
Nixon, F., *Managing to Achieve Quality and Reliability*, McGraw-Hill, London, 1971.
Jardine, A.K.S., editor, *Operational Research in Maintenance*, Barnes and Nobles Books, New York, 1970.

Blanchard, B.S., et al., *Maintainability*, McGraw-Hill, New York, 1969.
Jorgenson, D.W., et al., *Optimal Replacement Policy*, Rand McNally, Chicago 1967.
Goldman, A.S., et al., *Maintainability*, Wiley, New York, 1964.
Haviland, R.P., *Engineering Reliability and Long Life Design*, Van Nostrand, Princeton NJ, 1964.
Landers, R.R., *Reliability and Product Assurance*, Prentice-Hall, Englewood Cliffs NJ, 1963.
Calabro, S.R., *Reliability Principles and Practices*, McGraw-Hill, New York, 1962.
Lloyd, D.K., et al., *Reliability, Management Methods and Mathematics*, Prentice-Hall, Englewood Cliffs NJ, 1961.

Design and Application
Klaassen, K.B., *Reliability of Analogue Electronic Systems*, Elsevier, Amsterdam, 1984.
Jensen, F., et al, Burn-in: *An Engineering Approach to the Design and Analysis of Burn-in Procedures*, Wiley, Chichester, 1982.
Dhillon, B.S., et al., *Engineering Reliability*, Wiley, New York, 1981.
O'Connor, P.D.T., *Practical Reliability Engineering*, Heyden and Son, London, 1981.
Longbottom, R., *Computer System Reliability*, Wiley, New York, 1980.
Endrenyi, J., *Reliability Modelling in Electric Power Systems*, Wiley, New York, 1978.
Bennet, S.B., et al., *Failure Prevention and Reliability*, Society of Mech. Eng., New York, 1977.
Anderson, R.T., *Reliability Design Handbook*, IIT Research Institute (RAC), RADC, Griffins Air Force Base, New York, 1976.
Brown, D.B., *Fault Tree Analysis*. In: System Analysis and Design for Safety, Prentice-Hall, Englewood Cliffs, NJ, 1976.
Myers, J., *Software Reliability Principles and Practice*, Wiley Interscience, New York, 1976.
Smith, C.O., *Introduction to Reliability in Design*, McGraw-Hill, New York, 1976.
Hammer, W., *Fault Tree Analysis*. In: Product Safety Management and Engineering, Prentice-Hall, Englewood Cliffs, NJ, 1975.
Myers, J., *Reliable Software through Composite Design*, Petrocelli Books, New York, 1975.
Cluley, J.C., *Electronic Equipment Reliability*, Macmillan, London, 1974.
Bompass-Smith, J.H., *Mechanical Survival*, McGraw-Hill, London, 1973
Jowett, C.E., *Electronic and Environments*, Business Books, London, 1973.
Carter, A.D.S., *Mechanical Reliability*, Wiley, London, 1972.
Cunningham, C.E., et al, *Applied Maintainability Engineering*, Wiley, New York, 1972.
Moranda, P.B., et al, *Software Reliability Research*. In: Statistical Computer Performance Evaluation, W. Freiberger, editor, Academic Press, New York, 1972.
Jowett, C.E., *Reliable Electronic Assembly Production*, Tab Books, Blue Ridge Summit, 1971.
Kivenson, G., *Durability and Reliability in Engineering Design*, Hayden, New York, 1971.
Billinton, R., *Power System Reliability Evaluation*, Gordon and Breach, New York, 1970.
Jelen, F.C., *Cost and Optimization Engineering*, McGraw-Hill, New York, 1970.
Jowett, C.E., *Reliability of Electronic Components*, London Iliffe Book, England, 1966.
Pierce, W., *Failure Tolerant Computer Design*, Van Nostrand Reinhold, New York, 1966.
Meyer, R.H., et al., *Reliability Engineering for Electronic Systems*, Wiley, New York, 1964.
Ankenbrandt, F.L, *Maintainability Design*, Engineering Publishers Division of A.C. Book Co., Elisabeth NJ, 1963.
Landers, R.R., *Reliability and Product Assurance*, Prentice-Hall, Englewood Cliffs NJ, 1963.
Haugen, E.B., *Probabilistic Approaches to Design*, Wiley, New York, 1962.

Human Reliability
Park, K.S., *Human Reliability: Analysis, Prediction and Prevention of Human Error*, Elsevier Press, Amsterdam, 1987.
Dhillon, B.S., *Human Reliability, with Human Factors*, Pergamon Press, New York, 1986.
Smith, H.T., et al, *Human Interaction with Computers*, Acadamic Press, New York, 1980.
Drury, C.G., et al, *Human Reliability in Quality Control*, Halstead, New York, 1976.
Henley, E.J., et al, *Generic Techniques in Systems Reliability Assessment*, Noordhoff, Leiden, Holland, 1976.
Lees, F.P. *Man-Machine System Reliability*. In: Man and Computer in Process Control, E. Edwards editor, Institution of Chemical Engineers, London, 1973.
Cunningham, C.E., et al, *Human Factors in Maintainability: Applied Maintainability Engineering*, Wiley-Interscience, New York, 1972.
Meister, D., *Human Factors: Theory and Practice*, Wiley-Interscience, New York, 1971.
Chapanis, A., *Man-Machine Engineering*, Wadsworth, Belmont CA, 1965.
Meister, D., et al, *Human Factors Evaluation in System Development*, Wiley-Interscience, New York, 1965.

Index

a

accelerated life test 57, 61
accelerating the ageing process 17
acceleration factor 17, 24
accumulated processor time 201
activation energy 23
—, effective 24
active m-out-of-n redundancy 103
active maintenance time 33
active redundancy 89, 150, 156
active-redundant system 153
adaptive majority voting 104
administrative maintenance time 33
ageing process, accelerating the 17
aircraft attitude control 196
alpha particles 26
analysis method 111
analysis module 201
analysis, consequence 190
—, criticality 188
—, failure criticality 176
—, failure effect 176
—, failure mode 176
—, fault tree 178
—, FMEC 174
—, reliability 11
—, risk 187
—, statistical data 32
AND-gate 179
anti-causal evaluation 177
a posteriori reliability 12
a priori reliability 12
Arrhenius' model 23
assembler language 203
availability 15, 35, 139
—, intrinsic 35
—, long-term 139
—, mission 139
—, steady-state 139
—, system 158
availability of a repairable system 142
average life 37

b

backward method 173
BASIC 204
basic event 178
bath tub distribution 42
Bayes' theorem 114
bottom-up evaluation method 173
budget constrained optimisation 109
burn-in period 21
burn-in screen 29

c

CA 188
capacitor, electrolytic 13
case history 189
catastrophic failure 32, 33
catastrophic failure model 63
causal evaluation 173
CC-failure 98
central limit theorem 245
certification of software 200
chain structure 86
circuit, parallel 65
—, series 65
code walk-through 201
common-cause failure 86, 98, 116
common-cause failures, reliability of a system with 100
component, electronic 42
—, hi-rel 11, 93
computer 196
computer software program 196
computer test time 204
computer time model 201
condensation 13, 90
condition monitoring 132
condition-based maintenance 128, 132, 165
conditional event 179
conditional probability theorem 115
confidence level of life test 58
conformity 12
consequence analysis 190
cooperating repairmen 147
corrected software error 201
corrective maintenance 165
corrosion 27
cost of ownership 108
costs of scheduled maintenance 131
critical system 15
criticality analysis 188
cut 66
cut set 66
—, minimum 112, 184

d

decision theory 32
decomposition method 114
deductive method 173
degradation failure 32
delayed repair 158
delta-star transformation 111
dependent failure 115, 97
derating 11, 21, 71
derating factor 71
design error 21
design method, top-down 198
design, preliminary 88
detected software error 201
deterministic approach 17
deterministic reliability 23
diode 66, 83
direct maintenance 165
disjunct events 67, 211
distribution, bath tub 42
—, Erlang's 55
—, failure 36, 41
—, gamma 53
—, Gauss 48
—, life 36, 41
—, lognormal 50
—, negative-exponential 44, 55
—, normal 48, 245
—, repair time 42, 136, 141

252 Index

—, Student's 60
—, Weibull 51
distribution measurements, life 57
distribution of Student 60
documentation, software 199
dormant failure 197
dormant redundancy 90
down time 33

e

early failure 26
early failure period 42
effective activation energy 24
effective repair time 159
effectiveness, system 34
electrolytic capacitor 13
electromigration 28
electronic component 42
electronic system 134
end-of-life period 42
environment 23
—, specified 12, 13
environmental factor 70
Erlang's distribution 55
error, corrected software 201
—, detected software 201
—, human 190
estimation theory 11, 32
estimator, unbiased maximum likelihood 60
evaluation, anti-causal 177
—, backward 173
—, bottom-up 173
—, causal 173
—, deductive 173
—, forward 173
—, top-down 173
evaluation method 172
—, event-oriented 172
—, structure-oriented 172
event, conditional 179
—, resulting 179
—, trigger 179
—, undeveloped 179
event-oriented evaluation method 172
events, disjunct 67, 211
—, independent 211
—, multiple 77
—, stochastically independent 67
external stress 23, 68
externally applied voltage 26

f

factor, derating 71
—, environmental 70
—, safety 70
fail-safe system 127
failure 32
—, catastrophic 32, 33
—, CC- 86, 98, 116
—, common-cause 86, 98, 116
—, degradation 32
—, dependent 115
—, dormant 197
—, early 26
—, gradual 33
—, intermittent 32
—, multi-mode 66
—, partial 33

—, primary 86
—, safe 36
—, secondary 86
—, single mode 63
—, single-point 174, 181, 186
—, software 197, 244
—, 'system stuck' 163
—, total 33
—, unsafe 36
failure alarm 104
failure cautioning 104
failure criticality analysis 176
failure density of Weibull distribution 53
failure distribution 36, 41
failure distribution from life test 58
failure distribution of negative-exponential
    distribution 46
failure distribution of normal distribution 48
failure distribution of Weibull distribution 51
failure effect analysis 176
failure identification 104
failure isolation 161
failure mechanism 17, 23, 26
failure mode 23
—, effect analysis 188
—, effect and criticality analysis 174, 188
failure mode analysis 176
failure model, software 201
failure of components, statistical 41
failure probability density function 36
failure probability density of gamma distribution
    55
failure probability density of lognormal
    distribution 51
failure probability density of negative-exponential
    distribution 46
failure probability density of normal distribution
    48
failure rate 37, 141
failure reporting 104
failure-to-repair rate ratio 143
failures in software, number of 202
failures, dependent 97
fatigue fractures 13
fault tree 183
—, reduced 185
—, reducing a 184
fault tree analysis 178
fault tree symbols 178
FCA 176
FEA 176
final value theorem 245
FMA 176
FMEA 188
FMEC analysis 174
FMECA 188
forward method 173
FTA 178
functional description 199
functional model 63
functions, specified 12
Fussel-Vesely algorithm 184

g

gamma distribution 53
—, failure probability density 55
—, mean life 55
—, reliability 55

g

gamma function 53
Gauss distribution, *see normal distribution*
generalised costs 108
goto statement 198
gradual failure 33

h

hardware 196
hardware redundancy 89
hazard rate 37, 39, 207, 223
hazard rate of a parallel system 92
hazard rate of a series system 87
hazard rate of lognormal distribution 51
hazard rate of negative-exponential distribution 46
hazard rate of normal distribution 48
hazard rate of Weibull distribution 53
hi-rel component 11, 93
homogeneous Markov model, time- 76
homogeneous production 17
hot redundancy 89
human error 190
human intervention 35, 85
humans, reliability of 43
humidity, environmental 25

i

IC's 27
ideal replacement 136
increased stress 61
increasing reliability methods 19
independent events 211
—, stochastically 67
induction 173
inductive method 173
infancy failure period 42
information redundancy 89
inherent reliability 11
inhibit-gate 179
inhomogeneous Markov model, time- 76
inhomogeneous passive-redundant system 163
inhomogeneous system 159
initial value theorem 245
inspection method 107
instrumentation 201
intermittent failure 32
internal stress 23, 68
intrinsic availability 35
inventory list 106
invest now, save later 16

j

jump statement 199

l

Laplace transform 245
life cycle of a system 126
life distribution 36, 41
life distribution measurements 57
life test 22
—, accelerated 57, 61
—, confidence level 58
—, failure distribution 58
life variable 12, 13, 32
life, average 37
life-cycle cost, minimum 127
light bulb 17, 137
logic structure 199

logistic maintenance time 33
logistics 11
lognormal distribution, failure probability density of 51
—, hazard rate of 51
—, mean value of 51
lognormal failure distribution 50
long-term availability 139
longer-used software 198
low-maintenance 165

m

m-out-of-n system 90, 101
maintainability 35, 138
maintainable 11
maintainable system 11, 14, 85
maintained system 14, 35, 85, 126
maintenance 14, 126, 128, 164
—, condition-based 128, 132, 165
—, corrective 165
—, costs of scheduled 131
—, direct 165
—, low- 165
—, multi-stage 165
—, non-perfect scheduled 130
—, preventive 22, 89, 128, 165, 223
—, scheduled 128, 165, 223
maintenance friendliness 165
maintenance interval 130
maintenance operation 85
maintenance period 169, 223
maintenance strategy 164, 166
maintenance time 33, 85
—, active 33
—, administrative 33
—, logistic 33
maintenance to system with negative-exponential failure distribution 89
majority voting system 92, 103
majority voting, adaptive 104
management 11, 198
manual code walk-through 201
Markov diagram 121
—, simplifying 217
Markov model 73
—, time-homogeneous 76
—, time-inhomogeneous 76
Markov process 74, 115
Markov state 74
Markov chain model 73
material fatigue 27
mean life of a schedule-maintained system 130
mean life of a system 118
mean life of gamma distribution 55
mean life of negative-exponential distribution 46
mean life of Weibull distribution 53
mean time between failures 38, 140
mean time to failure 24, 37
mean time to first failure 38
mean time to repair 140
mean useful life of a series system 87
mean value of lognormal distribution 51
measurement system 196
measuring reliability 11
method, backward 173
—, bottom-up 173
—, decomposition 114
—, deductive 173

254 Index

—, evaluation 172
—, event-oriented evaluation 172
—, forward 173
—, inductive 173
—, state-space 115
—, structure-oriented 172
—, tie set 193
—, tie set and cut set 114, 184
—, top-down 173
—, top-down design 198
methods for increasing reliability 19
minimum cut set 112, 184
minimum life-cycle cost 127
minimum tie set 112
mission availability 139
mission reliability 34
misuse 13, 32
mixed system 106
model, Arrhenius' 23
    —, catastrophic failure 63
    —, computer time 201
    —, functional 63
    —, Markov 73
    —, reliability 63
    —, schematic 63
    —, software failure 201
    —, stress-strength 68
    —, time-homogeneous Markov 76
    —, time-inhomogeneous Markov 76
module 198
monitoring system 196
monitoring, condition 134
    —, trend 134
MTBF 38, 140
MTTF 24, 37, 118
MTTF of software 203
MTTFF 38, 150, 228
MTTR 140
multi-mode failure 66
multi-stage maintenance 165
multiple events 77

n

navigation 196
negative-exponential distribution 44, 55
    —, failure distribution 46
    —, failure probability density 46
    —, hazard rate of 46
    —, mean life 46
    —, reliability 46
negative-exponential failure distribution, maintenance to a system with 89
network reduction method 111
non-maintained system 14, 35, 85
non-perfect scheduled maintenance 130
normal distribution 48, 245
    —, failure distribution 48
    —, failure probability density 48
    —, hazard rate 48
normal life period 42
nuisance alarm 100
number of failures in software 202

o

operation time 13
operational readiness 35
operational state 85
operational time 33

optimisation 107
    —, budget constrained 109
    —, reliability constrained 109
OR-gate 179
organisation 11
overall cost of ownership 108
overdimensioned 71

p

pacemaker 15
parallel circuit 65
parallel system 92, 98, 181
    —, hazard rate 92
partial failure 33
partitioning, system 108
passive m-out-of-n redundancy 103
passive redundancy 90, 94, 150, 156
passive-redundant system 154
    —, inhomogeneous 163
period, burn-in 21
    —, early failure 42
    —, end-of-life 42
    —, infancy failure 42
    —, normal life 42
    —, useful life 42
    —, wear out 42
physical deterioration process 17
physics of failure 11
Poisson process 45
power distribution system 21
power supply 26
predicting approach 16
preliminary design 88
preventive maintenance 22, 85, 89, 128, 165, 223
primary failure 86
probability 12
    —, transition 75
probability theorem, conditional 115
process, Poisson 45
process control 196
processor time, accumulated 201
production, homogeneous 17
program, computer 196
    —, reliable 197
    —, sub- 198
program development 243
program failures 244
pure parallel system 90
purple plague 29

q

qualitative complexity 15
quality 12
quality circle 205
qualtity control 70
quantitative complexity 15
quantitative engineering 15
queueing theory 11

r

radiation, ionising 26
RAM 26
reduced fault tree 185
reducing a fault tree 184
redundancy 11, 22, 89
    —, active 89, 150, 156
    —, active m-out-of-n 103

Index 255

—, dormant 90
—, hardware 89
—, hot 89
—, information 89
—, passive 90, 94, 150, 156
—, passive m-out-of-n 103
—, repairable system with 150
—, signal 89
—, stand-by 90
—, structural 89
redundant system 85
—, active- 153
—, inhomogeneous passive- 163
—, passive- 154
reliability 12, 34, 139
—, a posteriori 12
—, a priori 12
—, deterministic 23
—, inherent 11
—, measuring 11
—, methods for increasing 19
—, mission 34
—, software 201
—, statistical 32
—, testing 11
reliability analysis 11
reliability constrained optimisation 109
reliability engineering 11, 15
—, software 196
reliability group 205
reliability management 198
reliability model 63
reliability of a m-out-of-n system 101
reliability of a schedule-maintained system 129
reliability of a system with common-cause failures 100
reliability of gamma distribution 55
reliability of humans 43
reliability of negative-exponential distribution 46
reliability of Weibull distribution 51
reliability theory 11
reliable software program 197
renewal density 136
renewal rate 143
renewal strategy 137
renewal theory 11
repair 127, 136, 138, 156
—, delayed 158
—, shared 141, 156
repair capacity 141, 166
repair channel 141, 158
repair policy 166
repair rate 141
repair strategy 141
repair time, effective 159
repair time distribution 42, 136, 141
repairability 35
repairable system 141
—, availability 142
repairable system with redundancy 150
repairmen, cooperating 147
replacement 127, 136
reserve state 90
resulting event 179
risk 36, 187
risk analysis 187
run-to-break strategy 132

s
safe failure 36
safety 36, 187
safety factor 70
sample 17
schedule-maintained system, mean life 130
—, reliability 129
scheduled maintenance 128, 165, 223
—, costs 131
schematic model 63
screening 26, 29
secondary failure 86
sensitivity coefficient 39
series circuit 65
series structure 86
series system 86, 90, 98, 181
—, hazard rate 87
—, mean useful life 87
shared repair 141, 156
signal redundancy 89
similarity voter 103
simplify Markov diagram 217
single mode failure 63
single-point failure 174, 181, 186
software 196
—, certification 200
—, computer 196
—, failure model 201
—, longer-used 198
—, MTTF 203
—, number of failures 202
—, reliability 200, 201
—, testing reliability 200
—, validation 200
—, writing 198
software development, 243
software documentation 199
software error, corrected 201
—, detected 201
software failure 197, 244
software reliability engineering 196
spare parts 165, 166
specified environment 12, 13
specified functions 12
stand-by redundancy 90
stand-by time 33
state, active 90
—, Markov 74
—, reserve 90
state-space method 115
statistical approach 16
statistical data analysis 32
statistical failure of components 41
statistical reliability 32
steady-state availability 139
step-stress test method 24, 61
stochastically independent events 67
strategy, repair 141
strength 68
stress 68
—, external 23, 68
—, increased 61
—, internal 23, 68
stress method, step- 24
stress quantity 23
stress-derating 71
stress-strength model 68
structural redundancy 89

structure-oriented evaluation method 172
structured programming 198
Student, distribution of 60
subprogram 198
success tree 183
symbols 246
—, fault tree 178
system 12
—, active-redundant 153
—, availability of a repairable 142
—, critical 15
—, electronic 134
—, fail-safe 127
—, hazard rate of a parallel 92
—, hazard rate of a series 87
—, inhomogeneous 159
—, inhomogeneous passive-redundant 163
—, life cycle 126
—, m-out-of-n 90, 101
—, maintainable 14, 85
—, maintained 14, 35, 85, 126
—, majority voting 92, 103
—, mean life 118
—, mean life of a schedule-maintained 130
—, mean useful life of a series 87
—, measurement 196
—, mixed 106
—, monitoring 196
—, MTTF 118
—, non-maintained 14, 35, 85
—, parallel 92, 98, 181
—, passive-redundant 154
—, power distribution 21
—, pure parallel 90
—, redundant 86
—, reliability of a m-out-of-n 101
—, reliability of a schedule-maintained 129
—, repairable 141
—, series 86, 90, 98, 181
system availability 158
system effectiveness 34
system partitioning 108
'system stuck' failure 163
system with common-cause failures, reliability 100
system with negative-exponential failure distribution, maintenance to 89
system with redundancy, repairable 150

t

temperature 23
test, accelerated life 57, 61
—, confidence level from 58
—, failure distribution from 58
—, life 22
—, step-stress 61

test time, computer 204
testing reliability 11
testing software reliability 200
theorem, Bayes' 114
—, central limit 245
—, conditional probability 115
—, final value 245
—, initial value 245
tie 66
tie set 66
—, minimum 112
tie set and cut set method 112, 184
tie set method 193
time, active maintenance 33
—, administrative maintenance 33
—, computer test 204
—, down 33
—, logistic maintenance 33
—, maintenance 33
—, operational 33
—, stand-by 33
—, up 33
—, waiting 33
time-homogeneous Markov model 76
time-inhomogeneous Markov model 76
tolerance 32
top-down design method 198
top-down method 173
total failure 33
transform, Laplace 245
transition probability 75
trend monitoring 134
trigger event 179

u

unbiased maximum likelihood estimator 60
undeveloped event 179
unreliability 36
unsafe failure 36
update 198
useful life period 42

v

validation of software 200
voltage, externally applied 26

w

waiting time 33
wear out period 42
Weibull distribution 51
—, failure density 53
—, failure distribution 51
—, hazard rate 53
—, mean life 53
—, reliability 51
writing software 198